Structural Health Monitoring of Long-Span Suspension Bridges

Long-span suspension bridges cost billions. In recent decades, structural health monitoring systems have been developed to measure the loading environment and responses of these bridges in order to assess serviceability and safety while tracking the symptoms of operational incidents and potential damage. This helps ensure the bridge functions properly during a long service life and guards against catastrophic failure under extreme events.

Although these systems have achieved some success, this cutting edge technology involves many complex topics which present challenges to students, researchers, and engineers alike. Systematically introducing the fundamentals and outlining the advanced technologies for achieving effective long-term monitoring, this book covers

- the design of structural health monitoring systems
- finite element modelling and system identification
- highway loading monitoring and effects
- railway loading monitoring and effects
- temperature monitoring and thermal behaviour
- wind monitoring and effects
- seismic monitoring and effects
- SHMS-based rating method for long-span bridge inspection and maintenance
- structural damage detection and test-bed establishment.

These are applied in a rigorous case study, using more than 10 year's worth of data, to the Tsing Ma suspension bridge in Hong Kong to examine their effectiveness in the operational performance of a real bridge. The Tsing Ma bridge is the world's longest suspension bridge to carry both a highway and railway, and is located in one of the world's most active typhoon regions. Bridging the gap between theory and practice, this is an ideal reference book for students, researchers and engineering practitioners.

You-Lin Xu is Chair Professor, Head of the Department of Civil and Structural Engineering, and founding Director of the Research Centre for Urban Hazards Mitigation at the Hong Kong Polytechnic University.

Yong Xia is an Assistant Professor at the Hong Kong Polytechnic University.

Structural Health Monitoring of Long-Span Suspension Bridges

You-Lin Xu and Yong Xia

CRC Press
Taylor & Francis Group
Boca Raton London New York

CRC Press is an imprint of the
Taylor & Francis Group, an **informa** business

A SPON PRESS BOOK

First published 2012
by Spon Press
2 Park Square, Milton Park, Abingdon, Oxfordshire OX14 4RN

Simultaneously published in the USA and Canada
by Spon Press
711 Third Avenue, New York, NY 10017

First issued in paperback 2017

Spon Press is an imprint of the Taylor & Francis Group, an informa business

British Library Cataloguing in Publication Data
A catalogue record for this book is available from the British Library

Library of Congress Cataloging in Publication Data
Xu, You-Lin, 1952-
Structural health monitoring of long-span suspension bridges / You-Lin Xu and Yong Xia.
p. cm.
Includes bibliographical references and index.
1. Suspension bridges -- Testing. 2. Structural health monitoring. I. Xia, Yong, 1971- II. Title.
TG400.X82 2012
624.2'30287 -- dc22

2011009700

ISBN 13: 978-1-138-07563-4 (pbk)
ISBN 13: 978-0-415-59793-7 (hbk)

This book has been prepared from camera-ready copy provided by the authors.

To Wei Jian, my wife

– Y.L. Xu

To Xiao Qing, my wife

– Y. Xia

Contents

Acknowledgements

The writing of this book has been a challenging and time-consuming task that could not have been completed without the help of many individuals. We are grateful to many people who helped in the preparation of this book.

A few PhD students, present and former, at The Hong Kong Polytechnic University participated in some research works presented in this book: Dr. Wai-Shan Chan, Dr. Bo Chen, Dr. Zhi-Wei Chen, Prof. Wen-Hau Guo, Dr. Chi-Lun Ng, Dr. Shun Weng, Ms. Xiao-Hua Zhang, Prof. Wen-Shou Zhang, and Mr.Yue Zheng.

Several colleagues and research staff, present and former, at The Hong Kong Polytechnic University made contributions to some research works described in this book: Dr. Jun Chen, Dr. Quan-Shun Ding, Prof. An-Xin Guo, Dr. Xiao-Jian Hong, Dr. Liang Hu, Dr. Qi Li, and Mr. Shan Zhan.

Prof. Jan-Ming Ko, former Vice President of The Hong Kong Polytechnic University, and Ir. Chai-Kwong Mak, former Director of the Highways Department of the Government of Hong Kong Special Administrative Region (HKSAR), initiated and approved the collaborative research project "Establishment of Bridge Rating System for Tsing Ma Bridge" between The Hong Kong Polytechnic University and the Highways Department in 2005.

Dr. Kai-Yuen Wong, a senior engineer from the Highways Department, and his team members assisted in numerous ways and played an important role in the collaborative research project "Establishment of Bridge Rating System for Tsing Ma Bridge".

Our work in this research area, some of which has been incorporated into the book, has been largely supported by the Hong Kong Research Grants Council, The Hong Kong Polytechnic University, The Natural Science Foundation of China, and the Highways Department over many years. All the support is gratefully acknowledged.

A vote of thanks must go to Mr Tony Moore, senior editor at Spon Press, for his patience and encouragement from the beginning and during the preparation of this book, and to Mr Simon Bates and Ms. Siobhán Poole for their patience and scrutiny in the editing of this book.

Finally, we are grateful to our families for their help, encouragement and endurance.

Preface

Structural health monitoring technology has been developed for decades, whereas its systematic application to long-span bridges started in the 1990s. The Tsing Ma Bridge in Hong Kong is one of the pioneering long-span bridges that were installed with an advanced health monitoring system. The first author has fortunately carried out long-term collaborative researches on the structural health monitoring of the Tsing Ma Bridge with The Hong Kong Highways Department since 1995. Although the definition of structural health monitoring is still controversial among some people and some early health monitoring systems were not as successful as expected, the authors have noticed that the clients of many newly constructed large-scale bridges have increased the investment in the health monitoring systems and, as a result, the functionality of the monitoring systems is being improved. With years' experience and lessons learned from the Tsing Ma Bridge project and others, the authors feel that it is the proper time to further promote and advance the structural health monitoring technology. However, there is no book on the market suitable for graduate students and practising engineers, although a few books on structural health monitoring have been published in recent years. This is the original incentive of the authors in writing the present book to bridge the gap between theory and practice.

Structural health monitoring systems have been designed for various purposes. It is the authors' opinion that, based on the current health monitoring technology, detecting damage of long-span bridges is very difficult. Therefore, this book does not focus too much on damage detection, but demonstrates other functions of structural health monitoring, for example, monitoring real loading conditions of a bridge, understanding the bridge performance on site, verifying the design rules or assumptions employed, and guiding bridge rating and maintenance. All of these functions will improve the safety and serviceability of the bridge and finally benefit society.

The book mainly presents the monitoring of common loading effects on long-span bridges from railway traffic, highway traffic, temperature, wind, and earthquake. For each loading effect, the sensing technology, theoretical background, and practical implementation are described, followed by the application to the Tsing Ma Bridge. A few other loading effects are also summarized. In consequence, each chapter can be read independently. In this regard, the notations and references are also listed at the end of each chapter. To achieve a comprehensive monitoring, numerical simulation is imperative as the number of sensors is always insufficient for such sizable structures. Consequently structural health monitoring oriented finite element modelling and model updating are introduced prior to describing the monitoring of loading effects. Before this, a general introduction to long-span bridges and structural health monitoring systems is provided at the beginning of this book so that the readers have a complete view on the characteristics of long-span bridges and components of health monitoring systems. Although successful applications of damage detection to large-scale bridges are very limited, various damage detection methods are introduced. A test-bed is then established for examining the feasibility of these damage detection methods in real long-span bridges. The first author mainly contributes to Chapters Five, Six, Eight, Nine, Twelve to Fourteen, and the second author mainly contributes to the others.

In writing the book, the authors are always reminded that the book mainly serves as a textbook for graduate students and practising engineers to understand structural health monitoring. The readers are assumed to have some background in structural analysis, structural dynamics, and mathematics. Although the applications in the book are focused on a long-span bridge, the readers can extend the techniques to high-rise buildings, large-scale spatial structures, flexible towers, offshore structures, dams, and others.

We would be very happy to receive constructive comments and suggestions from readers.

You-Lin Xu and **Yong Xia**
The Hong Kong Polytechnic University
ceylxu@polyu.edu.hk;ceyxia@polyu.edu.hk

February 2011

Foreword (1)

Structural Health Monitoring (SHM) is the integration of a Sensory System, a Data Acquisition System, a Data Processing and Archiving System, a Communication System, and a Damage Detection and Modelling System to acquire knowledge of the integrity of in-service structures on a continuous basis.

In the past, civil engineering professionals have maintained the integrity of structures, bridges, buildings, offshore platforms and power plants by means of manual systems of inspection, nondestructive evaluation (NDE) and interpretation of data using conventional technologies. The profession has relied heavily on evaluation parameters given in codes of practice that are conservative and often costly. This approach has led to the North American infrastructure of aged bridges whose health is not easy to monitor. For example, the structural health of existing nuclear power plant structures cannot be readily evaluated. Also, many bridges and buildings constructed in earthquake prone areas cannot be opened immediately after a seismic event due to the time and cost involved in performing extensive safety checks.

Engineers have long sought for a means to obtain information on how structures behave in service. By incorporating, at the time of construction or subsequently, sensing devices (such as structurally-integrated fibre optic sensors) that provide information on conditions such as strain, temperature, and humidity this can be accomplished. The development of these structurally integrated fibre optic sensors (FOSs) has led to the concept of smart structures.

Recently, methods have been sought for transmitting signals from intelligent structures over telephone lines to a central monitoring station where the signals are interpreted. This procedure is termed remote monitoring and is especially valuable for monitoring the condition of structures situated in remote locations, such as bridges and dams. However, to avoid transmitting a flood of sensing data, the imposition of some form of intelligent processing is required on-site to ensure that only condensed and meaningful information is sent to the central monitoring station.

At present, the discipline of Civil Engineering is too conservative. It is noteworthy that in Aerospace Engineering the risk of acceptance for failure of a structure is 24 per million per year whereas in civil engineering the risk factor is 0.1 per million per year. It would seem that with the ability to closely monitor structures, the aircraft industry has found higher risk factors to be acceptable. This could apply equally well to civil engineering structures now that the ability exists to monitor them.

Since 1990, long-term SHM systems have been implemented in major long-span bridges around the world, however, how to utilise these advanced systems is a practical challenge that many researchers and practitioners are facing.

The present book by Professor You-Lin Xu and Dr. Yong Xia at The Hong Kong Polytechnic University is a unique one that provides a systematic and comprehensive account for SHM in long-span suspension bridges and fills a gap between theory and practice. The book details the design of sensory system, SHM-oriented modelling, monitoring of various loading and environmental effects, damage detection methods, SHM-based rating system, and test-bed establishment. For each loading and environmental effect, the sensing requirement, theoretical background, and practical implementation are described for an efficient monitoring. Moreover, the application of these technologies to a just right example – the Tsing Ma Bridge – is detailed. In such a way, one can easily follow the procedures and apply the technologies to other large-scale bridge health monitoring exercises.

The first author is a well-known expert in the fields of wind engineering and structural health monitoring. The second author also has many years experience in structural health monitoring. The authors present the SHM from a new angle of understanding, that is, the structural health monitoring is not only used for identification of possible damage of a structure, but also can be employed for assessment of loadings and environmental factors, evaluation of structural performance, verification of design criteria, and guidance on inspection and maintenance. As stated by the authors, damage identification of long-span bridges is still not matured at the present stage, whereas other objectives can be realised through a good monitoring exercise, like the SHM of the Tsing Ma Bridge. The information presented in this book will be of special value to the engineers and researchers. The book can serve as a textbook for graduates of the new generation. I would like to extend my personal congratulations to the authors on this extraordinary achievement.

Aftab Mufti, C.M., PhD., FCAE, P.Eng.
President, ISIS Canada

Foreword (2)

Hong Kong has an extensive infrastructure development programme, and ten major infrastructure projects (around HK$250 billion investment) are under way. The construction activities in Mainland China are also booming and will last a few decades. Increasingly developed large-scale and complex civil infrastructures, particularly long-span cable-supported bridges, need huge investment and are critical to the economy of highly urbanised cities such as Hong Kong and special economic zones in Mainland China. The exposure of civil infrastructures to harsh environment and natural hazards thus requires innovations in structural design, construction, maintenance, and management.

Recently developed long-term structural health monitoring technology is one of the cutting-edge technologies for the better understanding of structural performance and the subsequent improvement of safety and serviceability of the structures. In the mid-1990s, I was involved in pioneering a Wind and Structural Health Monitoring System (WASHMS) for the Tsing Ma Bridge when I was working in the Highways Department of the Government of Hong Kong Special Administrative Region (HKSAR). A sophisticated WASHMS for all the three cable-supported bridges in the Tsing Ma Control Area (the Tsing Ma Bridge, the Kap Shui Mun Bridge, and the Ting Kau Bridge) has now been devised, installed, and operated by the Highways Department. To make best use of the WASHMS, the Highways Department has established a long-term collaboration with local universities, in which the Department of Civil and Structural Engineering of The Hong Kong Polytechnic University has carried out extensive studies on the structural health monitoring of the Tsing Ma Bridge. This book presents a systematic investigation and comprises the most important works that the authors and their research team have performed in the last 15 years. Their works demonstrate the pioneering and successful implementation of bridge health monitoring on a real long-span suspension bridge and the mutual respect and effective collaboration between professional engineers, academic researchers, government agencies, contractors and managers.

This book is a valuable contribution to the further development of the field of structural health monitoring. I believe that this book will benefit not only the students and academics but also engineers and other professionals in bridge engineering.

Ir **Chi-Sing Wai**, JP, BSc (Eng), MSc
Permanent Secretary for Development (Works)
The Development Bureau of HKSAR, Hong Kong

CHAPTER 1

Introduction

1.1 THE OBJECTIVE OF STRUCTURAL HEALTH MONITORING

Civil structures, including bridges, buildings, tunnels, pipelines, dams, and many others, provide very fundamental means for a society. Safety and serviceability of these civil structures are therefore basic elements of a civilized society and a productive economy, and also the ultimate goals of engineering, academic, and management communities.

However, many civil structures in-service are in fact defective owing to many factors. For example, they might not be properly designed or constructed according to modern standards, or they have deteriorated due to environmental corrosion and long term fatigue after many years in service. The American Society of Civil Engineers estimated that between one-third and one-half of infrastructure in the United States are structurally deficient, and the investment needs about US$1.6 trillion in a fiveyear period (ASCE, 2005). Due to the economic boom, a huge number of large-scale and complex civil structures such as long-span bridges, high-rise buildings, and large-space structures, have been constructed in China during the past twenty years. However, according to other countries' experiences, enormous cost and effort will be required for maintenance of these structures and for safeguarding them from damage in the next twenty years (Chang et al., 2009). It is crucial that a concerted effort be made to identify practical and effective methods for monitoring, evaluation, and maintenance of civil structures.

The recently developed structural health monitoring (SHM) technology provides a better solution for the problems concerned. The SHM technology is based on a comprehensive sensory system and a sophisticated data processing system implemented with advanced information technology and structural analysis algorithms. The main objectives of the SHM are to monitor the loading conditions of a structure, to assess its performance under various service loads, to verify or update the rules used in its design stage, to detect its damage or deterioration, and to guide its inspection and maintenance (Aktan et al., 2000; Ko and Ni, 2005; Jiang and Adeli, 2005; Brownjohn and Pan, 2008; Xu, 2008). These aspects are addressed as follows.

1. A large-scale civil structure, for example, a long-span suspension bridge, is often located in a unique environment condition. Design loads for a suspension bridge mainly include dead load, traffic load, temperature load, wind load, and seismic load. Except the dead load, which can be determined from the design with a higher accuracy, other loads are usually based on design standards or measured from scaled models in laboratories, which may not fully present the actual ones on

each unique bridge. In addition, idealistic laboratory conditions cannot offer the realistic loading environment that a bridge is located in. Moreover, the loading may vary from time to time (for example, the traffic conditions). With a well designed sensory system, various types of loadings on-site can be directly or indirectly measured and monitored.

2. Although numerical analysis and laboratory experimental techniques have been extensively developed and employed, their accuracy in predicting the structural responses is limited due to idealistic assumptions used in the mathematical models and the size effect of the scaled physical models. For example, scaled models in wind tunnels for a long-span suspension have a much smaller Reynolds number than the practical one so that the laboratory results must be used with great caution. By contrast, the SHM system can provide the prototype bridge responses which can be employed to accurately evaluate the bridge performance and physical condition.

3. Structural design usually adopts some assumptions and parameters, particularly for long-span bridges and skyscrapers whose geometry is beyond the design standards. These assumptions need to be verified through on-site structural monitoring. The results from SHM can also provide more accurate evidence for designing other structures in the future and updating earlier design specifications.

4. Civil structures are exposed to natural and man-made hazards, such as typhoons, strong earthquakes, flood, fire, and collisions. Failure of civil structures could be catastrophic not only in terms of losses of life and economy, but also subsequent social and psychological impacts. Through the online SHM system, structural responses are monitored continuously and the possible abnormality can be detected in the early stage for prevention of potential collapse of the structure and loss of lives.

5. Traditional structural maintenance and management strategies (such as maintenance methods and frequencies) are mainly based on experience. The SHM system can provide the holistic, realistic, and latest condition of the structure so that the management authority can make the effective and efficient inspection and maintenance measures and, finally the benefit to the clients is maximized in the life-cycle.

The above mentioned objectives of the SHM have been recognized and accepted in the community, but the writers would like to clarify a few viewpoints as follows.

First, although numerical analyses and laboratory experiments have some limitations, the SHM does not reduce or downplay their roles in structural health monitoring. For example, laboratory experiments can be well controlled so that a particular issue can be investigated with other factors controlled. It is quite difficult or sometimes impossible for one to perform in this way to a prototype structure. This is because the data measured by an SHM are often the combination of many factors and the dominant factor may be masked. Furthermore, the number of sensors in an SHM system is always limited due to budget and other limitations and some positions are not accessible to install sensors. On the contrary, the numerical analysis offers more detailed spatial information of the monitored parameters. Therefore, integration of numerical analyses, laboratory experiments, and field SHM can fulfil the objectives better.

Second, damage detection methods, especially vibration based methods, have been developed for decades and most dramatically from the 1990s to the early twenty-first century. However, their application in large civil structures is still limited at the present stage (Brownjohn, 2007). Many problems in damage detection of large civil structures have not been solved. For several reasons (for example, redundancy and safety margin), many new built structures have little damage. For sure, damage detection is one important objective of SHM, but not the sole one. More attention should be paid to other objectives of SHM as described above at present.

Third, an SHM system is not just a field measurement. Although one or more field measurements can fulfil some functions of an SHM system, they usually comprise limited types of sensors and carry out measurement in a short time. On the other hand, the SHM is a more systematic exercise that integrates sensing, data network and transmission, data processing, data mining, and structural analysis. Therefore, the SHM is a long-term system and all the parameters of interest are measured continuously. Consequently any changes in the loading, environmental factors, and responses can be captured in real time or near real time, which cannot be achieved by one single or a few periodical measurements.

Based on the above facts and in consideration that long-span suspension bridges are the flagships of our civil infrastructure, this book aims to provide a holistic framework of health monitoring of long-span suspension bridges and demonstrate how to achieve the above-mentioned objectives using the SHM system. The SHM practices in the Tsing Ma suspension bridge with a main span of 1,377 m in Hong Kong will be described in detail as a real example to demonstrate the procedures.

1.2 THE HISTORY OF STRUCTURAL HEALTH MONITORING OF BRIDGES

Major applications of SHM have been implemented in the oil industry, large dams and bridges and have acquired a great deal of research and practical efforts. Residential and commercial structures have received relatively little attention in comparison (Doebling *et al.*, 1996).

Offshore platforms are usually operated in depth water and subjected to extreme environmental loads. Due to the high cost of inspection, vibration-based inspection and damage detection have been employed in offshore structures since 1980s. For example, Duggan *et al.* (1980) used ambient vibration measurements taken during a seven-month period on three offshore platforms in the Gulf of Mexico as a means of structural integrity monitoring.

The majority of monitoring exercises on buildings and towers have been aimed at improving understanding of loading and response including those induced by earthquakes and strong winds. For example, three tall buildings in Chicago have been equipped with monitoring systems to compare their wind-induced responses (measured by GPS and accelerometers) with predictions made using wind tunnels and finite element models and discrepancies have been identified (Kijewski-Correa *et al.*, 2006). Brownjohn and Pan (2008) compared data from a decade of monitoring of a 280 m office tower in Singapore with design code requirements for both wind and seismic effects, showing that code provisions for both types of loadings are very conservative. Earlier studies have reported that more than 150 buildings in California, more than 100 buildings in Japan, and more than 40 buildings in Taiwan have had strong motion monitoring systems for seismic excitation/response measurement and post-earthquake damage assessment installed (Huang and Shakal, 2001; Lin *et al.*, 2003; Huang, 2006).

Bridge monitoring can be dated back to the construction of the Golden Gate and Bay Bridges in San Francisco in the 1930s in USA, in which the dynamic behaviour of the bridges was studied. The collapse of the Tacoma Narrows Bridge in Washington State, USA in 1940 led to the inspection and modification of other suspension bridges, including strengthening the Golden Gate Bridge. The widespread introduction of systematic bridge inspection programmes was directly attributed to the catastrophic bridge collapse at Point Pleasant, West Virginia, USA, in 1967 (Doebling *et al.*, 1996). At present, bridges are generally inspected every two years, largely using visual inspection techniques with the aid of some NDE methods such as acoustic or ultrasonic methods. There is the possibility that damage could not be detected between inspection intervals.

In the last three decades, many attempts have been made to detect structural damage using vibration data, in both time domain and frequency domain (Doebling *et al.*, 1996; Sohn *et al.*, 2003). The vibration-based methods have achieved some success in mechanical and aerospace engineering, whereas their successful applications to large-scale civil structures are very limited (Brownjohn, 2007) due to the uniqueness of civil structures, significant uncertainties of structures, complicated environmental factors, and so forth.

Since the 1990s, long-term monitoring systems have been implemented in major bridges in China, Japan, America, and Europe. In Hong Kong and mainland China alone, more than 40 long-span bridges had been equipped with long-term monitoring systems by 2005 (Sun *et al.*, 2007). Table 1.1 lists major bridges in the world instrumented with real-time monitoring systems (Ko and Ni, 2005; Ou and Li, 2005; Chang *et al.*, 2009).

Table 1.1 Major bridges equipped with health monitoring systems.

No.	Bridge name	Bridge type [1]	Location	Main span (m)	Sensors installed [2]
1	Akashi Kaikyo Bridge	Susp	Japan	1991	(1) (2) (4) (5) (6) (7) (16)
2	Great Belt East Bridge	Susp	Denmark	1624	(1) (2) (3) (4) (5) (9) (12) (13) (20) (21)
3	Runyang South Bridge	Susp	China	1490	(1) (2) (3) (4) (7)
4	Humber Bridge	Susp	UK	1410	(1) (2) (3) (4) (6) (9)
5	Jiangyin Bridge	Susp	China	1385	(1) (2) (3) (4) (5) (7) (10) (11) (14)
6	Tsing Ma Bridge	Susp	China	1377	(1) (2) (3) (4) (5) (7) (8) (13)
7	Golden Gate Bridge	Susp	USA	1280	(1) (4) (16)
8	Minami Bisan-Seto Bridge	Susp	Japan	1100	(4) (7) (9) (16)
9	Forth Road Bridge	Susp	UK	1006	(2) (3) (7) (9) (18)

10	Humen Bridge	Susp	China	888	(3) (7) (12) (13)
11	Ohnaruto Bridge	Susp	Japan	876	(1) (2) (3) (4) (5) (7) (16)
12	Hakucho Bridge	Susp	Japan	720	(1) (4) (16)
13	Gwangan Bridge	Susp	Korca	500	(1) (2) (3) (4) (12) (18) (19) (20)
14	Namhae Bridge	Susp	Korea	404	(1) (2) (3) (4) (12) (19)
15	Tamar Bridge	Susp	UK	335	(1) (2) (3) (13) (19) (20)
16	Youngjong Bridge	Susp	Korea	300	(1) (2) (3) (4) (12) (18) (19) (20)
17	Sutong Bridge	Cable	China	1088	(1) (2) (3) (4) (5) (7) (8) (9) (10) (11) (12) (21) (22)
18	Stonecutters Bridge	Cable	China	1018	(1) (2) (3) (4) (5) (7) (8) (9) (10) (11) (12) (21) (22)
19	Tatara Bridge	Cable	Japan	890	(4) (16)
20	Normandie Bridge	Cable	France	856	(1) (2) (3) (4) (7)
21	3rd Nanjing Yangtze River Bridge	Cable	China	648	(1) (2) (3) (4) (5) (8) (11) (12) (15) (18)
22	2nd Nanjing Yangtze River Bridge	Cable	China	628	(1) (2) (3) (4) (8) (10) (16)
23	Xupu Bridge	Cable	China	590	(2) (3) (4) (8) (13)
24	Rio-Antirio Bridge	Cable	Greece	560	(2) (3) (4) (5) (16) (22) (25)
25	Skarnsundet Bridge	Cable	Norway	530	(1) (2) (3) (4) (5) (12) (20)
26	Oresund Bridge	Cable	Sweden	490	(1) (2) (3) (4) (21)
27	Zhanjiang Bay Bridge	Cable	China	480	(1) (2) (3) (5) (7) (10) (12) (16) (21)
28	Ting Kau Bridge	Cable	China	475	(1) (2) (3) (4) (5) (7) (8) (13) (22)
29	Seohae Bridge	Cable	Korea	470	(1) (2) (3) (4) (12) (18) (19) (20)
30	Dafosi Bridge	Cable	China	450	(2) (3) (4) (5) (11) (13)
31	Rama IX Bridge	Cable	Thailand	450	(1) (2) (3) (4) (5) (7) (11) (12)
32	Kap Shui Mun Bridge	Cable	China	430	(1) (2) (3) (4) (5) (7) (8) (13) (22)
33	Tongling Yangtze River Bridge	Cable	China	432	(1) (2) (3) (4) (5) (11) (13)
34	Donghai Bridge	Cable	China	420	(1) (2) (3) (4) (5) (7) (9) (17) (18) (19) (21)

35	Hitsuishijima Bridge	Cable	Japan	420	(4) (16)
36	Runyang North Bridge	Cable	China	406	(1) (2) (3) (4)
37	Fred Hartman Bridge	Cable	USA	381	(1) (2) (3) (4) (5)
38	Sunshine Skyway Bridge	Cable	USA	366	(2) (3) (5) (7) (9)
39	Songhua River Bridge	Cable	China	365	(1) (2) (3) (4) (11)
40	Jindo Bridge	Cable	Korea	344	(1) (2) (3) (4) (11) (12) (16) (18) (19) (20)
41	Wuhu Bridge	Cable	China	312	(2) (3) (4) (5) (11) (13)
42	Binzhou Yellow River Bridge	Cable	China	300	(1) (2) (3) (7) (11)
43	Bayview Bridge	Cable	USA	274	(9) (18)
44	Samcheonpo Bridge	Cable	Korea	230	(1) (2) (3) (4) (12) (19) (20)
45	Pereria-Dos Quebradas Bridge	Cable	Columbia	211	(1) (2) (3) (4) (5) (9) (12) (19) (21)
46	Shenzhen Western Corridor	Cable	China	210	(1) (2) (3) (4) (5) (7) (8) (9) (21) (22)
47	Flintshire Bridge	Cable	UK	194	(1) (2) (3) (4) (5) (9)
48	New HaengJu Bridge	Cable	Korea	160	(1) (2) (3) (4) (12) (14) (19)
49	4th Qianjiang Bridge	Arch	China	580	(1) (2) (3) (4) (5) (8) (10) (15)
50	Lupu Bridge	Arch	China	550	(2) (3) (4) (13)
51	Banghwa Bridge	Arch	Korea	540	(1) (2) (3) (4) (5) (12)
52	Maocao Street Bridge	Arch	China	368	(1) (2) (3) (4)
53	Yonghe Bridge	Arch	China	338	(1) (2) (3) (4) (5) (7) (22)
54	Commodore Barry Bridge	Truss	USA	548	(1) (2) (3) (4) (5) (8) (12) (19) (21) (22)
55	Ironton-Russell Bridge	Truss	USA	241	(2) (3)
56	1st Nanjing Yangtze River Bridge	Truss	China	160	(1) (2) (3) (4) (5) (8) (16)
57	Confederation Bridge	Box	Canada	250	(1) (2) (3) (4) (5) (8) (9) (11) (12) (13) (22)
58	Foyle Bridge	Box	UK	234	(1) (2) (3) (4) (5) (23)
59	New Benicia Martinez Bridge	Box	USA	201	(2) (3) (4) (9) (12) (14)
60	Saint Anthony Falls I–35W Bridge	Box	USA	154	(2) (3) (4) (9) (11) (24)
61	North Halawa Valley Bridge	Box	USA	110	(2) (3) (5) (12)
62	Dongying Yellow River	Rigid	China	220	(2) (3) (4) (11)

	Bridge				
63	Guangyang Island Bridge	Rigid	China	210	(2) (3) (11)

Note [1]: Susp – Suspension bridge; Cable – Cable-stayed bridge; Arch – Arch bridge; Truss – Steel truss bridge; Box – Box girder bridge; Rigid – Continuous rigid-frame bridge
Note [?]: (1) – anemometer; (2) – temperature sensor; (3) – strain gauge; (4) – accelerometer; (5) – displacement transducer; (6) – velocimeter; (7) – global positioning system; (8) – weigh-in-motion sensor; (9) – corrosion sensor; (10) – elasto-magnetic sensor; (11) – optic fibre sensor; (12) – tiltmeter; (13) – level sensing station; (14) – dynamometer; (15) – total station; (16) – seismometer; (17) – fatigue meter; (18) – cable tension force; (19) – joint meter; (20) – laser displacement sensor; (21) – meteorological station; (22) – video camera; (23) – jacking pressure sensor; (24) – potentiometer; (25) – water-level sensor

Among these bridges with the SHM system, the following two projects are worthy of special attention.

One is the Wind And Structural Health Monitoring System (WASHMS) for the Tsing Ma Bridge, Kap Shui Mun Bridge, and Ting Kau Bridge in Hong Kong, China (Wong, 2004). This may be one of the pioneer systematic SHM systems on long-span bridges. It consisted in total of 819 sensors in nine major types: anemometers, temperature sensors, dynamic weigh-in-motion sensors, accelerometers, displacement transducers, level sensing stations, strain gauges, GPS, and CCTV video cameras. The GPS with 29 stations was installed on the bridges in 2001. The systems have operated very well since the beginning from 1997. The SHM system on the Tsing Ma Bridge will be described in detail in later chapters of this book.

Another is the SHM system for the Stonecutters Bridge in Hong Kong, which is composed of 1,571 sensors in 15 types (some types of sensors are merged in Table 1.1). This may be the most comprehensive SHM system on bridges in the world. The system was devised to monitor and evaluate four main categories of quantities including environmental loads, operation loads, bridge features, and bridge responses (Wong, 2010).

Although many SHM systems have been installed on long-span bridges, there is a lack of corresponding legal standards and codes for practitioners. Some practical guidelines, handbooks, and research books have been published by academics. These include *Guidelines for Structural Health Monitoring* (Mufti, 2001), *Development of a Model Health Monitoring Guide for Major Bridges* (Aktan *et al.*, 2002), *Structural Health Monitoring with Piezoelectric Wafer Active Sensors* (Giurgiutiu, 2008), *Encyclopedia of Structural Health Monitoring* (Boller *et al.*, 2009), *Structural Health Monitoring of Civil Infrastructure Systems* (Karbhari and Ansari, 2009), and *Health Monitoring of Bridges* (Wenzel, 2009).

The rapid development of the SHM has supported a few regional and international series conferences in the related areas. The first one may be *The International Workshop on Structural Health Monitoring* that is held at Stanford University every two years starting from 1997. The *European Workshop on Structural Health Monitoring* takes place every two years from 2002. Since 2003, *The International Conference on Structural Health Monitoring and Intelligent Infrastructure* has been organized every two years by the International Society for Structural Health Monitoring of Intelligent Infrastructure. The International

Association on Structural Control and Monitoring organizes *The World Conference on Structural Control and Monitoring* every four years. The Asian-Pacific Network of Centers for Research in Smart Structures Technology has *The International Workshop on Advanced Smart Materials and Smart Structures Technologies*. Besides, more and more conferences have included the topics on SHM, such as the conferences organized by the International Society for Optics and Photonics.

1.3 RECENT DEVELOPMENTS IN STRUCTURAL HEALTH MONITORING

Several recent trends in SHM practice for large-scale bridges are worth mentioning.

• The recently devised long-term health monitoring systems emphasize multi-purpose monitoring of the bridge integrity, durability, and reliability. For example, the monitoring system for the Stonecutters Bridge includes sensors measuring environment, such as corrosion sensors, rainfall gauges, barometers, hygrometers, and pluviometers to facilitate bridge safety/reliability assessment (Ni *et al.*, 2011).
• For some recent bridges such as the Shenzhen Western Corridor and the Stonecutters Bridge, the design of a monitoring system is required in the tender as part of the bridge design. Integration of bridge design and monitoring system design ensures that design engineers' important concerns are reflected in the monitoring system while civil provisions for implementing a monitoring system are considered in the bridge design.
• The implementation of long-term monitoring systems on new bridges, such as the Stonecutters Bridge and the Sutong Bridge, is accomplished in synchronism with the construction progress. In this way some specific types of sensors, e.g., corrosion sensors, strain gauges, and fibre optic sensors, can be embedded into the bridge during certain bridge erection stages. In addition, the construction monitoring results provide initial values of parameters. This is especially important for the strain gauges such that the absolute strain rather than relative strain can be obtained (Xia *et al.*, 2011). Consequently the health condition of the structural components and the impact of extreme events (typhoons, earthquakes, man-made disasters, etc.) on the structural performance can be evaluated realistically. Moreover, the construction monitoring can provide a complete 'birth' archive of the bridge as the in-construction state may be different from the as-design state. Integration of the construction monitoring ensures the later long-term monitoring will be more accurate and effective.

1.4 ORGANIZATION OF THE BOOK

The objectives of health monitoring of long-span bridges have been defined and relevant background materials have been provided in this chapter. Chapter Two introduces the long-span suspension bridges including the historic development, structural configuration, and design criteria. Chapter Three introduces the components and design of the structural health monitoring system. Finite element modelling of long-span bridges is described in Chapter Four. Chapter Five to

Chapter Nine respectively describe the monitoring of railway traffic, highway traffic, temperature effect, wind effect, and seismic effect. In each chapter, the sensing techniques, theoretical background, data analysis, and real application to the Tsing Ma suspension bridge are detailed. Consequently these chapters are rather independent of the others and can be used individually. Except the above loading effects, monitoring of other loading effects is summarized in Chapter Ten. In Chapter Eleven, damage detection methods are described, followed by the bridge rating with application to the Tsing Ma Bridge. Due to the current difficulties encountered in application of damage detection methods to long-span bridges, a test-bed is established and presented in Chapter Thirteen. Finally challenges and prospects of SHM of long-span suspension bridges are highlighted.

1.5 REFERENCES

Aktan A.E., Catbas F.N., Grimmelsman K.A. and Tsikos C.J., 2000, Issues in infrastructure health monitoring for management. *Journal of Engineering Mechanics, ASCE*, **126**, pp. 711-724.

Aktan, A.E., Catbas, F.N., Grimmelsman, K.A. and Pervizpour, M., 2002, *Development of a Model Health Monitoring Guide for Major Bridges*, (Philadelphia: Drexel Intelligent Infrastructure and Transportation Safety Institute).

ASCE, (2005), *2005 Report Card for America's Infrastructure*, (Reston, VA: America Society of Civil Engineers).

Boller C., Chang F.K. and Fujino Y., 2009, *Encyclopedia of Structural Health Monitoring*, (Chichester: John Wiley & Sons).

Brownjohn, J.M.W., 2007, Structural health monitoring of civil infrastructure. *Philosophical Transactions of the Royal Society A*, **365(1851)**, pp. 589-622.

Brownjohn, J.M.W. and Pan, T.C., 2008, Identifying loading and response mechanisms from ten years of performance monitoring of a tall building. *Journal of Performance of Constructed Facilities, ASCE*, **22(1)**, pp. 24-34.

Chang, S.P., Yee, J.Y. and Lee, J., 2009, Necessity of the bridge health monitoring system to mitigate natural and man-made disasters. *Structure and Infrastructure Engineering*, **5(3)**, pp. 173-197.

Doebling, S.W., Farrar, C.R., Prime, M.B. and Shevitz, D.W., 1996, *Damage Identification and Health Monitoring of Structural and Mechanical Systems from Changes in Their Vibration Characteristics: A Literature Review,* (Los Alamos: Los Alamos National Laboratory).

Duggan, D.M., Wallace E.R. and Caldwell S.R., 1980, Measured and predicted vibrational behavior of Gulf of Mexico platforms. In *Proceedings of the 12th Annual Offshore Technology Conference*, Houston, Texas, pp. 92-100.

Giurgiutiu, V., 2008, *Structural Health Monitoring with Piezoelectric Wafer Active Sensors*, (Oxford: Elsevier Academic Press).

Huang, M.J. and Shakal, A.F., 2001, Structure instrumentation in the California strong motion instrumentation program. *Strong Motion Instrumentation for Civil Engineering Structures*, **373**, pp. 17-31.

Huang, M.J., 2006, Utilization of strong-motion records for post-earthquake damage assessment of buildings. In *Proceedings of the International Workshop on Structural Health Monitoring and Damage Assessment*, Taichung, Taiwan, pp. IV1-IV29.

Jiang, X. and Adeli, H., 2005, Dynamic wavelet neural network for nonlinear identification of high-rise buildings. *Computer-Aided Civil and Infrastructure Engineering*, **20(5)**, pp. 316-330.

Karbhari, V.M. and Ansari, F., 2009, *Structural Health Monitoring of Civil Infrastructure Systems*. (Cambridge: Woodhead Publishing Limited).

Kijewski-Correa, T., Kilpatrick, J., Kareem, A., Kwon, D.K., Bashor, R., Kochly, M., Young, B.S., Abdelrazaq, A., Galsworthy, J., Isyumov, N., Morrish, D., Sinn, R.C. and Baker, M.F., 2006, Validating wind-induced response of tall buildings: synopsis of the Chicago full-scale monitoring program. *Journal of Structural Engineering, ASCE*, **132(10)**, pp. 1509-1523.

Ko J.M. and Ni Y.Q., 2005, Technology developments in structural health monitoring of large-scale bridges. *Engineering Structures, ASCE*, **27**, pp. 1715-1725.

Lin, C.C., Wang, C.E. and Wang, J.F., 2003, On-line building damage assessment based on earthquake records. In *Proceedings of the First International Conference on Structural Health Monitoring and Intelligent Infrastructure*, Tokyo, Japan, pp. 551-559.

Mufti, A., 2001, *Guidelines for Structural Health Monitoring*, (Winnipeg, Canada: Intelligent Sensing for Innovative Structures).

Ni., Y.Q., Wong, K.Y. and Xia, Y., 2011, Health checks through landmark bridges to sky-high structures. *Advances in Structural Engineering*, **14(1)**, pp. 103-119.

Ou, J.P. and Li, H., 2005, The state-of-the art and practice of structural health monitoring for civil infrastructures in the mainland of China. In *Proceedings of the 2nd International Conference on Structural Health Monitoring of Intelligent Infrastructure*, Shenzhen, China, pp. 69-94.

Sohn, H., Farrar, C.R., Hemez, F.M., Shunk, D.D., Stinemates, S.W., Nadler, B.R. and Czarnecki, J.J., 2003, *A Review of Structural Health Monitoring Literature form 1996-2001*, (Los Alamos: Los Alamos National Laboratory).

Sun, L.M., Sun, Z., Dan, D.H. and Zhang, Q.W., 2007, Large-span bridges and their health monitoring systems in China. In *Proceedings of 2007 International Symposium on Integrated Life-cycle Design and Management of Infrastructure*, Shanghai, China, edited by Fan, L.C., pp. 79-95.

Xia, Y., Ni, Y.Q., Zhang, P, Liao, W.Y. and Ko, J.M., 2011, Stress development of a super-tall structure during construction: Numerical analysis and field monitoring verification. *Computer-Aided Civil and Infrastructure Engineering*, accepted.

Xu, Y.L., 2008, Making good use of structural health monitoring systems: Hong Kong's Experience. In *Proceedings of the Second International Forum on Advances in Structural Engineering, Structural Disaster Prevention, Monitoring and Control*, Dalian, China, edited by Li, H.N., (Beijing, China: China Architecture & Building Press), pp. 159-198.

Wenzel, H., 2009, *Health Monitoring of Bridges*, (New York: John Wiley & Sons).

Wong, K.Y., 2004, Instrumentation and health monitoring of cable-supported bridges. *Structural Control and Health Monitoring*, **11(2)**, pp. 91-124.

Wong, K.Y., 2010, Structural health monitoring and safety evaluation of Stonecutters bridge under the in-service condition. In *Proceedings of The 5th*

International Conference on Bridge Maintenance, Safety and Management, edited by Frangopol, D.M, Sause, R. and Kusko, C.S., (Balkema: CRC Press), pp. 2641-2648.

CHAPTER 2

Long-span Suspension Bridges

2.1 PREVIEW

Bridge engineering has been developed for centuries. According to the structural configuration, bridges can be mainly categorized into five types: beam bridges, cantilever bridges, arch bridges, truss bridges, and cable-supported bridges. Cable-supported bridges include cable-stayed bridges and suspension bridges, which are competitive for long spans. The longest cable-stayed bridge in the world is currently the Sutong Bridge with a main span of 1,088 m, located in Jiangsu Province, China and opened to the public in 2008. The longest suspension bridge in the world as of 2010 is the Akashi-Kaikyo Bridge with a main span of 1,991 m, located in Kobe, Japan and completed in 1998. The suspension bridge is best suited for extending span lengths even farther.

This chapter will overview the historic development of long-span suspension bridges, and describe the common types of loadings that long-span suspension bridges are subjected to. Structural configuration, design aspects, modern construction methods, and maintenance measures of long-span suspension bridges are subsequently outlined, followed by an introduction to the Tsing Ma suspension bridge in Hong Kong.

2.2 THE HISTORY OF LONG-SPAN SUSPENSION BRIDGES

The ancient Chinese employed ropes or iron chains to construct structures over rivers or obstacles 2000 years ago. It was reported that the first suspension bridge was built in the United States in 1796 (Xanthakos, 1993). In the 19th century, many suspension bridges were designed and constructed. Thomas Telford designed the Menai Bridge connecting the island of Anglesey and mainland of Wales with a main span of 176 m, which was opened in 1826. Brunel designed the Clifton Bridge with a main span of 214 m, which was opened in 1864 after his death. It should be noted that the cables in this bridge consisted of iron eye chains and the tower was made of stone. Ellet designed and built the Wheeling Bridge over the Ohio River in 1849 with the main span of 308 m.

Another notable bridge engineer, John A. Roebling, designed the Cincinnati-Covington Bridge over the Ohio River, which was completed in 1866 and had the main span of 322 m. He also designed the famous Brooklyn Bridge across the East River in New York City with three spans of 286 + 486 + 286 m, which was completed in 1883. The Brooklyn Bridge is believed to be the ancestor of the modern suspension bridges. In it, steel wires were used for the first time and the aerial spinning method was invented for constructing parallel cables. This method is still widely used in construction of modern suspension bridges. Moreover, stay cables were employed to strengthen the suspension. The stay cables not only bore external forces, but also increased the stability of the bridge deck. The second bridge over the East River is the Williamsburg Bridge with spans of 284 + 488 + 284 m, opened in 1903. This was the first bridge to use steel towers instead of masonry. Different from the modern suspension bridges, the main cables of both bridges passed beneath the top chord of the stiffening truss and went down to the bottom chord of the truss at the centre of the main span (Gimsing, 1997). In this stage, the bridges were analysed based on the elastic theory.

The third bridge across the East River is the Manhattan Bridge, designed by L.S. Moisseiff. This was the first bridge to use the deflection theory in its design, which considered the deflection of the main cables under traffic load in calculating the bending moment of the stiffening girder (Gimsing, 1997). It resulted in smaller bending moment in the stiffening girder, and thus a lighter and shallower stiffening girder and less materials in construction. Development of the deflection theory led to the breakthrough of suspension bridges. The first suspension bridge over 1,000 m came in 1931: the George Washington Bridge over the Hudson River with a main span of 1,067 m. It nearly doubled the previous record. Later the San Francisco-Oakland Bay Bridge and Golden Gate Bridge were constructed in 1936 and 1937, respectively. In particular, the Golden Gate Bridge, as shown in Figure 2.1 with a main span of 1,280 m, is widely recognized. It is regarded as one of the most beautiful bridges, from the aspects of both structural design and aesthetic appearance.

In the design of these bridges, the wind load was considered a static force on the bridges. The importance of torsional rigidity to the aerodynamic stability had not been realized yet and the flexural rigidity of the stiffening girder was small. This concept was extended further to the Tacoma Narrows Bridge that was designed by L. S. Moisseiff by adopting a very shallow and narrow plate girder. The main span of the bridge collapsed four months after its opening in 1940 under a wind speed of 19 m/s. The failure of the bridge boosted the research in the field of bridge aerodynamics. This has influenced the design of the long-span bridges since the 1940s, including strengthening the Golden Gate Bridge.

In the subsequent twenty years, the truss-type stiffening girders with large flexural rigidity and torsional rigidity dominated long-span suspension bridges especially in the United States. The depth-to-span ratio usually ranged from 1/100 to 1/150. In 1964, the upper level of the double-decked Verrazano Narrows Bridge was opened and took over the world record for length (1,298 m) from the Golden Gate Bridge.

In the 1960s, the Severn Bridge over the River Severn in UK with the main span of 988 m was designed. A new box girder was employed with a depth-to-span ratio of 1/324 only. The shallow wide steel box girder had a streamlined shape to reduce the wind load and had a sufficient torsional rigidity. After that, the streamline box girder was used popularly, for example, Little Belt Bridge in Denmark with the main span of 600 m, and the Bosphorus Bridge connecting Europe and Asia with a 1,074 m main span. In 1981, the longest suspension bridge, Humber Bridge with the main span of 1,410 m, was built in England. The tower was made of concrete, instead of steel used in the preceding bridges.

Figure 2.1 The Golden Gate Bridge (Photo courtesy of Rich Niewiroski Jr., from Wikipedia)

Rapid development of computational techniques and hardware allow more detailed, accurate, and realistic analysis of large-scale bridges. In particular, with the finite element approach, global bridges can be modelled and nonlinearity can be considered. In addition, wind tunnel tests and computational fluid dynamics (CFD) have helped the design to ensure the flutter stability of long-span suspension bridges.

The 1990s saw the construction of quite a few long-span suspension bridges around the world. In Europe, the 1,624 m long main span Great Belt East Bridge was completed in 1997. In China, the 1,385 m long main span Jiangyin Bridge across the Yangtze River was opened in 1999. The Akashi Kaikyo Bridge (Figure 2.2), the longest suspension bridge and also the longest of any type, was completed in 1998 in Japan with the main span of 1,991 m. In the new century, two more long-span suspension bridges, the 1,490 m long main span Runyang Bridge and the 1,650 m long main span Xihoumen Bridge, were completed respectively in 2005 and 2009, both in China. It is worth noting that the Messina Strait Bridge with a center span of 3,300 m has been proposed in Italy (Diana, 1993). The bridge girder is 60 m wide and made of three oval box girders which support the highway and railway traffic.

The ten longest suspension bridges in the world as of 2010 are listed in Table 2.1, according to the length of the main span between the bridge towers. The development of the suspension bridges in terms of their main span lengths is illustrated in Figure 2.3.

Figure 2.2 The Akashi Kaikyo Bridge (photo courtesy of Kim Rötzel, from Wikipedia)

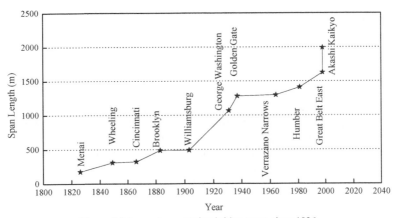

Figure 2.3 Longest suspension bridge spans since 1826

Table 2.1 List of ten longest suspension bridges as of 2010.

Rank	Name	Country	Main span (m)	Year opened
1	Akashi Kaikyo Bridge	Japan	1,991	1998
2	Xihoumen Bridge	China	1,650	2009
3	Great Belt East Bridge	Denmark	1,624	1997
4	Runyang Bridge	China	1,490	2005
5	Humber Bridge	UK	1,410	1981
6	Jiangyin Bridge	China	1,385	1999
7	Tsing Ma Bridge	China	1,377	1997

8	Verrazano Narrows Bridge	USA	1,298	1964
9	Golden Gate Bridge	USA	1,280	1937
10	Yangluo Bridge	China	1,280	2007

2.3 THE BUILT ENVIRONMENT AND LOADING CONDITIONS

A long-span suspension bridge is often located in a unique environment condition. Design loads for a suspension bridge mainly includes dead load, traffic load, wind load, seismic load, and temperature load. Others like erection load, impact load, and support movement may also be considered in some cases. A brief description of loading conditions is given in this section.

2.3.1 Dead Load

Dead load typically dominates the forces on the main components of the bridge. It includes weights of all components of the structure, appurtenances, and utilities attached thereto, earth cover, wearing surface, future overlays, and planned widenings (AASHTO, 2005).

2.3.2 Railway Load

When a long suspension bridge carries railways, the bridge is subjected to moving loads of railway vehicles, which include vertical forces of railway vehicles, longitudinal forces from acceleration or deceleration of vehicles, lateral forces caused by irregularities at the wheel-to-rail interface, and centrifugal forces due to track curvature. Railway vehicles vary greatly with respect to weight, number of axles, and axle spacing. This variability requires a representative live load model for design that provides a safe and reliable estimation of characteristics of railway vehicles within the design life of the bridge (Unsworth, 2010). For both bridge safety and vehicle comfort assessment, the interaction between the railway vehicles and long-span suspension bridge is important (Frýba, 1996).

In design of railroad bridges, the Cooper E80 load (AREMA, 1997) is the most common design live load. It consists of a series of point loads simulating locomotive wheels, followed by a uniformly distributed load of 8 kip per linear foot, equivalent to 14.6 kN/m. Monitoring of railway effects will be described in Chapter 5.

2.3.3 Highway Load

When a long-span suspension bridge carries a highway, the bridge is therefore subjected to a variety of non-stationary loads due to motorcycles, cars, buses, trucks, and heavy goods vehicles. Highway loadings are rather complicated. The effect of highway loadings on a suspension bridge is a function of several parameters, such as the axle loads, axle configuration, gross vehicle weight, number of vehicles, speed of vehicles, and the bridge configuration.

The bridge responses under highway loadings can be analysed using the moving load model (Timoshenko *et al.*, 1974), the moving mass model (Blejwas *et al.*, 1979), and the advanced vehicle-bridge interaction models with consideration of the road roughness (Yang and Lin, 1995; Cheung *et al.*, 1999).

In *American Association of State Highway and Transportation Officials* (AASHTO) specification (AASHTO, 2005), Highway Load '93' or HL93 is adopted as the vehicular live loading of highway bridges, which is a combination of design truck or design tandem and design lane load. The AASHTO design truck is shown in Figure 2.4. The axle spacing between the two 145 kN loads can be varied between 4.3 m and 9.0 m to create a critical condition for the design of each location in the structure. The AASHTO design tandem consists of two 110 kN axles spaced at 1.2 m apart. The transverse spacing of wheels shall be taken as 1.8 m. The design lane load is equal to a load of 9.3 kN/m uniformly distributed over a 3 m width. In Eurocode 1 (2003), the normal highway load model comprises a tandem axle system acting in conjunction with a uniformly distributed load.

Traffic running on bridges produces a stress spectrum which may cause fatigue. Fatigue load models of vertical forces are defined in Eurocode 1 (2003) and AASHTO specification (AASHTO, 2005). Monitoring of highway effects will be described in Chapter 6.

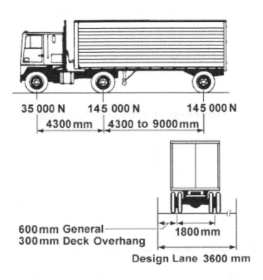

Figure 2.4 Design truck in AASHTO load and resistance factor design

2.3.4 Temperature Effects

Bridges are subjected to daily and seasonal environmental temperature effects (or thermal effects) induced by solar radiation and ambient air temperature. Variation in temperature of bridge components will cause movements and, usually, thermal stress due to indeterminacy and non-uniform distribution of temperature.

The temperature effects on a structure are dependent on the temperature distribution, structural configuration and boundary conditions, and material mechanical properties of the structure. The local climatic conditions, structural orientation, structural configuration and material thermal properties will affect the structural temperature distribution, which may be divided into three components:

- a uniform temperature component;
- a linearly varying temperature gradient component; and
- a nonlinear temperature gradient component.

A uniform temperature change will result in a change in length for an unrestrained structure. The linearly varying temperature gradient component will produce a curvature of the element. The nonlinear temperature gradient component results in self-equilibrated stresses which have no net load effect on the element. The uniform temperature ranges are determined in design specifications according to the climatic conditions and material types of the bridge. Monitoring of temperature effects will be described in Chapter 7.

2.3.5 Wind Load

Wind load is particularly important for long-span suspension bridges. Strong wind may induce instability and excessive vibration in bridges. Wind effects on a long-span suspension bridge are mainly due to static forces induced by mean winds, buffeting excitation, flutter instabilities, and vortex shedding excitation. These types of instability and vibration may occur alone or in combination (Cai and Montens, 2000).

Buffeting action on a long suspension bridge is caused by fluctuating winds that appear within a wide range of wind speeds. In wind resistance design of a long-span suspension bridge, the buffeting responses are normally dominant to determine the size of structural members. In addition to buffeting action, the self-excited forces induced by wind-structure interaction are also important because the additional energy injected into the oscillating structure by self-excited forces increases the magnitude of vibrations. The buffeting response prediction can be performed in both the frequency domain (Davenport, 1962; Scanlan, 1978) and the time domain (Bucher and Lin, 1988; Chen *et al.*, 2000). Flutter instabilities may occur in several types that occur at very high wind speeds as a result of the dominance of self-excited aerodynamic forces. The flutter instabilities always involve torsional motions, and the most common consideration of the flutter for a long-span suspension bridge is the coupled translational-torsional form of instability. In the design stage, a critical flutter speed of the bridge shall be determined and shall exceed, by a substantial margin, the design wind speed of the bridge site at the deck height (Holmes, 2007). Vortex shedding excitation can induce significant, but limited, amplitude of vibration of a long-span suspension bridge in low wind speed and low turbulence conditions. Scanlan's model can be used for calculating the vortex-shedding force (Simiu and Scanlan, 1996).

Design codes usually provide wind loads as a function of design wind velocity at a reference height above the ground or sea level. The mean wind velocity profile within the atmospheric boundary layer is approximated by a logarithmic or power law, while the former is preferred in the new codes (AASHTO, 2005; Eurocode 1, 2005). Terrain conditions should be considered in determination of the design wind velocity.

For large-scale cable-supported bridges (including cable-stayed bridges), appropriate wind tunnel tests are usually required to simulate the wind environment, determine the wind characteristics, and examine the responses of the bridge to various winds. Nevertheless, the on-site wind monitoring can provide a more realistic wind environment and behaviours of the bridge. Monitoring of wind effects will be described in Chapter 8.

2.3.6 Seismic Effects

In a region prone to earthquakes, the seismic design is also important. As the fundamental frequency of long-span suspension bridges is generally low, the seismic load is relatively small. During the construction of the Akashi Kaikyo Bridge, the Kobe Earthquake which occurred in 1995 caused a new fault in the seabed below the bridge and the towers moved by 1 metre. Fortunately, no damage was reported to the bridge itself.

According to the AASHTO *Load and Resistance Factor Design* specification (AASHTO, 2005), earthquake loads are specified as the horizontal force effects and are given by the product of the elastic seismic response coefficient and the equivalent weight of the superstructure, and divided by a response modification factor. The elastic seismic response coefficient is a function of the acceleration coefficient determined from the contour map of the region or nation, period of vibration, and site coefficient. Monitoring of seismic effects will be described in Chapter 9.

2.3.7 Other Effects

Besides the common types of loading described above, some bridges may also be subject to other loading effects, for example, ship collision and ice load, depending on the particular environmental conditions of the bridge. A few of them are outlined as follows.

2.3.7.1 Corrosion

Corrosion is the deterioration of a metal that results from a reaction with the environment. This reaction is an electrochemical oxidation process that usually produces rust. In bridges, corrosion may occur in structural steels, reinforcing bars, and strands in cables. The corrosion of the reinforcing steel is considered to be the primary contributor to the deterioration of highway bridge decks (Mark, 1977).

For protection against corrosion, structural steels must be self-protecting or have a coating system or cathodic protection. Reinforcing bars in concrete components must be protected by epoxy, galvanized coating, concrete cover, or painting. Prestressing strands in cable ducts are usually grouted against corrosion. Chapter 10 has more on corrosion monitoring.

2.3.7.2 Vessel Collision

According to the AASHTO *Load and Resistance Factor Design* (AASHTO, 2005), all bridge components in a navigable waterway crossing, located in design water depths not less than 600 mm, must be designed for vessel impact. The vessel collision loads should be determined on the basis of the bridge importance and characteristics of the bridge, vessels and waterway.

In AASHTO's specifications (AASHTO, 1991; 2005), the head-on ship collision impact force on a pier is taken as a static equivalent force. In Eurocode 1 (2006), a dynamic analysis or an equivalent static analysis is suggested for inland waterways and sea waterways. This topic will be detailed in Chapter 10 as well.

2.3.7.3 Hydraulics

The design of hydraulics involves the hydrology study, hydraulic analysis, drainage design, and bridge scour evaluation. Scour is actually not a force effect, but changes the conditions of the substructure of the bridge and consequently changes the force effects. It was reported that the most frequent causes of bridge failures were attributed to hydraulics (flood and scour) and collisions (Harik *et al.*, 1990; Wardhana and Hadipriono, 2003). Scour monitoring will be described in Chapter 10.

2.3.7.4 Ice Load

Ice load should be considered in cold regions where the sea is covered with ice during winter. Ice loads were the dominant loading condition for the design of the piers of the Confederation Bridge in Canada (Brown *et al.*, 2010). Dynamic forces occur when a moving ice floe strikes a bridge pier. The forces are dependent on the size of floe, and the strength and thickness of the floe. AASHTO (2005) specification suggests the horizontal force on the basis of field measurement, and vertical forces due to ice adhesion.

2.4 STRUCTURAL DESIGN, CONSTRUCTION AND MAINTENANCE

2.4.1 Structural System

A suspension bridge basically consists of stiffening girders/trusses, main cables and hangers (or suspenders), towers, and anchorages (Gimsing, 1997). Figure 2.5 illustrates a typical suspension bridge.

The main cables and hangers support the bridge girders, and the towers in turn support the main cables and transfer the loads to foundations. Flexible-type towers have predominated among main towers in recent long-span suspension bridges. They can be portal or diagonally braced types. The Xihoumen Bridge, Great Belt East Bridge, Runyang Bridge, and Tsing Ma Bridge use the former; the Akashi Kaikyo Bridge uses the latter; and main towers of the Golden Gate Bridge have the combined type. Towers are generally made of steel or concrete. The Akashi Kaikyo Bridge, Golden Gate Bridge, and Verrazano Narrows Bridge consist of steel towers. Examples of concrete towers include the Great Belt East Bridge, Tsing Ma Bridge, and Humber Bridge.

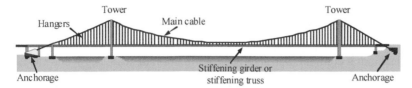

Figure 2.5 Components of a suspension bridge

Main cables are made of a group of parallel-wire bundled cables to support stiffening girders/trusses through hangers. In the modern suspension bridges, galvanized steel wires are usually bundled into a circle and protected by a Polyethylene tube outside. For example, the main cables of the Akashi Kaikyo Bridge are 1.12 m in diameter and each consists of 36,830 wires of 5.23 mm in diameter. Cable saddles support the main cable at the top of the main towers, known as the tower saddle, and at the splay bents in the anchorages, known as the splay saddle. Hanger ropes, vertical or slightly inclined, are fixed to the main cable with the cable band and clamp.

Stiffening girders/trusses are longitudinal components which support the operating loads such as traffic and provide aerodynamic stability of the bridge. Trusses or box girders are usually adopted in long-span suspension bridges. The former generally has higher section than the latter. The stiffening girders can be continuous over the towers or pin-pin connected with the towers.

Anchorages act as end supports of main cables. It can be classified into gravity and tunnel anchorage systems. The former resists the tension of the main cables via the mass of the anchorage itself. The latter takes the tension of the main cables into the ground.

2.4.2 Design Consideration and Criteria of Suspension Bridges

2.4.2.1 General

The classic design method of suspension bridges is based on the elastic theory or deflection theory; both are in-plane analytical methods. In both theories, the entire bridge is regarded as a continuous body, the cable is completely flexible, and the dead load of the stiffening girder and the cables is uniform. The difference between the two theories is that cable deflection resulting from the live load is considered in the deflection theory, which results in smaller bending moment in the stiffening girder. Rapid development of computational techniques allows more detailed and accurate analysis of modern suspension bridges with the finite displacement method.

Design of a suspension bridge should consider many factors, such as geology condition, wind environment, water condition, navigation, climatic conditions, bridge function, and design life. These factors affect the loading effects of the bridge and thus the configuration, material, construction, and maintenance of the bridge. Various types of loading in bridges have been described in the last section. Generally speaking, dead load, traffic load, wind load, and seismic load are most important for most suspension bridges.

General design procedure for a suspension bridge is as follows (Harazaki *et al.*, 2000):

1. Select initial configuration such as determination of span length and cable sag and so on.
2. Structural global analysis: A two-dimensional model can be used for in-plane analysis and a three-dimensional model can be used for out-of-plane analysis. Seismic responses are calculated using response spectrum or time history analysis
3. Member design: The towers, cables, and girders are designed from the previous analyses.
4. Verification of aerodynamic stability: Aerodynamic stability is investigated through buffeting analysis and wind tunnel tests.

In most countries, limit state design is adopted in national standards of structural engineering, which requires structures (including bridges) to satisfy a set of performance criteria when the structure is subject to loads. The ultimate limit state and serviceability limit state are two principal criteria. To satisfy the ultimate limit state, the structure must not collapse when subjected to the peak design load. A structure is deemed to satisfy the ultimate limit state criteria if all factored bending, shear, and tensile or compressive stresses are below the factored resistance for the section under consideration. To satisfy the serviceability limit state criteria, a structure must remain functional for its intended use subject to routine loading.

2.4.2.2 Seismic Design

Seismic design criteria must be established for different limit states, for example, serviceability limit state, ultimate limit state, and structural integrity limit state. Each limit state corresponds to a specific return period. A probabilistic seismic hazard assessment should be carried out to determine the ground motion level for each return period.

According to the performance based seismic design, performance requirements can be made for the bridge components under each limit state. In the serviceability limit state, bridges must behave elastically during frequently occurring or minor earthquakes and are expected to be serviceable immediately without the need for any repair after the earthquake. In the ultimate limit state, certain components may undergo large deformation without potential reduction in strength during a moderate earthquake and the damage is still considered as feasible to repair. In the structural integrity limit state, the deformation and damage of the bridge during a severe earthquake must not be such as to endanger emergency traffic or cause loss of structural integrity.

The seismic performance is assessed by comparing structural demands with capacities of components and the criterion is the limitation of the force demand/capacity ratio to 1.0. Demands on structural components are determined using global three dimensional computer models in terms of load-type quantities (forces and moments) or displacement-type quantities (displacements and rotations). For all seismic events, seismic demands are determined by elastic response spectrum analysis. For structural integrity limit state seismic event, a nonlinear time history analysis may be required.

2.4.2.3 Wind Resistant Design

Wind-resistant design is performed through integration of analysis and wind tunnel tests. Bridge response includes the combination of the static response to mean wind and dynamic response to buffeting wind. To conduct a buffeting analysis, mean wind speed profile, turbulence intensity profile, spectral distribution of turbulence velocity fluctuations and their coherence must be known. Wind tunnel tests serve to derive the drag, lift, and moment coefficients of bridge components, and to verify the dynamic stability in terms of flutter derivatives. In general, wind tunnel tests include two types of tests: one is the two-dimensional test on a rigid model (the section model test) to obtain the aerodynamic coefficients and flutter derivatives of a bridge deck; and the other is the global three-dimensional test (the aeroelastic model test) to examine the aerodynamic response and behaviour of the entire bridge. In the wind resistant design, a critical flutter speed of the bridge shall also be determined through integration of analysis and wind tunnel tests. The critical flutter wind speed predicted must exceed, by a substantial margin, the design wind speed of the bridge site at the deck height (Holmes, 2007).

2.4.2.4 Traffic Load Design

Traffic load analysis should consider the spatial distribution of the vehicles in different lanes and different spans with the load distribution coefficients. This usually leads to a large number of possible locations of vehicles.

For each load case, the load models on one lane should be applied on such a length longitudinally that the most adverse effect is obtained. On the remaining area, the associated load model should be applied on such lengths and widths in order to obtain the most adverse effect. For fatigue models, the location and the number of the lanes should be selected depending on the traffic to be expected in normal conditions.

Finally each traffic load case should be combined with non-traffic loads to determine the most adverse values.

2.4.3 Construction of Suspension Bridges

Construction of a suspension bridge involves sequential construction of the following components: the towers and cable anchorages, main cables, and deck structure.

2.4.3.1 Main Towers and Anchorages

Tower construction begins with foundations founded on sufficiently firm rock. The towers may stand on dry land or in water with a caisson. The caisson, a steel and concrete cylinder, is lowered to the ground underneath the water, emptied of water, and filled with concrete in preparation for the tower.

Erection accuracy of tower shafts is particularly important to a suspension bridge. The allowable inclination of towers of the Akashi Kaikyo Bridge was specified to be 1/5,000, about 6 cm at the top of the 297 m high tower. The shaft is vertically divided into 30 sections. The sections were prefabricated and barged to the site. The base plate and the first section were erected using a floating crane. The remainder was erected using a tower crane supported on the tower pier (Harazaki *et al.*, 2000). During construction, towers are cantilevered and thus easily vibrate due to wind. In the Akashi Kaikyo Bridge, tuned mass dampers were installed inside the tower to control wind-induced vibration.

The Tsing Ma Bridge used concrete towers each having 206 m in height, 6.0 m in width transversely, and tapered from 18.0 m at the bottom to 9.0 m at the top longitudinally. The tower shafts are hollow. Each main tower was slip-formed in a continuous around-the-clock operation, using two tower cranes and concrete buckets (Figure 2.6).

Figure 2.6 Main tower construction of the Tsing Ma Bridge

Anchorages are the structures to which the ends of the main cables are secured. They are massive concrete blocks securely attached to strong rock formations. During construction of the anchorages, strong eyebars (steel bars with a circular hole at one end) are embedded in the concrete. A spray saddle is mounted in front of the anchorage and supports the cable at the point where its individual wire bundles fan out. Each wire bundle will be secured to one of the anchorage's eyebars. The two main anchorages of the Tsing Ma Bridge are gravity structures constructed in the aforementioned way.

2.4.3.2 Cables

A majority of the modern suspension bridges use the aerial spinning method, invented by John A. Roebling, to install cables. By this method, the individual wires are pulled across from one anchorage to the other over the pylon saddles. A catwalk is constructed along the bridge's entire length, about 1 m below the pilot line, used by workers for cable formation. After erection of all wires, the cable is compacted and then wrapped by a galvanized steel wire. The aerial spinning method was used in the construction of the main cables of the Tsing Ma Bridge (Figure 2.7).

A prefabricated parallel-wire strand method has been introduced to reduce the labour involved and speed erection of main cables of suspension bridges. This method differs from the aerial spinning method by pulling across the prefabricated strands in bundles, whereas it costs more than the aerial spinning method. The prefabricated parallel-wire strand method was used in the Akashi Kaikyo Bridge and the Second Bosporus Bridge.

Cable spinning   Compacted main cable before wrapping

Figure 2.7 Cable installation in the Tsing Ma Bridge

2.4.3.3 Stiffening Girders/Trusses

Stiffening girders/trusses are typically installed using either the girder-section method or cantilevering from the towers or the anchorages.

In the girder-section method, stiffening girder sections are fabricated at the shop and placed on a barge below each erection point. Then they are hoisted up into position using lifting beams and secured to hanger ropes. This method was employed in the Tsing Ma Bridge (Figure 2.8).

Fabrication of deck unit in mainland China   Installed deck unit

Transportation of a completed unit to site   Lifting of a unit

Figure 2.8 Girder construction of the Tsing Ma Bridge

In the cantilevering method, preassembled section panels are erected by extending the stiffening girders as a cantilever from the towers and anchorages. This avoids disrupting marine traffic, which is required for the girder-section method.

2.4.4 Life-cycle Maintenance and Inspection of Suspension Bridges

Long-span suspension bridges are vulnerable to and are constantly subjected to aggressive environments. Deterioration begins after bridge construction is completed. For example, after the Tsing Ma Bridge was completed, about 40 workers immediately began painting, checking, and replacing corroding rivets. Maintenance and inspection of bridges is important to ensure the safety of bridges during their specified lifetime.

From the 1960s, interest in the inspection and maintenance rose considerably. In the United States, the National Bridge Inspection Standard (NBIS) sets the national policy regarding bridge inspection procedures, inspection frequency, inspector qualifications, reporting format, and rating procedures. The Federal Highway Agency, AASHTO, and Department of Transportation of each state have developed bridge inspection manuals and programs. Over the past three decades, the inspection program evolved into sophisticated bridge management systems such as Pontis (Thompson *et al.*, 1998) and BRIDGIT (Hawk and Small, 1998).

The frequency, scope, and depth of the inspection of bridges can be determined by bridge owners according to age, traffic, known deficiencies, and so forth. A suspension bridge requires a field inspection team led by a qualified professional engineer or certified inspector. The regular inspection interval should not exceed two years. For concrete members, common inspection defects include cracking, scaling, delamination, corrosion, and scour. For steel and metal members, common defects include corrosion and cracks, especially fatigue cracking. Both visual and physical examination are commonly employed for inspection. Destructive and nondestructive testing techniques, such as acoustic emissions, computer tomography, ultrasonic, and infrared, are also available.

Inspection reports are required to establish and maintain a bridge history file. Findings and results of a bridge inspection must be recorded, preferably on standard inspection forms. In the Pontis bridge management system (Thompson *et al.*, 1998), approximately 120 basic elements are required for inspection. Each element is characterized by discrete condition states, describing the type and severity of element deterioration. Pontis uses a Markovian deterioration model to predict the probability of transitions among the condition states each year. For each state, available actions and associated cost are defined. Network optimization is performed to minimize expected life-cycle costs while keeping the element beyond risk of failure.

Maintenance measures include preventive procedures and corrective procedures (NYDOT, 1997). Planned preventive procedures at appropriate regular intervals can significantly reduce the rate of deterioration of critical bridge elements. Cyclical preventive maintenance procedures include cleaning, sealing cracks, sealing the concrete deck and concrete substructures, replacing the asphalt wearing surface, lubricating bearings, and painting steel. Corrective procedures are performed to remedy existing problems. These mainly include repairing the asphalt wearing surface, repairing the concrete deck, repairing or replacing joints, repairing or replacing joints and steel members, repairing or replacing bearings, repairing or replacing concrete substructures, and repairing erosion or scour.

A life-cycle cost consists of not only the initial design and construction costs, but also those due to operation, inspection, maintenance, repair, and damage consequences during a specified lifetime (Frangopol and Messervey, 2007). However, it is noted that as the aging and deterioration of existing bridges dramatically increase, the current and future needs for bridge maintenance face a major and difficult challenge due to insufficient available funds (Frangopol and Messervey, 2007; Frangopol and Liu, 2007).

In a resource constrained environment, bridge managers will need to see how the initial cost of an SHM results in total life-cycle cost savings as well as improved structural safety. Combination of SHM and life-cycle management will be a tendency in near future. Chapter 12 in this book presents a preliminary work in this respect.

2.5 THE TSING MA SUSPENSION BRIDGE IN HONG KONG

2.5.1 Background

Hong Kong Special Administrative Region (HKSAR) consists of Hong Kong Island, Kowloon, and New Territories with a total land area of 1,104 km^2. As the financial and transportation hub of South China, air and marine traffic have expanded rapidly during the past decades. The new airport and container port have been constructed on Lantau Island. In addition, new highway and railway systems must be developed to provide access to the established urban areas of Hong Kong. These include the Tsing Ma Bridge, Kap Shui Man Bridge (cable-stayed bridge with a total length of 770 m, completed in 1997), Ting Kau Bridge (cable-stayed bridge with a total length of 1,177 m, completed in 1998), and Stonecutters Bridge (cable-stayed bridge with a total length of 1,596 m, completed in 2009).

2.5.2 Structural System

Located at latitude N22.2° and longitude E114.1°, the Tsing Ma Suspension Bridge has a total span of 2,132 m and carries a dual three-lane highway on the upper level of the bridge deck and two railway tracks and two protected carriageways on the lower level within the bridge deck. The main span across the Tsing Yi Island and the Ma Wan Island is 1,377 m long, as shown in Figure 2.9, making this bridge the longest of its type. The angle between the bridge longitudinal axis and the south is 73°. The bridge was designed by Mott MacDonald of Croydon (Mott MacDonald Group, 1990; Beard, 1995).

2.5.2.1 The Bridge Towers and Piers

Two reinforced concrete bridge towers, the Tsing Yi tower and the Ma Wan tower, are located in shallow water and stand on a ship impact protection island. The two towers are of almost identical geometric and structural configuration except that the topmost portal beam in the Ma Wan tower is 150 mm higher than the counterpart in the Tsing Yi tower but the top level of both towers is the same: 206.4 m with respect to the base level (Figure 2.10). The breadth of each tower leg changes from 9 m at its top to 18 m at the base level with a constant width of 6 m.

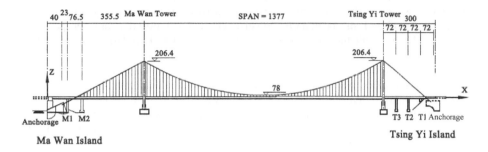

Figure 2.9 Configuration of the Tsing Ma Bridge (unit: m)

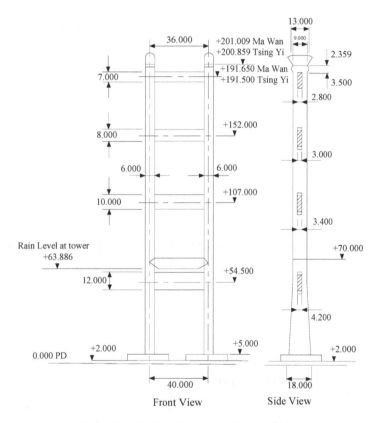

Figure 2.10 Configuration of the bridge towers (unit: m)

Each tower leg is designed with two rectangular hollow shafts that start from around 15 m above the base level to the topmost portal beam. The two legs in each tower are inclined towards each other at a slope of 1 in 100 and connected by four deep pre-stressed concrete portal beams. The dimensions of the beams vary from locations and they are the smallest of 7×2.8 m² (height × width) at the top and the largest of 12×4.2 m² at the bottom. The four portal beams are centrally hollow with a steel truss cast in the concrete enclosing a narrow corridor. The narrow corridor of the size of 2×1 m² is equally constructed in each portal beam for passage purpose. Underneath the tower legs are two heavily RC foundation bases which are laid on suitable rock.

The two side spans on the Ma Wan side and Tsing Yi side are supported by two and three piers, respectively, as seen in Figure 2.9. The two piers supporting the Ma Wan approach span, namely piers M1 and M2, are at a distance of 355.5 m and 432 m, respectively, measured from the Ma Wan tower. For the Tsing Yi approach span, the three piers are located at an identical distance of 72 m from each other starting from the Tsing Yi tower in a sequence of piers T3, T2 and T1. All supporting piers in the side spans are reinforced concrete structures. Piers M1, T2 and T3 are free-standing piers of similar design. Pier M2 provides lateral restraint to the bridge deck against lateral loads and carries two saddles at its top above the deck. These two saddles deflect the main cables through a small angle. Pier T1 is part of the approach road and slip road structure on the Tsing Yi side. It also provides lateral restraint to the bridge deck. Piers M1, M2, T2 and T3 are hollow for most of their height whereas pier T1 is almost a solid wall. Out of the five piers, piers M2 and T1 are in different structural forms whereas the remaining piers (M1, T2 and T3) share a common structural layout except that the height of the three piers is unequal.

2.5.2.2 The Bridge Deck

The bridge deck is a hybrid steel structure consisting of Vierendeel cross-frames supported on two longitudinal trusses acting compositely with stiffened steel plates that carry the upper and lower highways. The stiffened plates acting with two longitudinal diagonally braced trusses at 26 m centre provide the vertical bending stiffness of the bridge deck. Transverse shear forces are carried by the steel plates together with the plane bracing systems that join the plates at both upper and lower flanges and span vent openings. The mixed plane bracing-plate systems enable the longitudinal trusses to provide lateral bending stiffness. At the main span and Ma Wan side span, the deck is suspended from the main cables at an 18 m interval. Near the Ma Wan and Tsing Yi bridge towers, the bridge deck changes to incorporate two additional inner longitudinal trusses to share forces with the main trusses, and the deck plates extend over the centre to cover the full width of the bridge without vent openings. The cross section of the bridge deck also changes in the Tsing Yi side span and in the area near the Ma Wan abutment, where the deck is supported by piers rather than suspenders.

The bridge deck at the main span is a suspended deck and the structural configuration is typical for every 18 m segment. Figure 2.11 illustrates the configuration of the suspended deck module consisting of mainly longitudinal trusses, cross frames, highway decks, railway tracks, and bracings. Two outer longitudinal trusses, spacing at 26 m, link up the cross frames along the bridge longitudinal axis and act as the main girders of the bridge.

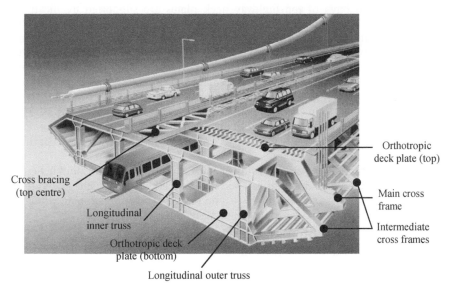

(a) Isometric view of the deck module

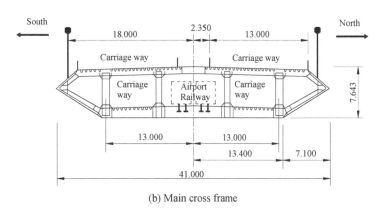

(b) Main cross frame

Figure 2.11 Configuration of the deck at the main span (unit: m)

Each longitudinal truss is comprised of upper and lower chords and vertical and diagonal members. In the 18 m deck module, there are one main cross frame and four neighbouring intermediate cross frames with two on each side of the main cross frame. The five cross frames are 4.5 m apart from each other and connected by the two outer longitudinal trusses. Each cross frame (see Figure 2.11b) is comprised of upper and lower chords, inner struts, outer struts, and upper and lower inclined edge members. Through suspender units connected to the intersections of edge members of the main cross frame, this deck module is suspended to the main cable. Two pairs of sway bracings are connected from the suspension points at the main cross frame to the outer ends of the upper chords of the two adjacent intermediate cross frames to strengthen the structural stability.

Two symmetrical bays of top highway deck plates are supported by the upper chords of cross frames and longitudinal trusses. Between them is a row of central cross bracing stretching from neighbouring cross frames. Regarding the bottom deck, there are two railway tracks laid on the central bay with one row of central cross bracing and two rows of outer cross bracings added to brace the bottom chords of cross frames. Two bays of deck plates in the bottom deck on two outer sides are supported on the bottom chords of cross frames and longitudinal trusses. This deck module is symmetric to the middle vertical plane, with a width of 41 m (= 2×20.5), a lateral distance between the two suspension points at the main cross frame of 36 m (= 2×18), a height of 7.643 m and an inner clearance of 5.35 m in the middle.

The structural translational movements at the Ma Wan abutment are restrained in three translational directions. At the Tsing Yi abutment, the vertical (z-axis) and lateral (y-axis) movements of the bridge deck are restrained while the deck can move freely along the longitudinal direction (x-axis).

2.5.2.3 The Bridge Cables and Suspenders

The cable system of the Tsing Ma Bridge consists of two main cables, 95 pairs of suspender units and 95 pairs of cable bands.

The two main cables 36 m apart in the north and south are accommodated by the four saddles located at the top of the tower legs. Each main cable consists of 91 strands of parallel galvanized steel wires in the main span and 97 strands in the side spans. The number of wires per strand is 360 or 368 with the wire diameter of 5.38 mm. Each cable has a cross sectional area of 0.759 m^2 consisting of 33,400 wires in the main span and 0.800 m^2 consisting of 35,224 wires in the side span. The main cables were formed by the traditional aerial spinning technique. The resultant cable has an overall diameter of approximately 1.1 m after compacting.

The two parallel main cables are seen as horizontal sagged cables fixed at the tower saddles and moving together with the towers and transferring the loadings into the anchorages. The Tsing Yi side span cables are seen as inclined sagged cables with the top ends fixed at the tower saddles and the lower ends fixed at the main anchorage. The Ma Wan side span cables are also inclined sagged cables with the top ends fixed at the tower saddles and with the lower ends fixed at the main anchorage through the saddles on pier M2. On each main cable, there are 19 suspender units within the Ma Wan approach span and 76 suspender units in the main span. In the Tsing Yi approach span, there are no suspender units. Each suspender unit is made of a pair of wire ropes of 76 mm diameter that passes over the cable bands on the main cables and is then attached to the chords of the bridge deck by steel sockets. There are thus four strands in each suspender unit, held apart by spacer clamps. The distance between two suspender units is 18 m along the longitudinal axis of the bridge deck.

2.5.2.4 The Bridge Anchorages and Foundations

On the Tsing Yi side, the main cables are extended from the tower saddles to the main anchorage on the ground. On the Ma Wan side, the main cables extended from the Ma Wan Tower are fixed first by pier saddles at the deck level and then by the main anchorage saddles on the ground. The two anchorages are gravity structures which are integral with the deck abutments. At Tsing Yi the anchorage is largely below ground in a 290,000 m³ rock excavation. In contrast, the Ma Wan anchorage is only partially buried. Figure 2.12 shows the Tsing Yi anchorage.

The foundations are generally simple spread footings laid on suitable rock, expect the Ma Wan tower is laid on caissons in approximately 12 metres of water.

Figure 2.12 The Tsing Yi anchorage

2.5.3 Design Considerations and Criteria

The design criteria adopted by the designers are generally incorporated in the Design Requirements and Loading Specification document (Mott MacDonald Group, 1990), which makes reference to other design standards, notably BS5400 and the Hong Kong Civil Engineering Manual (Hong Kong Government, 1983). Certain criteria have been modified to reflect the bridge's location and the exceptionally long span. In particular, specific highway, railway, and wind loading criteria have been developed.

2.5.3.1 Dead Load

The dead weights of all structural components were estimated from the details given on the drawings, which were typically about 204 kN/m in the global analysis. The superimposed dead weights of non-structural elements were 88 kN/m.

2.5.3.2 Railway Load Effects

The Tsing Ma Bridge consists of highway traffic and railway traffic, whereas the latter plays a more significant role in fatigue life of the bridge deck. Airport railway design loading is based on an eight-car-four-axle train. The maximum axle loading has been taken as 17 tonnes for the static design. For fatigue effects, a 13 tonnes axle load has been adopted with an annual tonnage per track of 51 million tonnes (Beard, 1995).

The design criteria for the Airport railway are generally based on passenger comfort requirements. These are defined in terms of horizontal and vertical accelerations at the track level, the maximum values being 0.05 g horizontally and 0.03 g vertically. In addition, the maximum angular deflections of the rail are 0.042^{o} horizontally and 0.24^{o} vertically to avoid unacceptable high accelerations due to abrupt changes in rail alignment (Beard and Young, 1995).

In order to satisfy the stringent deflection criteria imposed by the railway the bridge deck is continuous between anchorages. Specially designed rail movement joints are located at each end of the deck to accommodate all longitudinal and angular movements. Short supported side spans, continuous with the main deck, are introduced immediately in front of the anchorage abutments to control and minimise the rate of change of vertical accelerations. Lateral bearings are provided at the towers, abutments, and piers to control horizontal displacements.

2.5.3.3 Highway Load Effects

At the maximum loaded length of 2 km on the highway, the predicted lane loading is 14.85 kN/m. In parallel with lane load intensities, coexistent lane loading has been considered for the specific planned operational patterns of this bridge, including combinations of lane loading on the upper and lower deck carriageways.

2.5.3.4 Wind Load Effects

The wind data most relevant to the Tsing Ma Bridge were those obtained from the Hong Kong Royal Observatory from its station at Waglan Island, which was used to determine the design wind speed. The hourly mean wind speed for the site at 10 m above mean sea level, having a return period of 120 years, was taken as 35 m/s. The variation of hourly mean wind speed with height follows a power law variation with an exponent of 0.19. Three-second gust speeds have been taken as 44 and 50 m/s with normal highway and railway operations respectively. Due to the higher dispersion of wind speeds in Hong Kong than in the United Kingdom, a partial factor on wind loading of 1.9 was adopted rather than 1.4. In the extreme wind conditions, highway transport on the Tsing Ma Bridge ceases to function. Highway loading is therefore not considered for maximum wind speeds.

The bridge deck must be stable in high speed winds and have a low drag factor. A section was therefore developed using double-deck box construction with truss stiffening and non-structural edge fairings. Longitudinal air vents were provided in the upper and lower surfaces to enhance stability. The aerodynamic characteristics of the bridge deck were determined from wind tunnel tests carried out by British Maritime Technology. Aerodynamic tests confirmed that the absence of divergent oscillations and vortex shedding oscillations occurred at relatively low wind speeds with acceptable magnitude and frequency.

2.5.3.5 Seismic Load Effects

Based on the *Hong Kong Civil Engineering Manual*, a seismic peak ground acceleration of 0.05 g was used for the bridge design. Although the seismic coefficient in Hong Kong is low, multimode spectral method was used due to the complexity and importance of this large-scale bridge. A global finite element model was employed for seismic response analysis via the complete quadratic combination rule in which the lowest 30 modes were used.

2.5.3.6 Temperature Load Effects

The minimum effective temperature of the bridge at a return period of 120 years is suggested as −2°C, and the maximum values are 46°C for the deck (with 40 mm surfacing), 50°C for the main cables and hangers, and 36°C for the towers, respectively. Range of bridge effective temperature is taken as 23°C in calculating expansion and contraction due to temperature restraint. Temperature difference profile within the bridge deck was derived from the *Hong Kong Civil Engineering Manual*. The maximum positive temperature differences were considered to coexist with the maximum bridge effective temperatures. The maximum reverse temperature differences were considered to coexist with the minimum bridge effective temperatures and also with effective temperatures 8°C below the maximum for steel elements and 2°C below the maximum for concrete elements.

2.5.3.7 Other Effects

The Tsing Ma Bridge is in a largely marine environment and has concrete towers, piers, and anchorages. In order to maximize its design life a stringent specification has therefore been adopted, which contains quality control requirements relating to both the concrete constitutes and the mix designs. These include long term strength and bulk chloride diffusion durability tests. It also requires temperature tests and strict controls for concrete pours in large members. The design has also adopted both epoxy coated reinforcement and increased cover for parts of foundations, piers, and towers.

The foundations of the bridge are laid on suitable rock or caissons. The governing design criterion is the settlement of individual foundations with differential settlement between the legs of a tower limited to 10 mm.

Other loading criteria include protection against the impact forces resulting from the collision of ships of 220,000 dead weight tonnes.

2.5.3.8 Load Effects Combination

The dead load, highway load, railway load, wind load, seismic load, temperature load, earth pressure, and settlement were applied to a three-dimensional spine finite element model of the entire bridge individually. The calculated load effects arising from each of the loading types were combined with partial factors. Envelopes of maximum and minimum values were produced for each load effect under each combination. Finally the governing values were determined from the combination results. The global static and dynamic analyses produced sets of global load effects in the bridge deck, cables, suspenders, towers, and piers.

As the global model cannot capture the loading effects in the local individual elements, local analysis was carried out using detailed local models, which included deck grillage, typical suspended deck girder section, suspenders, towers and piers, anchorages, and saddles. In these local models, finer beam elements or shell elements were used. The critical load cases identified from the global analysis were then analysed on these local models in order to obtain the maximum load effects for use in design.

The maximum combined load effects in each component were derived from the global and local analyses. The ultimate limit state and serviceability limit state criteria were taken on all parts. In general, the reinforced concrete parts including the towers, side span piers, anchorages, and approaches are governed by the ultimate limit state criteria, and the prestressed parts including the tower portal beams and the deck elements in the slip roads at the Tsing Yi approaches are governed by the serviceability limit state criteria. The cable system including main cables, suspenders, cable clamps and bands, handstrands, and saddles is governed by the serviceability limit state criteria.

2.6 REFERENCES

AASHTO, 1991, *Guide Specification and Commentary for Vessel Collision Design of Highway Bridges*, (Washington, D.C.: American Association of State Highway and Transportation Officials).

AASHTO, 2005, *LRFD Bridge Design Specifications*, 3rd ed., (Washington, D.C.: American Association of State Highway and Transportation Officials).

AREMA, 1997, *Manual for Railway Engineering*, (Landover, MD: American Railway Engineering Maintenance of Way Association).

Beard, A.S., 1995, Tsing Ma Bridge, Hong Kong. *Structural Engineering International*, **5(3)**, pp. 138-140.

Beard, A.S. and Young, J.S., 1995, Aspects of the design of the Tsing Ma Bridge, In *Proceedings of International Conference on Bridges into the 21st Century*, Hong Kong, edited by Felber, A.J., (Seattle, Washington: Impressions Design & Print Ltd.), pp. 93-100.

Blejwas, T.E., Feng, C.C., and Ayre, R.S., 1979, Dynamic interaction of moving vehicles and structures. *Journal of Sound and Vibration,* **67**, pp. 513-521.

Brown, T.G., Tibbo, J.S., Tripathi, D., Obert, K. and Shrestha, N., 2010, Extreme ice load events on the Confederation bridge. *Cold Regions Science and Technology*, **60(1)**, pp. 1-14.

BS5400, *British Standard: Steel, Concrete and Composite Bridges*, (London: British Standards Institution), Parts 1-10.

Bucher, G.C. and Lin Y.K., 1988, Stochastic stability of bridges considering coupled modes. *Journal of Engineering Mechanics, ASCE*, **114(12)**, pp. 2055-2071.

Cai, C.S. and Montens, S., 2000, Wind effects on long-span bridges. In *Bridge Engineering Handbook*, edited by Chen, W.F. and Duan, L., (Boca Raton: CRC Press).

Chen, X.Z., Matsumoto, M., and Kareem, A., 2000, Time domain flutter and buffeting response analysis of bridges. *Journal of Engineering Mechanics, ASCE*, **126(1)**, pp. 7-16.

Cheung, M.S., Tadros, G.S., Brown, T., Dilger, W.H., Ghali, A. and Lau, D.T., 1997, Field monitoring and research on performance of the Confederation bridge. *Canadian Journal of Civil Engineering*, **24(6)**, pp. 951-962.

Cheung, Y.K., Au, F.T.K., Zheng, D.Y. and Cheng, Y.S., 1999, Vibration of multi-span non-uniform bridges under moving vehicles and trains by using modified beam vibration functions. *Journal of Sound and Vibration*, **228**, pp. 611-628.

Davenport, A.G., 1962, Buffeting of a suspension bridge by storm winds. *Journal of Structural Division, ASCE*, **88(3)**, pp. 233-268.

Diana, G., 1993, Aeroelastic study of long span suspension bridges, the Messina Crossing. In *ASCE Structures Congress 93*, Irvine, CA, (New York: ASCE), pp. 484-489.

Eurocode 1, 2003, *Actions on Structures-Part 2: Traffic Loads on Bridges*, EN 1991-2:2003, (Brussels, Belgium: European Committee for Standardization).

Eurocode 1, 2005, *Actions on Structures-Part 1-4: General Actions-Wind Actions*, EN 1991-1-4:2005, (Brussels, Belgium: European Committee for Standardization).

Eurocode 1, 2006, *Actions on Structures-Part 1-7: General Actions-Accidental Actions*, EN 1991-1-7:2006, (Brussels, Belgium: European Committee for Standardization).

Frangopol, D.M. and Liu, M., 2007, Maintenance and management of civil infrastructure based on condition, safety, optimization, and life cycle cost. *Structural and Infrastructure Engineering*, **3**(1), pp. 29-41.

Frangopol, D.M. and Messervey, T.B., 2007, Integrated life-cycle health monitoring, maintenance, management and cost of civil infrastructure. In *Proceeding of 2007 International Symposium on Integrated Life-cycle Design and Management of Infrastructure*, Shanghai, edited by Fan, L.C., Sun, L.M., and Sun, Z., (Shanghai: Tongji Universitty Press), pp. 3-20.

Frýba, L. 1996, *Dynamics of Railway Bridges*, (London: Thomas Telford).

Gimsing, N.J., 1997, *Cable Supported Bridges: Concept and Design*, 2nd ed., (New York: Chichester).

Harazaki, I., Suzuki, S. and Okukawa, A., 2000, Suspension bridges. In *Bridge Engineering Handbook*, edited by Chen, W.F. and Duan, L., (Boca Raton: CRC Press).

Harik, I.E., Shaaban, A.M., Gesund, H., Valli, G.Y.S. and Wang, S.T., 1990, United States bridge failures, 1951-1988. *Journal of Performance of Constructed Facilities, ASCE*, **4**, pp. 272-277.

Hawk, H. and Small, E.P., 1998, The BRIDGIT bridge management system. *Structural Engineering International*, **4**, pp. 309-314.

Holmes, J.D., 2007, *Wind Loading of Structures*, 2nd ed., (London: Taylor & Francis).

Hong Kong Government, 1983, Design of highway structures and railway bridges. In *Hong Kong Civil Engineering Manual, Volume V Roads*, (Hong Kong: Hong Kong Government).

Mark, V.J., 1977, *Detection of Steel Corrosion in Bridge Decks and Reinforced Concrete Pavement*. Report HR-156, (Ames, Iowa: Iowa Department of Transportation).

Mott MacDonald Group, 1990, Engineering, operating and maintenance requirements: Part B - Design requirements and loading specification. In *Lantau Fixed Crossing Project Brief*, (Surrey, UK: Mott MacDonald Group Limited).

NYDOT, 1997, *Fundamentals of Bridge Maintenance and Inspection*, (New York: New York State Department of Transportation).

Scanlan, R.H., 1978, The action of flexible bridge under wind, II: buffeting theory. *Journal of Sound and Vibration*, **60(2)**, pp. 201-211.

Simiu, E. and Scanlan, R., 1996, *Wind Effects on Structures*, 3rd ed., (New York: Wiley).

Thompson, P.D., Small, E.P., Johnson, M. and Marshall, A.R., 1998, The Pontis bridge management system. *Structural Engineering International*, **4**, pp. 303-308.

Timoshenko, S, Young, D.H. and Weaver, W., 1974, *Vibration Problems in Engineering*, 4th ed., (New York: Wiley).

Wardhana, K. and Hadipriono, F.C., 2003, Analysis of recent bridge failures in the United States. *Journal of Performance of Constructed Facilities, ASCE*, **17(3)**, pp. 144-150.

Unsworth, J.F., 2010, *Design of Modern Steel Railway Bridges*, (CRC Press, Taylor & Francis Group).

Xanthakos, P.P., 1993, *Theory and Design of Bridges*, (New York: John Wiley & Sons).

Yang, Y.B. and Lin, H.B, 1995, Vehicle-bridge interaction analysis by dynamic condensation method. *Journal of Structural Engineering ASCE*, **121**, pp. 1636-1643.

CHAPTER 3

Structural Health Monitoring Systems

3.1 PREVIEW

An on-line SHM system generally consists of the following modules, namely, sensory system (SS), data acquisition and transmission system (DATS), data processing and control system (DPCS), data management system (DMS), and structural evaluation system (SES). The first two are embedded on the structures, whereas the other three are usually located in the control office of the bridge administrative departments.

Designing an SHM system is a systematic work integrating various expertise. This chapter will first outline the design criteria of an SHM for a large-scale bridge. The commonly used types of sensors, data acquisition systems, basic signal processing techniques, and data management systems will then be introduced. Various structural evaluation and analysis methods are related to different loading effects, which will be detailed in the later chapters of the book. As a new and potentially promising sensing, transmission, and monitoring system method, wireless monitoring is specially described in this chapter. Finally, the Wind And Structural Health Monitoring System (WASHMS) for the Tsing Ma Bridge will be introduced.

3.2 DESIGN OF HEALTH MONITORING SYSTEMS

The components of a complete SHM system are as illustrated in Figure 3.1. The sensory system is composed of various types of sensors that are distributed along the bridge to capture different signals of interest. The data acquisition and transmission system is responsible for collecting signals from the sensors and transmitting the signals to the central database server. The data processing and control system is devised to control the data acquisition and transmission, process and store the data, and display the data. The data management system comprises the database system for temporal and spatial data management. In accordance with monitoring objectives, the structural evaluation system may have different applications. It may include an on-line structural condition evaluation system and/or an off-line structural health and safety assessment system. The former (on-line) is mainly to compare the measurement data with the design values, analysis results, and pre-determined thresholds and patterns to provide a prompt evaluation on the structural condition. The latter (off-line) incorporates varieties of model-based and data-driven algorithms, for example, loading identification, modal identification and model updating, bridge rating system, and damage diagnosis and prognosis.

Some large-scale SHM systems, such as those of the Tsing Ma Bridge and Stonecutters Bridge (Wong, 2004; Wong and Ni, 2009), have a portable data acquisition system (PDAS) and portable inspection and maintenance system (PIMS). The inspection and maintenance system is a portable system for inspecting and maintaining sensors, data acquisition units, and cabling networks.

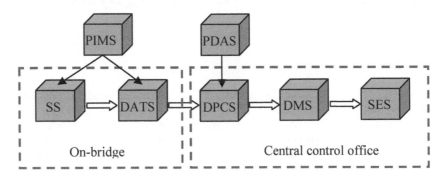

Figure 3.1 Architecture of a SHM system

Note: SS: sensory system, DATS: data acquisition and transmission system, DPCS: data processing and control system, DMS: data management system, SES: structural evaluation system, PDAS: portable data acquisition system, PIMS: portable inspection and maintenance system.

It can be seen from Figure 3.1 that each module is a stand-alone system. This implies that failure of an individual module will have no detrimental effect on the operation of the remaining parts of the entire system (Wong, 2004). This concept must be considered during the design of SHM systems. For example, if the data server in the control office does not work, this should not affect the data acquisition on the bridge. This requires that the data acquisition system has the capacity of on-board data storage for a specific duration, and the data management system has the data retrieval function after the server recovers.

Design of an SHM system requires the designer to understand the needs of monitoring, characteristics of the structure, environment condition, hardware performance, and economic considerations.

First of all, the designer should clearly know the bridge itself and the objectives of the monitoring. Different bridges have different characteristics and unique demands. Designers of the SHM system should work together with the designers of the bridge and know their main concerns. Next, the monitoring items and the corresponding information should be identified, which may include (Aktan *et al.*, 2002):

• The parameters to be monitored, such as temperature, wind, displacement, and corrosion;
• The nominal value and expected ranges of the parameters;
• The spatial and temporal properties of the parameters, for example, variation speed of the measurands, location of the measurands;
• The accuracy requirement;

- The environment condition of the monitoring;
- The duration of the monitoring.

After the monitoring parameters are identified, the number of sensors should be determined according to the size and complexity of the structure and the monitoring objectives. Then types of sensors are carefully selected such that their performance can meet the requirement of the monitoring. Important sensor performance characteristics include measurement range, sampling rate, sensitivity, resolution, linearity, stability, accuracy, repeatability, frequency response, durability, and so forth. In addition, sensors must be compatible with the monitoring environment, such as temperature range, humidity range, size, packaging, isolation, and thermal effect. Due to the limitation of length, this book will not consider in too much detail the optimal sensor placement, measurement hardware or signal processing. The interested readers may refer to some references on sensors, data acquisition, communication, and signal processing: for example, the guidelines on bridge health monitoring edited by Aktan *et al.* (2002) and Mufti (2001).

The data acquisition units (DAUs) should be compatible with the sensors as well. Location and number of DAUs should be determined to trade-off the distance from the sensors to the DAUs and number of channels of the DAUs. Sampling rate, resolution, accuracy, and working environment should be taken into account for selection of hardware.

Besides, the designer should consider the budget of the project, availability of hardware, wiring, and installation and protection of the hardware. In practical monitoring projects, wires or cables are more easily damaged artificially or naturally than the sensors themselves. Special protection of sensors and wires is worth the effort. Maintenance is also a factor to be considered during the design stage. Important sensors and DAUs should be accessible for check and repair after installation.

3.3 SENSORS AND SENSING TECHNOLOGY

In a bridge monitoring system, the sensors are mainly employed for monitoring three types of parameters: (i) loading sources such as wind, seismic, and traffic loading; (ii) structural responses such as strain, displacement, inclination, and acceleration; and (iii) environmental effects including temperature, humidity, rain, and corrosion. This section will introduce the commonly used sensors, categorized by the monitoring parameters. It is noted that many new sensor technologies are being developed, which may not be included in the book.

3.3.1 Wind Measurement Sensors

The traditional sensors widely used for wind monitoring of a long suspension bridge include anemometers for measuring wind velocity. Occasionally, pressure transducers are installed to measure wind pressures and pressure distribution over a particular part of the bridge envelope. In recent years, Doppler radar, GPS

(Global Positioning System) drop-sonde, and Doppler sodar have become powerful devices for measuring boundary layer wind profiles.

Propeller and ultrasonic anemometers are the most commonly used instruments for measuring wind velocity and direction on site. The propeller anemometer directly records wind speed and direction. The propeller anemometer is convenient and relatively reliable and sustainable in harsh environments but it is not sensitive enough to capture the nature of turbulent winds at higher frequencies. This is particularly true in situations when wind speed or direction changes rapidly. The ultrasonic anemometer measures wind velocity through its two or three orthogonal components. The ultrasonic anemometer is quite sensitive but it is not sustainable in harsh environments. The accuracy requirement for wind velocity measurement of the anemometers must be maintained under heavy rain, that is, no occurrence of spikes during heavy rainstorms. For a long suspension bridge, the anemometers are often installed at a few bridge deck sections on both sides and along the height of the towers so that not only wind characteristics at key points can be measured but also the correlation of wind velocity in both horizontal and vertical direction can be determined. The positions of the anemometers must be selected so as to minimize the effect of the adjacent edges of the bridge deck and towers on the airflow towards them. To meet this requirement, anemometer booms or masts are often needed so that the anemometer can be installed a few metres away from the bridge edges. The boom or mast must be equipped with a retrievable device to enable retracting of the anemometer in an unrestricted and safe manner for inspection and maintenance.

Wind pressure transducers sense differential pressure and convert this pressure difference to a proportional electrical output for either unidirectional or bidirectional pressure ranges. To measure the pressure difference accurately, the location of the reference pressure transducer needs to be selected appropriately to avoid possible disturbances from surrounding environment.

3.3.2 Seismic Sensors

Seismometers are instruments that measure motions of the ground, including those of seismic waves generated by earthquakes, nuclear explosions, and other sources. For short-period seismometers, the inertial force produced by a seismic ground motion deflects the mass from its equilibrium position, and the displacement or velocity of the mass is then converted into an electric signal as the output proportional to the seismic ground motion. Long-period or broadband seismometers are built according to the force-balanced principle, in which the inertial force is compensated with an electrically generated force so that the mass moves as little as possible. The feedback force is generated with an electromagnetic force transducer through a servo loop circuit. The feedback force is strictly proportional to the seismic ground acceleration and is converted into an electric signal as the output.

A strong-motion seismometer usually measures acceleration, which is also built on the force-balanced principle and can be integrated to obtain ground velocities and displacements.

3.3.3 Weigh-in-motion Stations

Weigh-in-motion (WIM) devices can measure the axle weight of the passing vehicles and thus the sum of the weight of the vehicles, velocity of the vehicles, and distance between the axles. These data can be used to evaluate the traffic load on bridges. Unlike older static weigh stations, current WIM systems are capable of measuring weight at normal traffic speeds and do not require the vehicle to stop, making them much more efficient.

There are a few available WIM devices, such as a bending plate WIM system with strain gauge bonded underside of the plate scale (McCall and Vodrazka, 1997), a piezoelectric WIM system using piezoelectric sensors embedded in the pavement which produce a charge when the tyres induce the deformation on the pavement, a load cell WIM system utilising a load cell with two scales to weigh both right and left sides of the axle simultaneously, a capacitive-based WIM system with two or more metal plates, and a fibre optic sensor-based WIM system.

A dynamic WIM station mainly consists of a metal housing with lightning protection, bending plate sensors and processing board, induction loop detection and loop processor board, central processing unit, and power supply. The metal housing must have sufficient room to house an inner cabinet of suitable size where the control electronics with power supply and maintenance-free backup batteries can be located. Proprietary plug-in circuit boards are designed with individual lightning protection for the bending sensors monitoring, induction loops monitoring, data storage memory and additional serial interfaces.

As compared with static weight stations, WIM systems can record dynamic axle load information. However, WIM systems are less accurate than static scales.

3.3.4 Thermometers

Temperature, including structural temperature and ambient air temperature, is frequently measured in many monitoring systems. It is widely recognised that changes in temperature significantly influence the overall deflection and deformation of bridges. Restraint of movement can induce stresses within a bridge. Excessive thermal stresses can damage bridges. It is noted the temperature is usually non-uniformly distributed over the entire structure and is different from the ambient temperature, due to heat transfer.

The most often used temperature sensors include thermocouples, thermistors, and resistance temperature detectors. Thermocouples have a wide measurement range, are inexpensive but less stable than the other two kinds. Thermistors and resistance temperature detectors are based on the principle that resistance of a material increases when temperature goes up. They are generally more accurate and stable than thermocouples. Resistance temperature detectors are usually made of platinum. The most common resistance temperature detectors used in industry have a nominal resistance of 100 ohms at 0°C, and are called PT100 sensors, which are often employed in bridge monitoring exercises as well. Thermistors differ from the resistance temperature detectors in that the material used in a thermistor is generally a ceramic or polymer. Thermistors have a smaller temperature range (−90°C to 130°C) but typically have a higher precision than resistance temperature detectors.

3.3.5 Strain Gauges

Foil strain gauges, fibre optic strain gauges, and vibrating wire strain gauges are commonly used sensors measuring strain in civil structures. Fibre optic strain gauges will be described later.

3.3.5.1 Foil Strain Gauge

Foil strain gauges are the most common type of strain gauge consisting of a thin insulting backing which supports a fine metallic foil. The gauge is attached to the object by a suitable adhesive. As the object is deformed, the foil is stretched or shortened causing the change in its electrical resistance in proportion to the amount of strain, which is usually measured using a Wheatstone bridge.

The physical size of most foil strain gauges is about a few millimetres to centimetres in length. Its full measurement range is about a few milli-strain. The foil strain gauges are economical and can measure dynamic strains. However, their long-term performance (for example, zero-stability) is not as good as other alternatives, particularly in a harsh environment. For example, the presence of moisture may result in electrical noise in the measurement and zero-drift.

3.3.5.2 Vibrating Wire Strain Gauge

The vibrating wire strain gauge consists of a thin steel wire held in tension between two end anchorages. The wire vibrates due to an excitation with a short pulse and the resonant frequency is measured. When the distance between the anchorages changes, the tension of the wire changes, so does the natural frequency. The change in the vibration frequency of the wire is transferred into the change in strain. Consequently the captured strain can be transmitted over a relatively long distance (a few hundred metres to kilometres) without much degradation. This is one advantage of vibrating wire gauges over foil gauges.

The cost of the vibrating wire gauges is between the foil strain gauges and fibre optic strain gauges. Vibrating wire gauges are easy to install on the surface or embed in concrete. A typical vibrating wire gauge is about $100 \sim 200$ mm long and has a measurement range of 3000 $\mu\varepsilon$ with a resolution of 1.0 $\mu\varepsilon$, which is suitable for monitoring of civil structures. A main drawback of vibrating wire gauges is that they can measure the static strain only as it takes seconds to obtain the frequency of the vibrating wire.

3.3.6 Displacement Measurement Sensors

Displacement of bridge structures serves as an effective indicator of its structural performance condition. Large displacements or deformations may create hazardous conditions for traffic actually on the bridge, and excessive displacements may affect the bridge's structural integrity. Displacement monitoring is thus needed. Equipment measuring the displacement includes linear variable differential transformer, level sensing station, Global Positioning System (GPS), and so forth.

3.3.6.1 Linear Variable Differential Transformer

The linear variable differential transformer (LVDT) is a commonly used electro-mechanical facility for measuring relative displacements based on the principle of mutual inductance. It consists of a hollow metallic tube containing a primary and two secondary coils and a separate movable ferromagnetic core. The coils produce an electrical signal that is in proportion to the position of the moving core. The frictionless movement of the core leads to a very long mechanical life, very high resolution, good zero repeatability, and long-term stability of LVDTs. LVDTs are available in a wide range of linear stroke, ranging from micro-metres to 0.5 metre.

3.3.6.2 Level Sensing Station

The measurement of vertical displacement by the level sensors is in principle based on the pressure difference. The system basically consists of two or more interconnected fluid (usually water) filled cells. Relative vertical movement of the cells causes movement of water and variation in the water level is measured. The conventional level sensing system can detect the elevation difference of about 0.5 mm.

3.3.6.3 Global Positioning System

The Global Positioning System (GPS) developed by the US Department of Defense in 1973 was originally designed to assist soldiers, military vehicles, planes, and ships (Sahin *et al.*, 1999). It consists of three parts: the space segment, the control segment, and the user segment. The space segment is composed of 32 satellites in six orbital planes. Each satellite operates in circular 20,200 km orbits at an inclination angle of 55°, and each satellite completes an orbit in approximately 11 hours and 57.96 minutes (Hofmann-Wellenhof *et al.*, 2008). The spacing of satellites in orbits is arranged so that at least six satellites are within line of sight from any location on the Earth's surface at all times (Hofmann-Wellenhof *et al.*, 2001). The control segment is composed of a master control station, an alternate master control station, and shared ground antennas and monitor stations. The user segment is composed of thousands of military users of the secure precise positioning service, and millions of civil, commercial, and scientific users of the standard positioning service.

Basically, a GPS receiver receives the signals sent by the GPS satellites high above the Earth, determines the transit time of each message, computes the distances to each satellite, and calculates the position of the receiver. However, even a very small clock error multiplied by the very large speed of light (299,792,458 m/s) results in a large positional error. Therefore receivers use four or more satellites to improve the accuracy of the positioning. Nevertheless, this accuracy is in the order of a metre and cannot be used for displacement monitoring of bridges which is about in the order of a centimetre.

In the practical survey, the Real Time Kinematic (RTK) technique is used on the basis of carrier phase measurements of the GPS where a reference station provides the real-time corrections. A RTK system usually consists of a base station receiver and a number of mobile units. The base station re-broadcasts the phase of the carrier that it measures, and the mobile units compare their own phase measurements with the ones received from the base station. This system can achieve a nominal accuracy of 1 cm ± 2 parts-per-million (ppm) horizontally and 2 cm ± 2 ppm vertically. In the foreseeable future, this can be improved further.

A few factors affect the accuracy of GPS measurement, in particular the atmospheric conditions and multi-path effects. Inconsistencies of atmospheric conditions affect the speed of the GPS signals as they pass through the Earth's atmosphere. The GPS signals are also reflected by the surrounding obstacles, causing delay of signals.

3.3.7 Accelerometers

Although vibration can be measured in terms of velocity and dynamic displacement as well, acceleration can be measured more accurately. Accelerometers are widely used to measure acceleration of structures induced by force excitation or ambient excitation. The acceleration responses of a bridge are closely related to the serviceability and functionality of the bridge. In addition, a vibration testing can be employed to obtain the natural frequencies, damping ratios, and mode shapes of the global structure, which are directly associated with the mass, stiffness, and damping characteristics.

Basically an accelerometer is a mass-spring-damper system that produces electrical signals in proportion to the acceleration of the base where the sensor is mounted. Selection of accelerometers should consider the following parameters: usable frequency response, sensitivity, base strain sensitivity, dynamic range, and thermal transient sensitivity. Installation of accelerometers and cables is also critical for a good vibration measurement. There are four main types of accelerometers available: piezoelectric type, piezoresistive type, capacitive type, and servo force balance type.

The piezoelectric type accelerometers are very robust and stable in long-term use. However, the major drawback is that they are not capable of a true DC (0 Hz) response, which makes it not appropriate for some civil structures with very low frequency, for example, around 0.1 Hz. Actually the lower frequency limit of piezoelectric accelerometers is generally above 1 Hz. Piezoresistive and capacitive accelerometers are adequate for civil flexible structures as they can measure accelerations from DC level. Capacitive type accelerometers are very accurate and appropriate for low frequency and low level vibration measurement such as micro-g (gravity acceleration = 9.80 m/s^2). The force balance sensors are suitable for DC and low frequency measurement providing milli-g measurement capability.

3.3.8 Weather Stations

In some applications, it is desirable to measure the environmental conditions such as ambient temperature, humidity, rainfall, air pressure, and solar irradiation. A typical weather station usually integrates a few types of sensors and can measure the above-mentioned parameters besides the wind speed and direction. Solar irradiation intensity, air temperature, and wind are important parameters for deriving the temperature distribution of structures. With temperature distribution, the thermal effect on the structural responses can be evaluated quantitatively.

3.3.9 Fibre Optic Sensors

Optical fibres can be used as sensors to measure strain, temperature, pressure, and other quantities. Consequently fibre optic sensors measuring different parameters are summarised here together. The sensors modify a fibre so that the quantity to be measured modulates the intensity, phase, polarization, wavelength of light in the fibre. Accordingly the fibre optic sensors can be classified into four categories as: intensity modulated sensors, interferometric sensors, polarimetric sensors, and spectrometric sensors (Casas and Cruz, 2003).

A significant advantage of the fibre optic sensors is multiplexing, that is, several fibre optic sensors can be written at the same optical fibre and interrogated at the same time via one channel. In addition, fibre optic sensors are very small in size and immune to electromagnetic interferences. They are also suitable for both static and dynamics measurements with a frequency from hundreds to thousands Hertz. The major drawback of fibre optic sensors is the high cost of both sensors and the acquisition unit (or readout unit). In addition, the fibres are rather fragile and should be handled very carefully in the field installation.

In bridge monitoring, Fibre Bragg grating (FBG) sensors are commonly used for strain measurement (Seim *et al.*, 1999; Ni *et al.*, 2009). Its principle is that the strain variation causes the shift in the central Bragg wavelength. Consequently FBG strain sensors monitor changes in the wavelength of the light. Commercially available white light sources have a spectral width around 40 to 60 nm. An FBG sensor with the measurement range of 3000 με takes a wavelength of 3 nm. Counting the spectral space between the sensors, one optical fibre can accommodate six to ten FBG sensors.

Other applications of fibre optic sensors include monitoring fatigue crack (Austin *et al.*, 1999), wear of bearings (Cohen *et al.*, 1999), corrosion (Lo and Shaw, 1998), displacement (Vurpillot *et al.*, 1997), temperature (Stewart *et al.*, 2005), and acceleration (Sun *et al.*, 2006).

3.3.10 Others

A laser Doppler vibrometer (LDV) is an instrument that is used to make non-contact vibration measurements of a surface. The LDV basically uses the Doppler principle to measure velocity at a point to which its laser beam is directed. The reflected laser light is compared with the incident light in an interferometer to give the Doppler-shifted wavelength. This shifted wavelength provides information on surface velocity in the direction of the incident laser beam. Some advantages of an LDV over similar measurement devices such as an accelerometer are that the LDV can be directed at targets that are difficult to access, or that may be too small or too hot to attach a physical transducer. Also, the LDV makes the vibration measurement without mass-loading the target, which is especially important for very tiny devices such as micro-electro-mechanical systems.

Abe *et al.* (2001) and Kaito *et al.* (2001) have applied an LDV to measure vibration of bridge deck and stay cables. When the measurement grid was pre-determined, the LDV automatically scanned 45 points at a high frequency such that one LDV could measure the vibration of all points at the same time. With one reference, the modal properties can be extracted. Yan *et al.* (2008) applied an LDV to measure the vibration of hard disk drives.

Other non-contact measurement techniques, such as the photogrammetry and videogrammetry techniques, have been developed with the advance of inexpensive and high-performance charge-coupled-device cameras and associated image techniques. Bales (1985) applied a close-range photogrammetric technique to several bridges for estimation of crack sizes and deflection measurement. Li and Yuan (1988) developed a 3D photogrammetric vision system consisting of video cameras and 3D control points for measuring bridge deformation. Olaszek (1999) incorporated the photogrammetric principle with the computer vision technique to investigate the dynamic characteristics of bridges. Others applications include Patsias and Staszewski (2002). Ji and Chang (2008) and Zhou *et al.* (2010) employed the techniques for cable vibration measurement.

3.4 DATA ACQUISITION AND TRANSMISSION SYSTEM

3.4.1 Configuration of DATS

Sensors generate analogue or digital signals that represent the physical parameters being monitored. Data acquisition is an intermediate device between the sensors and computers, which collects the signals generated by the sensors, converts them, and transmits the signals to the computers for processing. For a small laboratory based experiment, the above function can be achieved with a card based data acquisition unit in a PC. However, configuration of a data acquisition and transmission system (DATS) in a long-term bridge monitoring system is generally much more complicated. It usually consists of local cabling network, stand-alone data acquisition units (DAUs) or substations, and global cabling network, as illustrated in Figure 3.2. The local cabling network refers to the cables connecting the distributed sensors to the individual DAUs, and the global cabling network refers to the cables connecting the DAUs to central database servers.

Appropriate deployment of DAUs plays a significant role in assuring the quality and fidelity of the acquired data in long-span bridges. Long distance wires cause noise and significant loss in analogue signals (especially for voltage signals) because the distributed sensors are far from the central control office (usually hundreds or thousands of metres away). It is also inefficient to wire all sensors to one central server. Therefore, DAUs are assigned at a few cross-sections of the bridge to collect the signals from surrounding sensors, condition the signals, and transmit the digital data into the central database server.

It is noted that some proprietary sensors such as GPS, video cameras, fibre optic sensors, and corrosion sensors have their specific data acquisition systems. These united systems capture corresponding information, transform the information into digital data, and directly connect to the central data server for processing (Figure 3.2).

For such a system as illustrated in Figure 3.2, it is desirable to have a uniform platform to assure the scalability, functionality, and durability of the system.

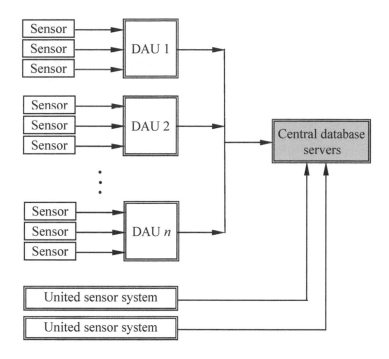

Figure 3.2 Configuration of a data acquisition and transmission system

3.4.2 Hardware of DAUs

A DAU generally comprises a number of electronic components including signal conditioners, memory and data storage unit, microcontroller, communication device, uninterruptible power supply, fan/air conditioner, lightning conductor, and GPS time synchronizer. All these components are integrated in a waterproof, rugged enclosure or cabinet for the long-term monitoring purpose.

It is common that different types of sensors with different output signals and different sampling rates are included in one DAU. Consequently a DAU can facilitate this flexibility and may have more than one signal conditioner. A signal conditioner manipulates an analogue signal such that it meets the requirements of further processing. Signal conditioning usually includes amplification, filtering, analogue-to-digital (A/D) conversion, and isolation. Amplification serves to amplify the analogue signal before A/D conversion to utilize the full range of the A/D converter, thus increases the signal-to-noise ratio and resolution of the input signal. For example, acceleration in an ambient vibration may be very low and the output may be in milli-volt level. An amplifier can multiply the voltage signal to up to that required by the A/D converter (0–10 V, for instance). The amplifier is preferably placed near to the sensor, but sometimes this is difficult and it is integrated inside the DAU. Filtering is used to remove the unwanted frequency components, for example, alternating current (AC). Most signal conditioners employ low-pass filters. Isolation is used to isolate the possible ground loop and protect the hardware from damage. As the sampling rate in the measurement data of bridges is usually low, a single A/D converter can perform A/D conversion by switching between several channels (i.e., multiplexing). This is much less expensive than having a separate A/D converter for each channel, and thus adopted in most practical SHM systems.

The internal memory serves as a temporary buffer of data for transmission. It is usually integrated with the microcontroller. The data storage unit can save the measurement data relatively longer in case the global cabling network does not work appropriately. The data can be retrieved manually to the external storage devices or automatically to the database server when the global cabling network recovers.

The microcontroller consists of internal electronic circuitry to execute commands sent by the users and to control other hardware components. For example, the sampling rates and acquisition duration of the sensors can be changed by the users.

The communication device is responsible for communication between the DAU and the computer. Usually Ethernet interface is employed.

The power supply provides power to the data acquisition system and to some sensors that require external power supply. An uninterruptible power supply provides instantaneous or near-instantaneous protection from unexpected power interruption or unstable input voltage.

A fan or air conditioner is used to cool the temperature of the DAU. A lightning conductor can provide protection of the DAU from lightning damage.

Previously the DAUs were synchronised through a synchronisation signal sent from the central station regularly. Nowadays GPS time synchronisers have become more popular as they can provide an easy way to keep the DAUs and united sensor systems accurately synchronised.

3.4.3 Network and Communication

In the DATS, a uniform network communication is crucial to assure the data can be transmitted over the entire system. Various communication network technologies such as Ethernet, RS-232, RS-485, IEEE-1394 can be employed for the common network.

In an SHM system for a long-span bridge, the distance between the DAUs to the central control office may be as far as a few kilometres, and fibre optic cabling is desirable. In fibre optic communication, there are basically two types of fibre optic cables: single mode and multi-mode fibres. Multi-mode fibres generally have a larger core diameter, and are used for shorter distance communication. Single mode fibres are used for communication links longer than 600 metres, thus preferable for long-span bridges.

Wireless communication and networking have been rapidly developed and employed for data transmission, whereas its transmission speed and accuracy are still not comparable with the cable-based network at the moment. Nevertheless, the wireless network is promising in the near future. It has advantages in some situations, particularly for construction monitoring when the cable network is not ready.

3.4.4 Operation of Data Acquisition and Transmission

After the hardware has been installed, the DATS should be tested or verified through field tests, for example, controlled load tests. This is because the actual performance of the hardware is uncertain under long-term exposure to the harsh conditions. Moreover, it is difficult in practice to identify, repair, and change the damaged facilities after the bridge is put into service, which is usually operated by a different sector from the construction stage. The field verification can help identify problems in hardware, installation, cabling, and software, such that the problems can be fixed before normal operation.

During normal operation, data acquisition and transmission are carried out in a systematic and organised manner. Depending on the nature of the monitored parameters, some sensors may work continuously while others may work in the trigger mode, in which the sensor signals are collected only when the parameters are above certain threshold values, usually some events occur (for example, earthquakes or typhoons). These two modes can be operated simultaneously in one data acquisition system.

The sampling rate (or sampling frequency) is an important factor affecting the data acquisition speed. It relies on the variation speed of the monitored parameters and can be programmed by the users. For example, ambient temperature usually changes slowly with time and can be treated as a static measurand with a low sampling rate. Acceleration, on the other hand, varies more rapidly with time and is usually regarded as a dynamic parameter with a higher sampling rate. If a parameter is not sampled fast enough, which is known as under-sampling, the resulting digitised signal will not represent the actual signal accurately, and this error is called as aliasing error. To avoid this error, the Nyquist criterion requires that the sampling rate should be more than twice the highest frequency component of the original signal. In bridge monitoring exercises, the sampling rate of most signals is usually no higher than 100 Hz as the fundamental frequencies of long-span bridges are relatively low, unless some special measurements like acoustic methods or guided-wave methods are employed.

For the low sampling rate, multiplexed sampling, rather than simultaneous sampling can be employed. The multiplexed sampling allows different channels to share one A/D converter and to be sampled sequentially. This can reduce cost compared with the simultaneous sampling, in which each signal channel has an individual A/D converter.

After operation for a period in the adverse environment, DAUs are inevitably subject to error or malfunction. It is preferable to carry out periodical calibrations. As mentioned previously, the DAUs should be accessible for maintenance.

3.5 DATA PROCESSING AND CONTROL SYSTEM

The functions of the data processing and control system include: 1) control and display of the operation of the data acquisition system; 2) pre-processing of the raw signals received from the data acquisition system; 3) data archive into a database or storage media; 4) post-processing of the data; and 5) viewing the data.

3.5.1 Data Acquisition Control

A large-scale SHM system comprises various types of data acquisition hardware. Therefore, centralised data acquisition control is preferable. As described above (Section 3.3.4), the signals can be collected in a long-term or short-term mode. Therefore the data acquisition control unit should be flexible in handling both continuous monitoring mode and scheduled trigger modes. In practical SHM systems, the centralised control unit is located in the central control office and operated by the users for carrying out the communication with the local acquisition facilities via a graphical user's interface.

The graphical user's interface program is an interface between the data acquisition hardware and the hardware driver software. It controls the DATS's operation, such as how and when the DATS collect data, and where to transmit. It provides the users with an easy interface as the details of the hardware are very complicated for most users, for example, civil engineers.

3.5.2 Signal Pre-processing

The collected raw signals are pre-processed prior to permanent storage. The data pre-processing has two primary functions: 1) transforming the digital signals into the monitored physical data; and 2) removing abnormal or undesirable data. Signal transforming is simply done by multiplying the corresponding calibration factor or sensitivities. There are several data-elimination criteria for removal of typical abnormalities associated with various types of statistical data. The source of abnormal data is possibly derived from malfunction of the measurement instrument. There are a few criteria defining abnormal data.

3.5.2.1 Data with Abnormal Magnitude

Extremely large or small data are regarded as abnormal. For example, the ambient temperature of Hong Kong must be within a certain range, and an extremely high or low temperature outside this range recorded by the temperature sensor does not have any physical meaning.

3.5.2.2 Data with Significant Fluctuation

The second criterion to eliminate the abnormal data is set in terms of difference between the maximum and minimum values in a specific period. For example, the change in ambient temperature of one day is within a certain range and an extremely significant increase or drop of temperature is suspicious. This criterion is

$$|x_{max} - x_{min}| > \varepsilon \tag{3.1}$$

where | | means the absolute value, x_{max} and x_{min} denote the maximum and minimum value, respectively; ε is a real number which defined the limit of the difference between the maximum and minimum values. The adopted values of ε vary for different types of measurement data recorded at different locations.

3.5.2.3 Data with no Variation

It is also observed that some data from vibration measurements are associated with zero standard deviation. A zero standard deviation physically indicates a steady measurement without any fluctuation within the statistical period. As a result, the statistical values of mean, maximum and minimum have the same magnitude. Having a perfectly flat signal might be considered as an abnormal measurement. Correspondingly, the third criterion to eliminate the abnormal data is set in terms of the zero standard deviation of measured records and is given by

$$x_{std} = 0 \tag{3.2}$$

where x_{std} is the standard deviation of measurement in a specific period.

3.5.3 Signal Post-processing and Analysis

The pre-processed signal will be saved into a database system for future management or storage media like hard disks and tapes after proper packaging and tagging. The stored data are processed for various uses. Here we only exemplify a few basic data processing techniques. Other techniques and more advanced analysis can be found in the following chapters or other books.

3.5.3.1 Data Mining

In a long-term SHM system, a huge number of data are recorded from the sensor system. How to extract important features or information is critical to effective use of SHM system for structural condition evaluation. The data mining is a bridge between the data and specific patterns (or features) for decision. The data mining technologies have attracted a great deal of attention in the artificial intelligence community, in which a wider term is *knowledge discovery* (Fayyad *et al.*, 1996a). Actually data mining can be regarded as a knowledge discovery process.

The data mining is defined as "the nontrivial process of identifying valid, novel, potentially useful, and ultimately understandable patterns in data" (Fayyad *et al.*, 1996b). The data mining mainly has the following several functions (Fayyad *et al.*, 1996a; Duan and Zhang, 2006):

- *Regression*: identify the relationships between a set of variables;
- *Classification*: classify a data item into one of several predefined classes;
- *Clustering*: identify a finite set of categories or clusters to describe the data without predefined class labels;
- *Summarization*: find a compact description for a subset of data;
- *Outlier detection*: detect data which do not comply with the general behaviour or model of the data in a database.

A wide variety of data mining methods exist, from conventional statistical methods such as regression analysis, clustering analysis, and principal component analysis to more advanced machine learning methods such as Support Vector Machine, Genetic Algorithm, Bayes Belief theory, Artificial Neural Networks (ANNs), and Colony Algorithms. Sohn *et al.* (2003) reviewed some applications of these techniques to structural damage detection. Here only the regression analysis is introduced as it is widely used for preliminary data processing in the context. Interested readers may refer to many textbooks in the area, for example, Kottegoda and Rosso (1997).

The regression analysis investigates the relationship between one variable and one or more other variables. The simplest relation is the linear regression as

$$y = \beta_0 + \beta_x x + \varepsilon_y \tag{3.3}$$

where x represents the explanatory variable, y is the response variable, β_0 (intercept) and β_x (slope) are regression coefficients, and ε_y is the error. With least-squares fitting, the regression coefficients and confidence bounds can be obtained. To examine goodness of fit of the linear relation between x and y, the correlation coefficient, ρ, is defined by

$$\rho = \frac{Cov(x, y)}{\sigma_x \sigma_y} \tag{3.4}$$

where σ and Cov are standard deviation and covariance, respectively. A higher correlation coefficient implies a good linear relation between the two variables. The linear regression can be easily extended to the multiple linear regression where the equation contains more than one explanatory variable.

3.5.3.2 Frequency Domain Analysis

Frequency domain analysis allows one to examine the data in the frequency domain, rather than in the time domain. It presents the frequency components of a signal and the contributions from each frequency to the signal. Usually the signal can be converted between the time and frequency domains with a pair of transform, for example, Fourier transform and inverse Fourier transform.

Quite often in the SHM, one is interested in the frequency spectrum of loading signals and response signals of a bridge to view their frequency components. For example, acceleration responses of a bridge can reveal the natural frequencies of the bridge. Further, its important vibration characteristics (frequencies, damping, and mode shapes) can be obtained via modal analysis tools, which will be described in Chapter 4.

In practice, signals are captured at a discrete set of times, say, $1/f_s$, $2/f_s$, …, n/f_s where n is the total number of data and f_s is the sampling rate. Accordingly the discrete Fourier transform is used in signal processing, which transforms a series of signal $x(0)$, $x(1)$, …, $x(n-1)$ into n complex numbers as

$$F(j) = \frac{1}{n} \sum_{k=0}^{n-1} x(k) \exp\left(\frac{-i2\pi jk}{n}\right) \tag{3.5}$$

where i is the imaginary unit, and $j=0$, 1, 2, …, $n-1$, and the inverse discrete Fourier transform takes the form

$$x(k) = \sum_{j=0}^{n-1} F(j) \exp\left(\frac{i2\pi jk}{n}\right) \tag{3.6}$$

The above two equations indicate that complex numbers $F(j)$ represent the amplitude and phase of the different sinusoidal components of signal $x(k)$ while signal $x(k)$ is a sum of sinusoidal components. The amplitude or phase of $F(j)$ represents the spectrum of the time series $x(k)$. Due to the symmetric property, usually only the first half spectrum is of interest. It is noted that the jth item is associated with the physical frequency (in Hertz) of jf_s/n (or circular frequency of $2\pi jf_s/n$).

The squared amplitude of the Fourier transform, or power, can be obtained as:

$$P(j) = |F(j)|^2 \tag{3.7}$$

The resulting plot is referred to as a power spectrum, indicating the averaged power over the entire frequency range. More common in signal processing, one considers power spectrum density, i.e., the power component of a signal in an infinitesimal frequency band. According to the Wiener–Khinchin theorem, the power spectrum density is the Fourier transform of the autocorrelation function of the signal (theoretically a random signal does not obey the Dirichlet condition and its Fourier transform does not exist, whereas its autocorrelation function obeys the Dirichlet condition and the Fourier transform is valid). That is,

$$S_{xx}(j) = \frac{1}{n}\sum_{k=0}^{n-1} R_{xx}(k)\exp\left(\frac{-i2\pi jk}{n}\right) \tag{3.8}$$

where S_{xx} is the auto-power spectrum density and R_{xx} is the autocorrelation function taking the form of

$$R_{xx}(k) = E[x(j)x(j-k)] \tag{3.9}$$

Similarly, for two discrete signals x and f, their cross-power spectrum can be obtained as the Fourier transform of the cross-correlation function of the two signals. When f and x refer to the input force and output response, respectively, the commonly used frequency response functions (FRFs) can be obtained as

$$H(j) = \frac{S_{fx}(j)}{S_{ff}(j)} \tag{3.10}$$

where S_{ff} and S_{fx} refer to the auto-power spectrum density of the input force and cross-power spectrum density between the input force and response, respectively.

It is recommended that a window is used to minimise the leakage problem during the transform unless the signal is transient and dies away within the record length. The Hanning window function is commonly used while the exponential window function is suggested for an impact testing (Avitabile, 2001).

The above mentioned frequency analyses, including the Fourier transform, power spectrum, and frequency response functions, are standard techniques and available in spectral analysers.

3.5.3.3 Time-frequency Domain Analysis

An important assumption of the Fourier transform is that the signal is stationary. Some signals in real world are nonstationary. i.e., the signal statistical characteristics vary with time. Examples of this include wind speed signal and earthquake ground motions. Recent advances in the field of signal processing have allowed characterization of the time-frequency properties of nonstationary signals. There are a few well-known time-frequency distributions and analysis tools such as short-time Fourier transform, Wavelet transform, and Hilbert-Huang transform.

The short-time Fourier transform computes the time-dependent Fourier transform of a signal using a sliding window (Oppenheim and Schafer, 1989). The method splits the original signal into overlapping segments and applies the discrete-time Fourier transform to each segment to produce an estimate of the short-time frequency content of the signal over the given time period.

The Wavelet transform is a new tool in mathematics and has broad applications (Daubechies, 1992). Wavelet functions are composed of a family of basis functions that are capable of describing a signal in a localized time (or space) domain and frequency (or scale) domain. Therefore using wavelets can perform local analysis of a signal, i.e. zooming on any interval of time or space.

The Hilbert-Huang transform was proposed by Huang *et al.* (1998). It decomposes a signal into a series of intrinsic mode functions with the empirical mode decomposition, and then uses the Hilbert spectral analysis to obtain instantaneous frequency data. The Hilbert-Huang transform is particularly designed for nonstationary and nonlinear processes.

3.5.3.4 Data Fusion

An SHM system usually includes various types of sensors located in different spatial positions. Different types of sensors located in the same position may capture different signals. Spatially distributed sensors may also demonstrate different features of the structure. In addition, different methods and different users may reach different results. Therefore, integration of data from different sensors and integration of results made by different algorithms are important to a robust monitoring exercise (Chan *et al.*, 2006). Data fusion is an important data processing tool to achieve this.

Data fusion is a process that integrates data and information from multiple sources in order to achieve improved information than could be achieved by use of single information alone (Hall, 1992). Fusion processes are often categorized as low, intermediate, and high levels fusion depending on the processing stage at which fusion takes place (Hall and Llinas, 1997). The low level fusion combines raw data from multiple sensors to produce new data that is expected to be more informative and synthetic than the inputs. In the intermediate level fusion or feature level fusion, features are extracted from multiple sensors' raw data and various features are combined into a concatenated feature vector that may be used by further processing. The high level fusion, also called decision fusion, combines decisions coming from several experts to reach a consistent conclusion. Techniques involved in feature/decision-level data fusion include a wide range of areas such as artificial intelligence, pattern recognition, statistical estimation, information theory. Detailed description of these techniques such as Neural Networks, Bayesian inference, Dempster-Shafer's methods can be found in corresponding references.

3.6 DATA MANAGEMENT SYSTEM

The collected data and processed data or results in a SHM system should be stored and managed properly for display, query, and further analysis. In addition, relevant information on the SHM system, computational models, and design files needs to be documented as well. These tasks are completed by the data management system (DMS) via a standard database management system such as MySQL and ORACLE.

3.6.1 Components and Functions of Data Management System

A standard database management system allows users to store and retrieve data in a structured way so that the later assessment is more efficient and reliable. In a long-term SHM system for large-scale structures, the data size is large in different types. Therefore a large SHM system usually consists of a few databases, including the device database, measurement data database, structural analysis data database, health evaluation data database, and user data database. The DMS manages each database to fulfil their corresponding functions.

The device database records the information on all sensors and substations. For sensors, their identification (ID), label, substation, location, specifications, manufacturer, installation time, initial values, sampling rate, thresholds, working condition, and maintenance record should be recorded. For substations, their ID, label, sensors, location, specifications, manufacturer, installation time, working condition, and maintenance record are recorded. These data are necessary to examine the collected measurements in the long term. Various types of sensors should be labelled properly so that they can be easily identified by users. The DMS has the function of inserting and deleting sensors and substations, and monitoring their conditions.

The measurement data database records all the data collected from the sensor system, including the loads, structural responses, and environmental parameters that are detailed in Section 3.2. For efficiency, the measurement data database usually stores data for a limited period only, for example, one year. The historical data beyond this period are archived in storage media such as tapes. For data safety, all the measurement data should have a spare backup. A DMS has the function of automatic retrieval and output of the measurement data, and data query from authorised users.

The structural analysis data database records the finite element model data, input parameters of the models and major output data, design drawings, and basic design parameters. For large-scale bridges, there might be more than one finite element model for cross checking and for different applications. The models can be input into the corresponding analysis software and the output data are employed for comparison and evaluation.

The health evaluation data database records the evaluation time, parameters, objects, criteria, results, and reporting. The structural health evaluation may be performed regularly for normal operation. Once an extreme event occurs, for example, an earthquake, a specified evaluation should be carried out immediately. For each kind of evaluation, the pre-determined criteria should be provided.

The user data database manages the users' information such as username, ID, user group, and personal data and contact information. Different users have been authorised different rights by the DMS to log into the SHM system.

A DMS should provide security management, which may include network security, data protection, database backup, and user operation audit.

Finally a DMS should provide the alarming function. The alarming module can automatically generate warning messages when some pre-defined criteria are satisfied. Important alarms should be sent to relevant staff through e-mail and short messages until countermeasures are taken.

3.6.2 Maintenance of Data Management System

A large SHM system is operated by authorised staff who have received basic training in computer technology and structure engineering. When the databases start working, their functions and performance need to be tested. After a period of operation, increase in sizes of the databases may cause physical storage malfunctions and reduce the efficiency of the databases. Therefore maintenance of the DMS is necessary. The duties of the administrator of a DMS include backup and restoration of the databases, monitoring and improvement of the databases, reconstruction and reconfiguration of the databases.

3.7 WIRELESS SENSOR AND WIRELESS MONITORING

Advances in micro-electro-mechanical systems (MEMS) technology, wireless communications, and digital electronics have enabled the rapid development of the wireless sensor technology since the late twentieth century. We place this part here rather than in Section 3.2 because the wireless sensor is actually neither one kind of pure sensing technology nor just a new transmission method, but a new system that can carry out many tasks certainly including SHM. A wireless sensor network can comprise all of the components in a wire-based SHM system described previously such as SS, DATS, DPCS, DMS, and SES, whereas it has its unique characteristics as compared with the wire-based SHM systems.

3.7.1 Overview of Wireless Sensors

Development of wireless sensors is due to the fact that a robust system may require a dense network of sensors throughout the system. Traditional sensing systems usually attempt to develop more and more accurate sensors of limited quantity at optimized locations. A bio-system, however, usually comprises a huge number of distributed sensors each with limited functions. This philosophy inspires researchers to develop a network of low-cost, small size, large quantity sensors. In addition, the traditional sensing systems are usually wire-based which have high installation costs. Maintenance of such a monitoring system at a reliable operating level under the adverse environment for a long period of time is very difficult. Experience in monitoring civil structures shows that communication wires are more vulnerable to the environment than sensors themselves. The wireless transmission provides a more flexible communication manner and sensors can be deployed and scalable easily.

With the support of the US Defense Advanced Research Projects Agency, researchers at the University of California at Berkeley have developed the open platform, well known as *Berkeley Mote*, or *Smart Dust*, whose ultimate goal is to create a fully autonomous system within a cubic millimetre volume (Kahn *et al.*, 1999). Such a system may comprise hundreds or thousands of sensor nodes each costing as little as about one US dollar.

Berkeley Mote is the first open hardware/software research platform, which allows users to customize hardware/software for a particular application. Its first generation is COTS Dust (Hollar, 2000), followed by Rene developed in 2000.

The third generation of Mote, the Mica, was released in 2001. Subsequent improvements to the Mica platform resulted in Mica2, Mica2Dot, and MicaZ. Another commonly used wireless sensor unit is the Intel Mote platform Imote (Kling, 2003) and Imote 2 (Kling, 2005).

Berkeley Mote, Intel Mote, and quite a few others have been used for general purposes in military, environment, health, home, and other commercial areas (Akyildiz *et al.*, 2002; Yick *et al.*, 2008). These systems have been customized for SHM applications (Kurata *et al.*, 2003; Ruiz-Sandoval *et al.*, 2006; Rice and Spencer, 2008). A wireless monitoring was implemented in the Jindo Bridge, Korea in June 2009, in which 110 wireless nodes were used. In the structural discipline, researchers from Stanford University have developed their own wireless sensor unit for SHM (Lynch *et al.*, 2001; 2002).

3.7.2 Basic Architectures and Features of Wireless Sensors

A wireless sensor node usually consists of four basic components as shown in Figure 3.3: a sensing unit, a processing unit, a transceiver unit, and a power unit (Akyildiz *et al.*, 2002). The components are carefully selected to meet the specified functions and keep a total low cost.

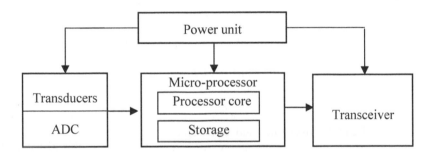

Figure 3.3 Structure of a wireless sensor node

The processing unit is a micro-processor (or micro-controller), which controls the sensing, data processing, computation, and communication with other sensor nodes or the central station. The on-board processor makes the wireless sensor node intelligent, which differs from a traditional sensor. The micro-processor has a small storage that stores internal programs and processed data.

The sensing unit is usually composed of a few sensors and the ADC. The analogue signals collected by the sensors are converted to digital signals by the ADC, and then sent to the processing unit. It is noted that the ADC in most general wireless sensor nodes is of only 8 or 10 bits. This is insufficient for vibration monitoring. In the customized Imote 2 (Rice and Spencer, 2008), a 16-bits ADC is embedded. The wireless sensor unit developed in Stanford University (Lynch *et al.*, 2001) has a 16-bits ADC as well.

A transceiver unit connects the node to the network. The transmission distance of most wireless nodes is about 50 m to 500 m in outdoor environment.

Consequently for large civil structures, this transmission range requires the sensor nodes communicate with the peers and send the data to the base station over the network. The wireless network has three kinds (Swartz and Lynch, 2009): star, peer-to-peer, and multi-tier, as shown in Figure 3.4. In the figure, the sensor nodes include generic nodes and gateway nodes. A gateway node, like the substation in the wired systems, gathers data from the adjacent generic nodes and transmits them to the base station. Most of the smart sensors to date adopt radio frequency for the wireless communication.

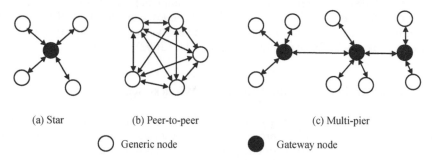

      (a) Star   (b) Peer-to-peer   (c) Multi-pier

         ○ Generic node   ● Gateway node

Figure 3.4 Wireless network topology

The power unit is an important component in wireless sensor node. Currently most of available smart sensors rely on the battery power supply, which has finite capacity and finite life. There have been several attempts to harvest energy at sensor nodes locally, for example, solar cell, wind turbine, mechanical vibration, fuel cells, and mobile supplier. Solar cell is the current mature technique and was used in wireless monitoring of the Jindo Bridge.

In wireless sensor nodes, communication consumes much more power than other operations including sensing and processing. Therefore, collected raw data are processed within the sensing unit to reduce the amount of the raw data transmitted wirelessly over the network. This also takes advantage of the computational characteristics of the processor board. Accordingly this distributed computation and monitoring make the wireless monitoring different from the tethered monitoring using the traditional wired system.

To facilitate this distributed monitoring, the micro-processor has two types of software: one is the operating system and the other is the engineering algorithms. The operating system controls the nodes and provides device drivers. One popular operating system is TinyOS (http://www.tinyos.net), an open-source operating system designed by the University of California at Berkeley. Both Berkeley Mote and Intel Mote run the TinyOS operating system.

Currently algorithms for distributed monitoring are relatively scarce and simple, mainly in modal analysis. The complicated monitoring algorithms used in the centralized monitoring usually need a large amount of memory, heavy computation, and data from multiple sensors. Consequently transplanting the available monitoring algorithms from the wired monitoring system directly is not feasible. It is imperative to develop appropriate algorithms for this distributed monitoring.

3.7.3 Challenges in Wireless Monitoring

Although wireless sensors and networks have been developed rapidly, at the moment the wireless monitoring is not mature for continuous health monitoring of large civil structures. Traditional wire-based systems still dominate practical SHM projects and wireless sensor nodes are mainly for research purposes or supplementary to the wired systems. Nevertheless, wireless sensors and networks might be a future direction for SHM. At present main challenges in wireless sensors are power supply, and communication bandwidth and range. As mentioned in the last section, lack of mature distributed monitoring algorithms is another big issue.

For many civil structures, AC power outlets are not available adjacent to the sensor nodes. (Even if AC power outlets are available, practitioners prefer to adopt the wired DAQ system adjacent to the power outlets and transmit the collected data to the base station via the wireless communication. In any case, wireless communication is just one communication method rather than wireless sensor networks.) For a battery-powered monitoring system, power consumption is a critical issue to maintain the operation of the sensor nodes in the long term. Currently used battery-powered sensor nodes can only operate for hours in full working state and weeks in standby state (or sleep mode). The power restraint requires the sensor components should be energy efficient. However, lower power consumption often comes with reduced functions such as lower resolution, shortened communication range, and reduced speed (Swartz and Lynch, 2009).

The majority of wireless sensor nodes operate with the unlicensed industrial, scientific, and medical radio band, in which the output power is limited, to 1 W, for example in the United States. The limited radio band limits the amount of data that can be reliably transmitted within the network during a given time period. In addition, the limitation of the output power restricts the effective communication range of the sensor nodes. For a large civil structure, a few sensor nodes may be insufficient and the sensor network should be designed carefully.

3.8 THE STRUCTURAL HEALTH MONITORING SYSTEM OF THE TSING MA BRIDGE

3.8.1 Overview of WASHMS

A Wind And Structural Health Monitoring System (WASHMS) for the Tsing Ma Bridge has been devised, installed, and operated by the Highways Department of the Government of Hong Kong Special Administrative Region. Actually the WASHMS also covers the Kap Shui Mun Bridge (cable-stayed) and Ting Kau Bridge (cable-stayed), and was operated from 1997 (Wong, 2004). The system architecture of the WASHMS is composed of six integrated modules: the sensory system (SS), data acquisition and transmission system (DATS), data processing and control system (DPCS), structural health evaluation system (SHES), portable data acquisition system (PDAS), and portable inspection and maintenance system (PIMS).

Table 3.1 Sensors deployed on the Tsing Ma Bridge.

Monitoring Item	Sensor Type	Sensor Code	No. of Sensor	Position
Wind speed and direction	Anemometer	WI	6	2: deck of main span; 2: deck of Ma Wan side span; 2: top of towers
Temperature	Thermometer	P1~P6, TC	115	6: ambient; 86: deck section; 23: main cables
Highway traffic	Weigh-in-motion station	WI	7	approach to Lantau Toll Plaza
Displacement	Displacement transducer	DS	2	1: lowest portal beam of the Ma Wan tower (lateral); 1: deck at the Tsing Yi abutment (longitudinal)
	GPS station	TM	14	4: top of towers; 2: middle of main cables; 2: middle of Ma Wan side span; 6: ¼, ½ and ¾ of main span
	Level sensing station	LV	9	1: abutment; 2: towers; 2: deck of Ma Wan side span; 4: deck of main span
Acceleration	Accelerometer	AS, AB, AT	13	4: uniaxial, deck; 7: biaxial, deck and main cables; 2: triaxial, main cables and Ma Wan abutment
Strain	Strain gauge	SP, SR, SS	110	29: Ma Wan side span 32: cross frame at Ma Wan tower 49: main span
Total			276	

The layout of the sensory system for the Tsing Ma Bridge is illustrated in Figure 3.5. The sensors were in seven major types: anemometers, temperature sensors, weigh-in-motion sensors, accelerometers, displacement transducers, level sensing stations, and strain gauges, as listed in Table 3.1. Tag numbers for the sensors except the GPS have the following format that is used in this book:

$$AA\text{-}TBX\text{-}YYC$$

where the letters and numbers have the following meanings: AA – sensor code as listed in Table 3.1, T – for Tsing Ma Bridge, B – section code as shown in Figure 3.5, X – location code ('A' for abutment, 'C' for main cable, 'N' for north side of deck, 'S' for south side of deck, 'T' for tower), YY – sequential number, and C – identifier. For some sensors, YYC is not shown in the book for brevity.

In 2001, a Global Positioning System (GPS) was installed in the Tsing Ma Bridge, Kap Shui Mun Bridge, and Ting Kau Bridge, adding 29 GPS stations which include 14 rover stations on the Tsing Ma Bridge, six on the Kap Shui Mun Bridge, seven on the Ting Kau Bridge, and two reference stations. Details of the various sensors are described in the following sections.

The DATS for Tsing Ma Bridge has three DAUs connected by a token ring fibre optic network. Position of the DAUs is shown in Figure 3.5. One DAU controls 64 to 128 data channels.

The DPCS comprises two UNIX-based 64-bit Alpha Servers and two 32-bit SGI Intel-based (Quad-CPU) Visual Workstations. The DPSC carries out the overall control of data acquisition, processing, transmission, filing, archiving, backup, display, and operation. The application software systems for WASHMS are customized MATLAB, customized GPS software, and MATLAB data analysis suite.

The SHES comprises one UNIX-based 64-bit (Quad-CPU) Alpha server and one UNIX-based 64-bit Alpha workstation. The server is used for structural health evaluation works based on a customized bridge rating system, together with advanced finite element solvers such as MSC/NASTRAN, ANSYS/Mulitiphysics, ANSYS/LS-DYNA, ANSYS/FE-SAFE, and MATLAB data analysis suite. The workstation is used for the preparation and display of graphical input and output files based on advanced graphical Input/Output tools such as MSC/PATRAN and ANSYS/7.1.

The PDAS comprises a 32-channel PC-based data-logger, 24 portable biaxial servo-type accelerometers, five portable uniaxial servo-type accelerometers and 16 cable drums. It is equipped with a customized LABVIEW software system for acquisition, processing, archiving, storage and display. It is used to measure the tensile forces in cables and to assist the fixed servo-type accelerometers to identify the global dynamic characteristics of the bridge.

The PIMS comprises three portable notebook computers. They are used to carry out the inspection and maintenance work on the SS and DATS. Another major function of the PIMS is to facilitate the system inspection and maintenance by storing and updating all the system design information including all drawings and all operation and maintenance manuals.

It is noted that the DPCS, SHES, PDAS, and PIMS mentioned above also serve the WASHMS of the Kap Shui Mun Bridge and Ting Kau Bridge. Except for SS and DATS deployed on three individual bridges, their DPCS and SHES are integrated and the central control office is located in the Tsing Yi administration building.

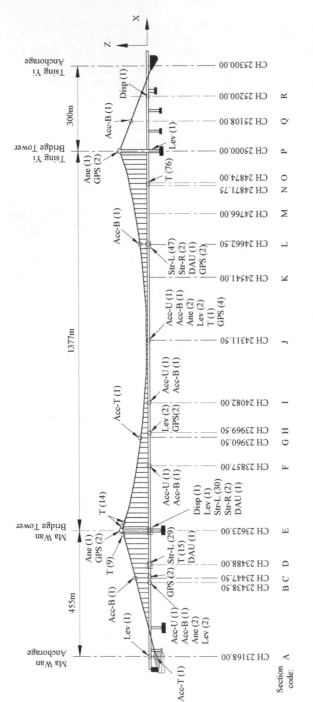

Figure 3.5 Layout of sensors and DAUs of the Tsing Ma Bridge

Note: 1) Numbers in parentheses are the numbers of sensors.

2) Lev: Level sensing (9); Ane: Anemometer (6); Acc–U: Uniaxial Accelerometer (4); Acc–B: Biaxial Accelerometer (7); Acc–T: Triaxial Accelerometer (2); Str–L: Linear strain gauge (106); Str–R: Rosette strain gauge (115); Disp: Displacement transducer (2); DAU: Data acquisition unit (3).

3) Weigh-in-motion sensors are not located on the bridge and not shown in the figure

3.8.2 Anemometers in WASHMS

The WASHMS of the bridge includes a total of six anemometers with two at the middle of the main span; two at the middle of the Ma Wan side span; and one of each on the Tsing Yi Tower and Ma Wan Tower (see Figure 3.6). To prevent disturbance from the bridge deck, the anemometers at the deck level were respectively installed on the north side and south side of the bridge deck via a boom 8.965 m long from the leading edge of the deck (see Figure 3.7).

The anemometers installed on the north side and south side of the bridge deck at the middle of the main span, respectively specified as WI-TJN-01 and WI-TJS-01, are the digital type Gill Wind Master ultrasonic anemometers. The anemometers located at the two sides of the bridge deck near the middle of the Ma Wan approach span, specified as WI-TBN-01 on the north side and WI-TBS-01 on the south side, are the analogue mechanical (propeller) anemometers. Each analogue anemometer consists of a horizontal component (RM Young 05106) with two channels, giving the horizontal resultant wind speed and its azimuth, and a vertical component (RM Young 27106) with one channel, providing the vertical wind speed. The other two anemometers arranged at 11 m above the top of the bridge towers, respectively specified as WI-TPT-01 for the Tsing Yi tower and WI-TET-01 for the Ma Wan tower, are analogue mechanical anemometers of a horizontal component only. The sampling frequency of measurement of wind speeds was set as 2.56 Hz.

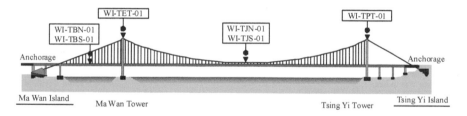

Figure 3.6 Distribution of anemometers in the Tsing Ma Bridge

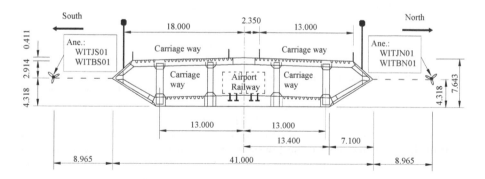

Figure 3.7 Deck cross-section and position of anemometers (unit: m)

3.8.3 Temperature Sensors in WASHMS

The total number of temperature sensors installed on the bridge is 115. Their positions are shown in Figure 3.8. The collected temperature data from WASHMS can be grouped into three categories: (1) ambient temperature (P3, 6 in number); (2) section temperature (P1, P2, P4, and P6, 86 in number); and (3) cable temperature (TC, 23 in number). The TC type is a thermocouple sensor, and P1 to P6 are PT100 Platinum resistance temperature sensors.

One air temperature sensor (P3-TJS) is approximately at the middle section of the main span and attached on a sign gantry which stands on the upper deck. The other five air temperature sensors (P3-TON and P3-TOS) measure the ambient temperature around the bridge deck section of the main span near the Tsing Yi tower. 23 thermocouples (TC-TEC) were embedded inside the main cables at three different locations to measure the cable temperature.

The deck section in the Ma Wan side span is equipped with 15 sensors (P1-TDS). Deck section 'O' close to Tsing Yi Tower is equipped with 71 sensors, in particular, 43 sensors (P1-TON, P1-TOS, P4-TOS, and P6-TOS) installed on the cross frame and 28 temperature sensors (P2-TNN and P2-TNS) mounted on the orthotropic deck plates 2.25 m away from the section. The sampling frequency of all of the temperature sensors is 0.07 Hz.

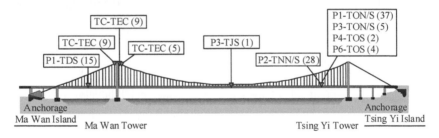

Figure 3.8 Distribution of temperature sensors in the Tsing Ma Bridge

3.8.4 Displacement Transducers in WASHMS

The displacement of Tsing Ma Bridge in the three orthogonal directions: (1) longitudinal (x-direction); (2) lateral (y-direction); and (3) vertical (z-direction), can be reflected by the measurement data recorded from displacement transducers, level sensing stations, and GPS stations together.

The lateral and longitudinal movements of the bridge deck are measured by two displacement transducers at two locations, as shown in Figure 3.9. One displacement transducer (DS-TEN-01), which indicates the lateral motion with a positive value if the deck sways to the north side, is installed in the north side of a bearing frame which sits on the lowest portal beam of the Ma Wan tower with bearing connection. The transducer is connected between the bearing frame and the tower leg of the Ma Wan tower, it thus measures the relative lateral movement of the deck to the Ma Wan tower. The longitudinal movement of the bridge deck at the Tsing Yi abutment is recorded by another displacement transducer (DS-TRA-

01). This transducer, which is exactly underneath the expansion joint bearing, is attached between the Tsing Yi abutment and the bottom chord of a cross frame next to the abutment in order to give the absolute displacement of the bridge deck. The measurements from this displacement transducer are directly in units of millimetre and positive if the deck moves toward the Tsing Yi Island.

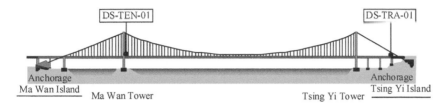

Figure 3.9 Distribution of displacement transducers in the Tsing Ma Bridge

3.8.5 Level Sensing Stations in WASHMS

The vertical motion of the bridge deck is monitored by the level sensing stations which give positive values when the deck moves downward. The level sensing stations are mounted at six locations along the bridge deck as displayed in Figure 3.10. There are in total three level sensing stations installed on one side of the bridge deck at the location of the Ma Wan abutment (LV-TAA-01), Ma Wan tower (LV-TEN-01) and Tsing Yi tower (LV-TPS-01). The vertical displacement at the Ma Wan approach span and the main span are monitored by one pair (LV-TBN/S-01) and two pairs (LV-THN/S-01 and LV-TJN/S-01) of level sensing stations, respectively. The installation positions of pairs of level sensing stations at the deck sections can be seen in Figure 3.11. The torsion of the deck sections is determined by the difference of vertical displacements between the two level sensing stations divided by the distance of separation (Xu *et al.*, 2007).

The level sensing stations, at a sampling rate of 2.56 Hz and cut-off frequency of 1.28 Hz, provide real-time monitoring of displacements, with a measurement accuracy of approximately 2 mm at typical deck sections at vertical planes only. They cannot measure the lateral and longitudinal displacements of the deck sections in the horizontal plane.

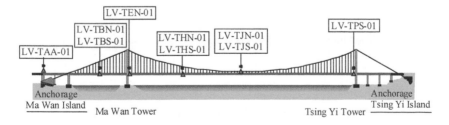

Figure 3.10 Distribution of level sensing stations in the Tsing Ma Bridge

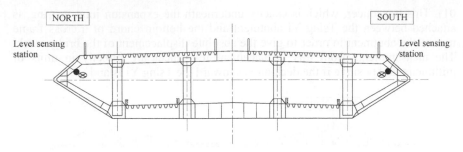

Figure 3.11 Mounting position of level sensing stations in the Tsing Ma Bridge

3.8.6 Global Positioning System in WASHMS

Real-time measurement accuracy of GPS has been improved to centimetre-level, making it well suited to monitor three-dimensional displacement of bridges in response to wind, temperature, and traffic loads. The commissioning of the GPS in January 2001 brought an additional 14 rover stations into the existing WASHMS, for improving the efficiency and accuracy of the bridge health monitoring system.

The components of the bridge implemented with the GPS receivers include bridge towers, main cables and bridge deck, as shown in Figure 3.12. Two base reference stations sit atop a storage building adjacent to the bridge monitoring room (see Figure 3.13). The Ma Wan tower and the Tsing Yi tower are respectively equipped with a pair of GPS receivers and they are mounted at the top of saddles on each of the tower legs. The displacement of the main cables is monitored through a pair of GPS receivers at the mid-span. The mid-span of the Ma Wan side span is equipped with a pair of receivers. Three pairs of GPS receivers are located at one quarter, one half, and three quarters of the main span of the bridge deck.

The displacements monitored by the GPS indicate positive movements in the corresponding three directions if the bridge component moves toward the Tsing Yi Island, sways to the north side, and goes upward, respectively.

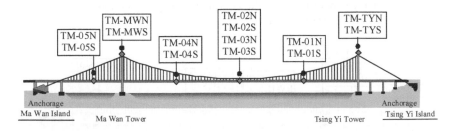

Figure 3.12 Distribution of GPS receivers in the Tsing Ma Bridge

Figure 3.13 A GPS reference station on the roof of a storage building

3.8.7 Strain Gauges in WASHMS

There are 110 strain gauges installed at three sections of the Tsing Ma Bridge with sampling rates of 51.2 Hz, as shown in Figure 2.5. The strain gauges have three types of configuration, that is, 44 of single gauges, 62 of pairs, and four of rosettes.

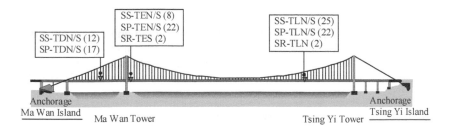

Figure 3.14 Distribution of strain gauges in the Tsing Ma Bridge

3.8.8 Accelerometers in WASHMS

The WASHMS of the bridge includes 13 Honeywell Q-Flex QA700 accelerometers: four uni-axial and four bi-axial accelerometers at four sections of the bridge deck, three bi-axial and one tri-axial on the main cables, and one tri-axial at the Ma Wan abutment, as shown in Figure 3.15. In each of the four deck sections, one bi-axial accelerometer (AB-TBS-01, AB-TFS-01, AB-TIS-01, and AB-TJS-01) is installed on the south side for measuring the accelerations in the vertical and lateral directions, while one uni-axial accelerometer (AS-TBN-01, AS-TFN-01, AS-TIN-01, and AS-TJN-01) was installed on the north side to measure the vertical acceleration. Three bi-axial accelerometers (AB-TCC-01, AB-TGC-01, and AB-TQC-01) were installed on the main cables for measuring the accelerations in the vertical and horizontal directions. Tri-axial accelerometers AT-TGC-01 and AT-TAA-01 respectively measure the acceleration of the main cable and Ma Wan abutment in the longitudinal, vertical, and lateral directions. Hourly acceleration time histories are recorded on tapes with the sampling frequency of 25.6 Hz before 2001 and 51.2 Hz from 2002 onward.

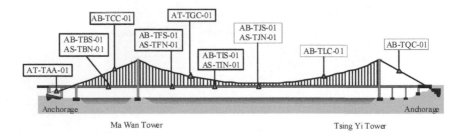

Figure 3.15 Distribution of accelerometers in the Tsing Ma Bridge

3.8.9 Weigh-in-motion Stations in WASHMS

The traffic loads on Tsing Ma Bridge are monitored by seven dynamic WIM stations which were respectively installed in seven lanes of the carriageways (three in the Airport bound and four in the Kowloon bound) at the approach to Lantau Toll Plaza near Lantau administration building, as shown in Figure 3.16. Each sensor is composed of two bending path pads and two magnetic loop detectors. The two bending path pads respectively placed on the left and right hand sides of the lanes were used for vehicle weight detection. The two magnetic loop detectors respectively installed in the front and rear of the carriageways at a defined distance were used for measuring axle numbers, axle length, as well as vehicle speed and height.

Figure 3.16 WIM stations in Lantau Toll Plaza

3.9 REFERENCES

Abe, M., Fujino, Y. and Kaito, K., 2001, Damage detection of civil concrete structures by laser Doppler vibrometry. In *Proceedings of the 19th International Modal Analysis Conference*, edited by Wicks, A.L. and Singhal R., (Bethel: Society for Experimental Mechanics), pp. 704-709.

Aktan, A.E., Catbas, F.N., Grimmelsman, K.A. and Pervizpour, M., 2002, *Development of a Model Health Monitoring Guide for Major Bridges*. Federal Report No. DTFH61-01-P-00347, (Philadelphia: Drexel Intelligent Infrastructure and Transportation Safety Institute).

Akyildiz, I.F., Su, W., Sankarasubramaniam, Y. and Cayirci, E., 2002, Wireless sensor networks: a survey. *Computer Networks*, **38(4)**, pp. 393-422.

Austin, T.S.P., Singh, M.M., Gregson, P.J., Dakin, J.P. and Powell, P.M., 1999, Damage assessment in hybrid laminates using an array of embedded fibre optic sensors. In *Proceedings of SPIE Smart Structures and Materials 1999: Smart Systems for Bridges, Structures, and Highways*, San Diego, California, edited by Liu, S.C., (Bellingham, Washington: SPIE), **3671**, pp. 281-288.

Avitabile, P., 2001, Modal space: back to basics. *Experimental Techniques*, **25(3)**, pp. 15-16.

Bales, F.B., 1985, Close-range photogrammetry for bridge measurement, *Transportation Research Record*, **950**, (Washington, D.C.: Transportation Research Board), pp. 39-44.

Casas, J.R. and Cruz, P.J.S., 2003, Fiber optic sensors for bridge monitoring. *Journal of Bridge Engineering, ASCE*, **8(6)**, pp. 362-373.

Chan, W.S., Xu, Y.L., Ding, X.L. and Dai, W.J., 2006, Integrated GPS-accelerometer data processing techniques for structural deformation monitoring. *Journal of Geodesy*, **80(12)**, pp. 705-719.

Cohen, E.I., Mastro, S.A., Nemarich, C.P., Korczynski, J.F., Jarrett, A.W. and Jones, W.C., 1999, Recent developments in the use of plastic optical fiber for an embedded wear sensor. In *Proceedings of SPIE Smart Structures and Materials 1999: Sensory Phenomena and Measurement Instrumentation for Smart Structures and Materials*, San Diego, California, edited by Claus, R.O. and Spillman, W.B., (Bellingham, Washington: SPIE), **3670**, pp. 256-267.

Daubechies, I., 1992, *Ten Lectures on Wavelets*, (Philadelphia: Society for Industrial and Applied Methematics).

Duan, Z.D. and Zhang, K., 2006, Data mining technology for structural health monitoring. *Pacific Science Review*, **8**, pp. 27-36.

Fayyad, U., Piatetsky-Shapiro, G. and Smyth, P., 1996a, From data mining to knowledge discovery in databases. *AI Magazine*, **17(3)**, pp. 37-54.

Fayyad, U., Piatetsky-Shapiro, G., Smyth, P. and Uthurusamy, R., 1996b, *Advances in Knowledge Discovery and Data Mining*, (Menlo Park: American Association for Artificial Intelligence Press).

Hall, D.L., 1992, *Mathematical Techniques in Multisensor Data Fusion*, (Norwood: Artech House Inc.).

Hall, D.L. and Llinas, J., 1997, An introduction to multisensor data fusion. In *Proceedings of IEEE*, **85(1)**, pp. 6-23.

Hofmann-Wellenhof, B., Lichtenegger, H. and Collins, J., 2001, *GPS: Theory and Practice*, 5th ed., (New York: Springer).

Hofmann-Wellenhof, B., Lichtenegger, H. and Wasle, E., 2008, *GNSS-Global Navigation Satellite Systems: GPS, GLONASS Galileo, and More*, (New York: Springer).

Hollar, S., 2000, *COTS Dust*. Master Thesis, (Berkeley: University of California at Berkeley).

Huang, N.E., Shen, Z., Long, S.R., Wu, M.C., Shih, H.H., Zheng, Q., Yen, N.C., Tung, C.C. and Liu, H.H., 1998, The empirical mode decomposition and the Hilbert spectrum for nonlinear and non-stationary time series analysis. In *Proceedings of Royal Society London A*, **454**, pp. 903-995.

Ji, Y.F. and Chang, C.C., 2008, Non-target image-based technique for bridge cable vibration measurement. *Journal of Bridge Engineering, ASCE*, **13**(1), pp. 34-42.

Kahn, J.M., Katz, R.H. and Pister, K.S.J., 1999, Mobile networking for smart dust. In *Proceedings of ACM/IEEE Internatioanl Conference on Mobile Computing and Networking*, Seattle, WA, (New York: IEEE Communications Society).

Kaito, K., Abe, M. and Fujino, Y., 2001, An experimental modal analysis for RC bridge decks based on non-contact vibration measurement. In *Proceedings of the 19th International Modal Analysis Conference*, edited by Wicks, A.L. and Singhal R., (Bethel: Society for Experimental Mechanics), pp. 1561-1567.

Kling, R.M., 2003, Intel Mote: an enhanced sensor network node. In *Proceedings of International Workshop on Advanced Sensors, Structural Health Monitoring, and Smart Structures*, Tokyo, Japan, [CD-ROM].

Kling, R.M., 2005, Intel Motes: advanced sensor network platforms and applications. In *Proceedings of IEEE MTT-S International Microwave Symposium Digest*, Long Beach, CA, USA, (Piscataway, New York: Institute of Electrical and Electronics Engineers), pp. 365-368.

Kottegoda, N.T. and Rosso, R., 1997, *Statistics, Probability, and Reliability for Civil and Environmental Engineers*, (New York: McGraw-Hill).

Kurata, N., Spencer Jr, B.F., Ruiz-Sandoval, M., Miyamoto, Y. and Sako, Y., 2003, A study on building risk monitoring using wireless sensor network MICA-Mote. In *Proceedings of First International Conference on Structural Health Monitoring and Intelligent Infrastructure*, Tokyo, Japan, [CD-ROM].

Li, J.C., and Yuan, B.Z., 1988, Using vision technique for bridge deformation detection. In *Proceedings of the International Conference on Acoustic, Speech and Signal Processing*, New York, (Piscataway, New York: Institute of Electrical and Electronics Engineers), pp. 912-915.

Lo, Y.L. and Shaw, F.Y., 1998, Development of corrosion sensors using a single-pitch Bragg grating fiber with temperature compensations. In *Proceedings of SPIE Smart Structures and Materials 1998: Smart Systems for Bridges, Structures, and Highways*, edited by Liu, S.C., (Bellingham, Washington: SPIE), **3325**, pp. 64-72.

Lynch, J.P., Law, K.H., Kiremidjian, A.S., Kenny, T.W., Carryer, E. and Partridge, A., 2001, The design of a wireless sensing unit for structural health monitoring. In *Proceedings of the 3rd International Workshop on Structural Health Monitoring*, Stanford, CA, edited by Livingston, R. A., (California: Stanford University).

Lynch, J.P., Sundararajan, A., Law, K.H., Kiremidjian, A.S., Kenny, T.W. and Carryer, E., 2002, Computational core design of a wireless structural health monitoring system. In *Proceedings of Advances in Structural Engineering and*

Mechanics Conference, Pusan, Korea, edited by Choi, C.K. and Schnobrich W.C., (Daejeon, Korea: Techno Press).

McCall, B. and Vodrazka, W.C., 1997, *States' Successful Practice Weigh-in-motion Handbook*, (Washington, D.C.: Department of Transportation, Federal Highway Administration).

Mufti, A., 2001, *Guidelines for Structural Health Monitoring*, (Winnipeg, Canada: Intelligent Sensing for Innovative Structures).

Ni, Y.Q., Xia, Y., Liao, W.Y. and Ko, J.M., 2009, Technology innovation in developing the structural health monitoring system for Guangzhou new TV tower. *Structural Control and Health Monitoring*, **16(1)**, pp. 73-98.

Olaszek, P., 1999, Investigation of the dynamic characteristic of bridge structures using a computer vision method. *Measurement*, **25**, pp. 227-236.

Oppenheim, A.V. and Schafer, R.W., 1989, *Discrete-Time Signal Processing*, (Englewood Cliffs: Prentice-Hall).

Patsias, S., and Staszewski, W.J., 2002, Damage detection using optical measurements and wavelets. *Structural Health Monitoring*, **1(1)**, pp. 7-22.

Rice, J.A. and Spencer Jr, B.F., 2008, Structural health monitoring sensor development for the Imote2 platform. In *Proceedings of SPIE Sensors and Smart Structures Technologies for Civil, Mechanical, and Aerospace Systems*, edited by Tomizuka, M., (Bellingham, Washington: SPIE).

Ruiz-Sandoval, M., Nagayama, T. and Spencer Jr, B.F., 2006, Sensor development using Berkeley Mote platform. *Journal of Earthquake Engineering*, **10(2)**, pp. 289-309.

Sahin, M., Tari, E. and Ince, C.D., 1999, Continuous earthquake monitoring with global positioning system (GPS). In *Proceedings of ITU-IAHS International Conference on Kocaeli Earthquake*, A Scientific Assessment and Recommendations for Re-Building, Istanbul, Turkey, edited by Karaca M. and Ural D. N., (Istanbul: Istanbul Technical University), pp. 231-238.

Seim, J.M., Udd, E., Schulz, W.L. and Laylor, H.M., 1999, Health monitoring of an Oregon historical bridge with fiber grating strain sensors. In *Proceedings of SPIE Smart Structures and Materials 1999: Smart Systems for Bridges, Structures, and Highways*, edited by Liu, S.C., (Bellingham, Washington: SPIE), **3671**, pp. 128-134.

Sohn, H., Farrar, C.R., Hemez, F.M., Czarnecki, J.J., Shunk, D.D., Stinemates, D.W. and Nadler, B.R., 2003, *A Review of Structural Health Monitoring Literature: 1996-2001*, Report LA-13976-MS, (Los Alamos: Los Alamos National Laboratory).

Stewart, A., Carman, G. and Richards, L., 2005, Health monitoring technique for composite materials utilizing embedded thermal fiber optic sensors. *Journal of Composite Material*, **39(3)**, pp. 199-213.

Sun, R.J., Sun, Z. and Sun, L.M., 2006, Design and performance tests of a FBG-based accelerometer for structural vibration monitoring. In *Proceedings of the 4th China-Japan-US Symposium on Structural Control and Monitoring*, edited by Xiang, Y.Q., (Hangzhou: Zhejiang University Press), pp. 450-451.

Swartz, R.A. and Lynch, J.P., 2009, Wireless sensors and networks for structural health monitoring of civil infrastructure systems. In *Structural Health Monitoring of Civil Infrastructure Systems*, edited by Karbhari, V.M. and Ansari, F., (Cambridge: Woodhead Publishing Limited).

Vurpillot, S., Casanova, N., Inaudi, D. and Kronenburg, P., 1997, Bridge spatial displacement monitoring with 100 fiber optic sensors deformations: sensors network and preliminary results. In *Proceedings of SPIE Smart Structures and Materials 1997: Smart Systems for Bridges, Structures, and Highways*, edited by Stubbs, N., (Bellingham, Washington: SPIE), **3043**, pp. 51-57.

Wong, K.Y., 2004, Instrumentation and health monitoring of cable-supported bridges. *Structural Control and Health Monitoring*, **11(2)**, pp. 91-124.

Wong, K.Y. and Ni, Y.Q., 2009, Chapter 123: Modular architecture of structural health monitoring system. In *Encyclopedia of Structural Health Monitoring*, edited by Boller C., Chang F.K. and Fujino Y, (Chichester: John Wiley & Sons), **5**, pp. 2089-2105.

Xu, Y.L., Guo, W.W., Chen, J., Shum, K.M. and Xia, H., 2007, Dynamic response of suspension bridge to typhoon and trains. I: Field measurement results. *Journal of Structural Engineering, ASCE*, **133(1)**, pp. 3-11.

Yan, T.H, Chen, X.D. and Lin, R.M., 2008, Feedback control of disk vibration and flutter by distributed self-sensing piezoceramic actuators. *Mechanics Based Design of Structures and Machines*, **36(3)**, pp. 283-305.

Yick, J., Mukherjee, B. and Ghosal, D., 2008, Wireless sensor network survey. *Computer Networks*, **52(12)**, pp. 2292-2330.

Zhou, X.Q., Xia, Y., Deng, Z.K., and Zhu, H.P., 2010, Experimental study on videogrammetric technique for vibration displacement measurement, In *Proceedings of the 11th International Symposium on Structural Engineering*, edited by Cui, J., Xing, F., Ru, J.P. and Teng, J.G., (Beijing: Science Press), pp. 1059-1063.

3.10 NOTATIONS

$Cov(\cdot)$	Covariance.
f	Input force.
$F(j)$	Fourier transform of signal, $j = 0, 1,\ldots, n$-1.
f_s	Sampling rate.
n	Total number of data.
$P(j)$	Power spectrum, $j = 0, 1,\ldots, n$-1.
R_{xx}	Auto-correlation function.
S_{ff}	Auto-power spectrum density of the input force.
S_{fx}	Cross-power spectrum density between the input force and response.
S_{xx}	Auto-power spectrum density.
x	Explanatory variable, Output response.
$x(k)$	A series of signal, $k = 0, 2,\ldots, n$-1.
x_{max}, x_{min}	Maximum and minimum absolute values of a series of numbers.
x_{std}	Standard deviation of measurement in a specific period.
y	Response variable.
ε	Real number defining the limit of the difference between the maximum and minimum values.

ρ	Correlation coefficient.
σ	Standard deviation.
β_0	Regression coefficient: intercept.
β_x	Regression coefficient: slope.
ε_y	Regression coefficients: error.

CHAPTER 4

Structural Health Monitoring Oriented Modelling

4.1 PREVIEW

The finite element (FE) method has become the widely accepted analysis tool in many disciplines. In bridge engineering, an accurate FE model of the bridge is often an essential tool to facilitate an effective assessment of bridge performance. This is particularly true of the prognosis of bridge performance because the performance predication of the bridge under a diversity of loadings and structural conditions could only be accomplished by an FE model based computational approach. To obtain an accurate FE model of the bridge, appropriate modelling is essential.

Due to the complexity of bridge configurations, construction variations, and uncertainties of boundary conditions, a highly idealized numerical model of a large-scale bridge usually differs from the as-built structure. The initial FE model needs to be fine tuned to match some references. Theses references are usually field measurement results or particularly vibration measurement results. The procedure is known as model updating.

This chapter will first introduce some commonly used FE modelling techniques for bridges, in particular, the spine beam type modelling and hybrid type modelling which integrates line elements, planar elements, and solid elements together. Next, two types of system identification techniques, the input-output methods and output-only methods, are described. The identified results are used in the vibration based FE model updating. Finally the SHM-oriented FE modelling, output-only field measurements, and the associated model updating of the Tsing Ma Bridge are described. The FE models will be applied for other purposes in later chapters, for example, bridge-vehicle interactions, wind-induced bridge responses, and so forth.

4.2 MODELLING TECHNOLOGY

4.2.1 Theoretical Background

Vibration of a linear structural system with N degrees of freedom (DOFs) is governed by the equation of motion as

$$[M]\{\ddot{x}\} + [C]\{\dot{x}\} + [K]\{x\} = \{f\} \tag{4.1}$$

where $[M]$ is the mass matrix, $[C]$ the viscous damping matrix, $[K]$ the stiffness matrix, $\{x\}$ is a vector containing nodal displacement, and $\{f\}$ the force vector. The matrices and vectors have an order of N in size, corresponding to the number of DOFs used in describing the displacements of the structure. It is noted that the linear viscous damping is commonly used because of the convenient form. Hysteretic damping, having a complex form of equation of motion, can lead to a more realistic result in terms of energy loss.

By the FE approach, the structure is assumed to be divided into a system of discrete elements which are interconnected only at a finite number of nodal points. The properties of the complete structure (the mass and stiffness matrices) are then found by evaluating the properties of the individual FE and superposing them appropriately (Clough and Penzien, 2003).

Within each element, each node holds specific DOFs depending on the problem described. Under external loads, the deflected shape of element follows a specific displacement function which satisfies nodal and internal continuity requirements. Based on the load-displacement relation, the element stiffness matrix can be established.

For mass property, the simplest method is to assume that the mass is concentrated at the nodes at which the translational displacements are defined and rotational inertia is null. This is referred to as lumped mass and the matrix has a diagonal form. On the other hand, following the same method and same displacement function in deriving the element stiffness matrix, the consistent mass matrix can be calculated.

Although the FE concept could also be used to define the damping coefficients of the system, the damping is generally expressed in terms of damping ratios which can be identified from experiments. If an explicit expression of the damping matrix is needed, it is generally calculated from the mass and stiffness matrices, which is called Rayleigh damping.

$$[C] = \alpha[M] + \beta[K] \tag{4.2}$$

Two Rayleigh damping factors, α and β, can be experimentally obtained from damping ratios of two modes (Clough and Penzien, 2003).

The concentrated load acting on the nodes can be directly applied. Other loads such as distributed forces can be similarly evaluated as deriving the consistent mass matrix with the same displacement function.

In engineering practice, the mass and stiffness matrices and load vector are automatically computed by structural analysis computer software. However, it is of the utmost importance for the users to understand the theories, assumptions, and limitations of numerical modelling using the FE method, as well as the limitations of the computation algorithms.

There have been numerous studies on the FE modelling of long-span cable supported bridges to facilitate static and dynamic analyses (Xu et al., 1997). Most of the studies are based on a simplified spine beam model of equivalent sectional properties to the actual structural components. Such a simplified model is effective to capture the dynamic characteristics and global structural behaviour of the bridge without heavy computational effort. However, local structural behaviour, such as stress and strain concentration at joints which is prone to cause local damage in static and/or dynamic loading conditions, could not be estimated directly. Apparently, the spine beam model is not the best option from the perspective of bridge health monitoring because, in general, the modelling of local geometric features is insufficient. In this regard, an FE model with finer details in highlighting local behaviours of the bridge components is needed. On the other hand, with the rapid development of information technology, the improvement of speed and memory capacity of personal computers has made it possible to establish a more detailed FE model for a long-span suspension bridge. Apparently, a finer FE model will cost more computational resources including computational time and storage memory.

In general, a simplified spine beam FE model of a bridge serves to understand the global behaviour of the structure, preliminary design, aerodynamic analysis, and so on. A fine model with solid and/or shell elements can be used for stress analysis so that the results can be directly compared with the measured ones (Fei *et al.*, 2007). To trade off the computational efficiency and capability, one can establish a hybrid model or multi-scale model to cater for the objectives. This section introduces some main issues in modelling a long-span bridge with the FE method. The theory of the FE method can be found in many references (Zienkiewicz and Taylor, 1994; Bathe, 1996) and will not be described here.

4.2.2 Spine Beam Model

In a spine beam model, components of a long-span bridge are modelled by line elements including beam elements, truss elements, and rigid links. Pylons and piers are usually modelled with beam elements based on the geometric properties. Each suspender is modelled by a truss element. A segment of main cables between two suspenders is modelled with a truss element and the geometric nonlinearity due to cable tension is taken into consideration. The static equilibrium profile of the main cable needs to be calculated iteratively based on the static horizontal tensions and the unit weight of both the cable and deck given in the design. A deck, however, is more challenging to model. Usually there are two simplified approaches to the deck modelling.

In the first approach, the deck is modelled as a central beam (the spine beam), suspended by suspenders through rigid links. The equivalent cross-sectional area of the deck is calculated by adding together all cross-sectional areas. In the case of a composite section, the areas should be converted to that of one single material, according to the modular ratio. The neutral axes and moments of inertia about the vertical and transverse axes are also determined in a similar way. The calculation of the torsional stiffness of the deck section should consider both pure and warping torsional constants. The mass moment of inertia of the deck should include those of all members according to their distances to the centroid of the section.

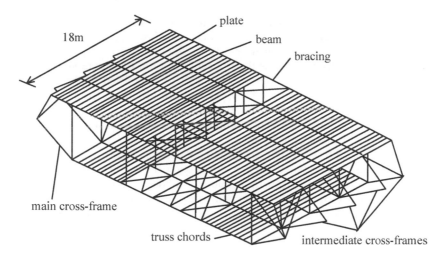

(a) The sectional model

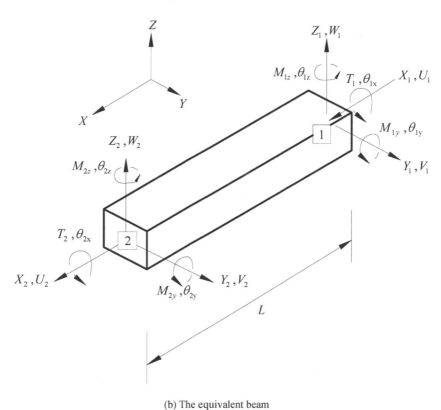

(b) The equivalent beam

Figure 4.1 One segment of the deck system of Tsing Ma Bridge

For a more complicated deck with a number of structural members especially with diagonally placed braces and irregular members, an alternative to the equivalent beam is to use the displacement method. As an example, Figure 4.1 shows an 18 m long segment of the deck of the Tsing Ma Bridge, represented by a fine three-dimensional model. The segment can be simplified to the equivalent beam of 18 m long and 12 DOFs, each corresponding to a generalised displacement and a generalised force. The resulting element stiffness has a size of 12×12. A positive unit displacement is imposed on one DOF of the segment while all other possible displacements in the three-dimensional model are prevented. The resultant generalized forces of the segment, representing the stiffness coefficients, are then calculated. It is noted that, in the section model, beam elements, shell elements and solid elements can be employed although Figure 4.1(a) has beam elements and shell elements only. The equivalent spine beam is connected to the suspender nodes through a series of transverse rigid links, as shown in Figure 4.2.

Constraints are necessary to connect different parts of the model together and to enforce certain types of rigid-body behaviour. For example, the nodes of the deck, bearings, and tower do not coincide with each other. Rigid links are used to restrain their motions in different directions, depending on the bearing types and boundary conditions.

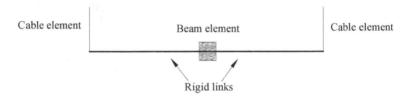

Figure 4.2 Connection between the spine beam element and the suspender cable element

4.2.3 Hybrid Model

To capture the local responses of some critical members and/or obtain more accurate results, the entire bridge can be modelled with different types of elements. Ever increasing capacity of computers makes this hybrid modelling more and more popular. In such a hybrid model, also known as a multi-scale model (Li *et al.*, 2007), the components of interest can be modelled with shell elements or solid elements and other components still with line elements.

The deck of a long-span suspension bridge is usually complicated in geometry and consists of deck plates and cross frames. The plate is generally modelled with shell elements and the cross frames with beam type, as shown in Figure 4.1(a). The towers and tiers are relatively simple in geometry, thus beam elements may be sufficient for most bridges. In this case the nodes of the cross frames do not coincide with the nodes of the shell elements which are located in the middle plane of the plate. Multi-point constraints (MPC) are needed to connect them to avoid displacement incompatibility among the nodes.

Where the joints or welded details are concerned, shell elements may be insufficient and solid elements are required. In a hybrid model or even finer model, solid elements can be employed to model most of members of a bridge. However, special care must be taken at the interface between a solid element and a shell element (or a beam element). This is because the former usually has three translational DOFs only at each node and does not have rotational DOFs as the latter has. MPCs must therefore be assigned to relate the shell element (or beam element) rotations to the solid element translations.

4.2.4 Modelling of Cables

Special attention should be paid to modelling of cables in considering the sag effect and the initial tension force under full dead loads. The sag effect of a cable is considered using the following equivalent modulus for the straight truss element to replace the actual modulus of the cable.

$$E_{eq} = \frac{E}{1 + \dfrac{(\rho A g \overline{l})^2 AE}{12T^3}} \tag{4.3}$$

where E_{eq} is the equivalent modulus of elasticity, E is the effective modulus of elasticity, ρ the density, g the gravity acceleration, $\overline{l}$ the horizontal projected length, A the effective cross sectional area, and T is the mean cable tension. The elastic stiffness matrix of the element in the three-dimensional local coordinate is given by

$$[K_e] = \frac{AE_{eq}}{L_c} \begin{bmatrix} 1 & 0 & 0 & -1 & 0 & 0 \\ 0 & 0 & 0 & 0 & 0 & 0 \\ 0 & 0 & 0 & 0 & 0 & 0 \\ -1 & 0 & 0 & 1 & 0 & 0 \\ 0 & 0 & 0 & 0 & 0 & 0 \\ 0 & 0 & 0 & 0 & 0 & 0 \end{bmatrix} \tag{4.4}$$

where L_c is the chord length of the cable. Furthermore, a geometric stiffness matrix of the element should be included for consideration of the cable tension force.

$$[K_g] = \frac{T}{L_c} \begin{bmatrix} 0 & 0 & 0 & 0 & 0 & 0 \\ 0 & 1 & 0 & 0 & -1 & 0 \\ 0 & 0 & 1 & 0 & 0 & -1 \\ 0 & 0 & 0 & 0 & 0 & 0 \\ 0 & 1 & 0 & 0 & -1 & 0 \\ 0 & 0 & 1 & 0 & 0 & -1 \end{bmatrix} \tag{4.5}$$

The tangent stiffness matrix of the element in the local coordinate is then obtained by

$$[K_t] = [K_e] + [K_g] \tag{4.6}$$

The tangent stiffness matrix of the cable system of the bridge in the global coordinate can then be assembled from the element tangent stiffness matrices.

4.3 SYSTEM IDENTIFICATION

An FE model of a large-scale bridge, based on design drawings, inevitably has errors to some extent due to several reasons, such as mesh discretisation, uncertainties in geometry and boundary conditions of the bridge, and deviation of the real material parameters from the design ones. Field measurement is usually carried out to validate or update the model. In particular, vibration measurement serves to extract bridge dynamic characteristics such as frequencies, damping, and mode shapes for model updating. How to extract dynamic characteristics is referred to as modal identification or experimental modal analysis.

In the 1970s and 1980s of the last century, a vast range of modal identification methods was developed. In general, according to the domain in which the data are treated, there are time domain and frequency domain methods. Maia *et al.* (1997) and Ewins (2000) extensively described these methods. When the excitation method is concerned, bridge tests can be generally divided into forced vibration tests and ambient vibration tests, although free vibration tests are employed occasionally. Correspondingly the modal identification methods are called as input-output methods and output-only methods, respectively.

4.3.1 Input-output Modal Identification Methods

In forced vibration tests, the bridge is excited by artificial means and both input and output need to be measured. The tests often require very heavy equipment and involve expensive resources in order to provide controlled excitation at a sufficient level (Okauchi *et al.*, 1997). Impulse hammers, drop weights, and electro-dynamic shakers are the excitation equipment usually used for relatively small structures.

Quite a few standard techniques of modal parameter estimation have been developed in mechanical and aerospace engineering. In brief, these modal parameter identification methods are based on the frequency response functions (FRFs) in the frequency domain or impulse response functions (IRFs) in the time domain. An FRF is the ratio of the output (or response) to the input (or force) in the frequency domain. In practice, it can be calculated as the ratio of the cross-power spectrum between the output and input to the auto-power spectrum of the input, although there are a few variants (Maia *et al.*, 1997).

To simply introduce the theoretical background of the identification methods, we start from the undamped free vibration equation:

$$[M]\{\ddot{x}\}+[K]\{x\} = \{0\} \tag{4.7}$$

It is known that the above equation has solutions of

$$\{x(t)\} = \{X\}e^{i\omega t} \tag{4.8}$$

where $\{X\}$ is an $N \times 1$ vector of time-independent amplitudes, i the imaginary unit, and ω the vibration circular frequency. Substituting the solution into Equation (4.7), we have

$$\left([K]-\omega^2[M]\right)\{X\}e^{i\omega t} = \{0\} \tag{4.9}$$

or,

$$\left([K]-\omega^2[M]\right)\{X\} = \{0\} \tag{4.10}$$

Equation (4.10) is a standard eigenvalue problem. As the trivial solution is of no interest, determinant of $\left([K] - \omega^2 [M] \right)$ must be zero, resulting in N possible positive real solutions ω_1^2, ω_2^2, ..., ω_N^2 known as the eigenvalues. ω_1, ω_2, ..., ω_N are the undamped natural frequencies of the system. For each natural frequency ω_r, Equation (4.10) has a set of nonzero vector solution of $\{X\}$, denoted as $\{\phi_r\}$ and known as the rth mode shape (or eigenvector) of the system. Each mode shape vector contains N elements that are relative values but not the absolute magnitude. All mode shape vectors form a mode shape matrix $[\Phi]$. Due to the important orthogonality property of the mode shape matrix, $[\Phi]$ is usually normalized with respect to the mass matrix as

$$[\Phi]^T [M][\Phi] = [I] \tag{4.11}$$

and

$$[\Phi]^T [K][\Phi] = [\Lambda] \tag{4.12}$$

where $[I]$ is a unit matrix and $[\Lambda]$ is a diagonal matrix consisting of the eigenvalues, ω_1^2, ω_2^2, ..., ω_N^2.

Considering a harmonic excitation force vector $\{f(t)\} = \{F\}e^{i\omega t}$, the responses are also harmonic with the same frequency (the transient vibration is neglected), i.e., $\{x(t)\} = \{X\}e^{i\omega t}$:

$$\left(-\omega^2 [M] + i\omega [C] + [K] \right) \{X\}e^{i\omega t} = \{F\}e^{i\omega t} \tag{4.13}$$

Therefore

$$\{X\} = \left(-\omega^2 [M] + i\omega [C] + [K] \right)^{-1} \{F\} = [H(\omega)]\{F\} \tag{4.14}$$

where $[H(\omega)]$ is the $N \times N$ FRF matrix. In the case of Rayleigh damping (Equation (4.2)), it has

$$[\Phi]^T [H]^{-1} [\Phi] = [\Phi]^T \left(-\omega^2 [M] + i\omega [C] + [K] \right)[\Phi] = -\omega^2 [I] + i\omega(\alpha [I] + \beta [\Lambda]) + [\Lambda] \tag{4.15}$$

The right-hand side of the above equation is a diagonal matrix as well and the rth item is $-\omega^2 + i\omega(\alpha + \beta\omega_r^2) + \omega_r^2 = -\omega^2 + 2i\omega\omega_r\xi_r + \omega_r^2$ where

$$\xi_r = \frac{\alpha}{2\omega_r} + \frac{\beta\omega_r}{2} \tag{4.16}$$

is defined as the modal damping ratio for the rth mode. Therefore, from Equation (4.15), the FRF matrix can be associated with the modal frequency, mode shapes, and damping as

$$[H] = [\Phi] \begin{bmatrix} \ddots & & \\ & \dfrac{1}{-\omega^2 + 2i\omega\omega_r\xi_r + \omega_r^2} & \\ & & \ddots \end{bmatrix} [\Phi]^T \tag{4.17}$$

The receptance h_{jk}, defined as the response displacement at coordinate j due to a harmonic excitation force at coordinate k with frequency of ω and all other forces are zero, is then given by

$$h_{jk}(\omega) = \frac{\{X\}_j}{\{F\}_k} = \sum_{r=1}^{N} \frac{\phi_{jr}\phi_{kr}}{-\omega^2 + 2i\omega\omega_r\xi_r + \omega_r^2} \tag{4.18}$$

Equation (4.18) is often expressed as

$$h_{jk}(\omega) = \sum_{r=1}^{N} \left(\frac{A_{jkr}}{i\omega - s_r} + \frac{A_{jkr}^*}{i\omega - s_r^*} \right) \tag{4.19}$$

where $s_r = -\omega_r \xi_r + i\omega_r \sqrt{1-\xi_r^2}$ is the pole and A_{jkr} is the residue for mode r, and superscript '*' is the complex conjugate. The pole is directly related to the frequency and damping and the residue is related to the mode shape. Therefore, the FRF (or poles and residues) can be employed to extract the modal parameters. The above equations show that:

1. The FRF matrix is a bridge connecting the modal parameters to the system matrices.
2. It is clear that the FRF matrix is symmetric as $h_{jk} = h_{kj}$, which is called reciprocity.
3. The FRF is the summation of the contributions of different vibration modes.
4. One column or one row of the FRF matrix includes all information of modal parameters, thus it is sufficient to extract the full modal parameters using one column or one row FRF vector only.

We will not give details of each modal parameter estimation method. Readers can refer to books such as Ewins (2000) and Maia *et al.* (1997).

Modal identification methods that can only handle a single FRF at a time are called single-input-single-output (SISO) methods. More common, a structure is tested with a set of sensors simultaneously collecting the responses to one excitation (position is fixed). Each pair of input-output forms an FRF and the total pairs are actually one column of the FRF matrix. The methods allowing for several FRFs to be analysed simultaneously are called global or single-input-multiple-output (SIMO) methods. These methods are based on the fact that the natural frequencies and damping ratios of a structure are global properties of the structure and do not vary from one FRF to another (Maia *et al.*, 1997). Finally, the methods that can simultaneously process a few columns of FRFs from the responses to various excitations are called multiple-input-multiple-output (MIMO) methods. This kind of methods can increase the spatial resolution (the physical measurement locations) of the test as one input may not excite some particular modes sufficiently.

Among many, two very common and useful methods, the peak amplitude method (Bishop and Gladwell, 1963) and the rational fraction polynomial method (Richardson and Formenti, 1982), are named here. The former is a simple, rapid, and useful modal identification method during the tests. The natural frequencies are simply taken from the observation of the peaks (or resonances) on the amplitude graphs of the FRFs. The mode shapes are estimated from the ratios of the magnitudes of the imaginary part (or amplitude) of the FRFs at the resonance points (Avitabile, 1999) without heavy computational efforts. In the rational fraction polynomial method, the FRF is formulated in a rational fraction form in terms of orthogonal Forsythe polynomials. When the error function between the formulated FRF and the measurement is minimized, the modal parameters can be estimated. The rational fraction polynomial method was expanded to the case of multiple FRFs, named as the global rational fraction polynomial method (Richardson and Formenti, 1985), such that the frequencies and damping ratios can be obtained from all FRFs consistently and mode shapes from each FRF. This method is now one of the most popular SIMO frequency domain methods and adopted by many commercial packages of modal analysis software.

4.3.2 Output-only Modal Identification Methods

In contrast, ambient vibrations, induced by traffic, winds, water waves, and pedestrians, are the natural or environmental excitations of bridges. An ambient vibration test has a few significant advantages over the forced vibration tests. First, it does not interrupt the operating condition of the bridge. Second, it is inexpensive as it does not require excitation equipment. Due to these merits, ambient vibration tests have been widely used in vibration tests of bridges such as the Golden Gate Bridge in USA (Abdel-Ghaffer and Scanlan, 1985), the Bosporus Suspension Bridge in Turkey (Brownjohn *et al.*, 1989), the Deer Isle Bridge in USA (Kumarasena *et al.*, 1989), the Quincy Bayview Bridge in Canada (Wilson and Liu, 1991), the Tsing Ma Suspension Bridge in Hong Kong (Xu *et al.*, 1997), the Kap Shui Mun Cable-Stayed Bridge in Hong Kong (Chang *et al.*, 2001), the Roebling Suspension Bridge in USA (Ren *et al.*, 2004), among many others.

In the case of ambient vibration testing, only output data (responses) are measured and actual loading conditions are unknown. Modal parameter identification methods using output-only measurements present a challenge that needs the use of special identification techniques which can deal with small magnitude ambient vibration contaminated by noise without knowledge of the input forces. Over the past decades, several output-only modal parameter identification techniques have been developed such as autoregressive moving-average (ARMA) method (Andersen *et al.*, 1996), natural excitation technique (NExT) (James *et al.*, 1995), and stochastic subspace identification (SSI) (Van Overschee and De Moor, 1996; Peeters, 2000).

It is noted that these methods are based on a fundamental assumption that the unknown excitation is taken as a white noise process. For example, it can be proved that the cross-correlation function between the responses from a multiple DOFs system excited by multiple white noise random inputs has the same form as the IRF of the system (Caicedo *et al.*, 2004). Consequently the time-domain modal parameter estimation methods such as Eigensystem Realisation Algorithm (ERA) (Juang and Pappa, 1985) can be employed to extract the modal parameters.

In practice, each set of measurements in ambient vibration tests is taken for a relatively long period, usually minutes to hours. One reason is to allow wider frequency components included in the excitation, which conforms to the stationary assumption of the input. In addition, more measurement data result in a better resolution in the frequency domain. Finally effect of measurement noise can be reduced by averaging data.

Reference points also need to be considered in practical ambient vibration tests as the number of available sensors is usually insufficient to measure all measurement points at one time. The points can be divided into a few groups and the measurement is taken in several sets, while the reference points should be kept unchanged throughout the entire test.

Using output-only modal identification methods, one cannot obtain an absolute scaling of the identified mode shapes (e.g. mass normalization) as the input is unknown. This is a big difference, sometimes also a disadvantage, from that in the forced vibration tests.

4.4 MODEL UPDATING

4.4.1 Overview

Accurate FE models are frequently required in a large number of applications, such as optimization design, damage identification, structural control, and SHM (Mottershead and Friswell, 1993). Due to the uncertainties in geometry, material properties, and boundary conditions, the dynamic responses of a structure predicted from a highly idealized numerical model usually differ from the measurements obtained from the as-built bridges. For example, Brownjohn and Xia (2000) reported that the difference between the experimental and the numerical modal frequencies of a curved cable-stayed bridge exceeded 10% for most modes and even reached 40%. In another study (Brownjohn *et al.*, 2003), 18% difference was found between the analytical and measured frequencies. Jaishi and Ren (2005) discovered that the difference between the natural frequencies of a steel arch bridge differed from the FE model predictions by up to 20%, and the modal assurance criterion (MAC) values could be 62%. Similarly, Zivanovic *et al.* (2007) found the differences of the natural frequencies for footbridges predicted by a very reasonable FE model before updating could be as large as 29.8%, as compared with the experimental counterparts. Therefore, an effective model updating is necessary to obtain a more accurate FE model for other purposes such as prediction of responses, damage identification, and so forth. Although the measurement noise is also unavoidable, the measurement data obtained from a well designed test are usually regarded as accurate and used as the reference for updating.

Theoretically many types of measurement data can be employed for model updating, while most exercises in past decades are based on modal data or their variants, such as frequencies, mode shapes, FRFs, modal flexibility, modal curvatures, and modal strain energy. Doebling *et al.* (1996) reviewed these methods extensively and compared their pros and cons. Other types of data used include time history (Ge and Soong, 1998) and static strain data (Sanayei and Saletnik, 1996). In this regard, this book focuses on the modal data based model updating.

The eigenvalue equation of an *N*-DOFs undamped FE model is given as

$$\left(-\lambda_i\left[M\right]+\left[K\right]\right)\left\{\phi_i\right\} = \left\{0\right\} \tag{4.20}$$

where λ_i is the ith eigenvalue ($\lambda_i=\omega_i^2$). When there is change or inaccuracy in the mass or/and stiffness, the vibration characteristics such as natural frequencies and mode shapes will change accordingly. Model updating is to find the change or inaccuracy in the structural properties based on the measured data.

Model updating methods are usually classified into one-step methods (or direct methods) and iterative methods (Friswell and Mottershead, 1995; Brownjohn *et al.*, 2001).

The one-step methods directly reconstruct the stiffness matrix and the mass matrix of the analytical model to reproduce the measured modal data. The pioneer work was done by Baruch (1978) and Berman and Nagy (1983). The former considered the mass matrix to be exact and only the stiffness matrix was updated. The latter updated the mass matrix and stiffness matrix sequentially by minimizing the norm of a weighted mass error matrix and stiffness error matrix. The advantage of the one-step methods, as the name implies, is that they do not require iteration and heavy computation. However, the main drawback of the methods is that the updated mass and stiffness matrices have little physical meaning, that is, they cannot be related to the changes in the parameters of the original model.

The iterative methods modify the physical parameters of the FE model repeatedly to minimize the discrepancy between the measurement data and the analytical counterparts. Other than changing matrices directly, this approach adjusts the physical parameters in elemental or substructural level. Then the system stiffness matrix and mass matrix are assembled from all elements in the discrete FE model. Thereby, (1) the matrix properties of symmetry, sparseness, and positive-definiteness are guaranteed; (2) the structural connectivity is preserved; (3) the changes in the updated global matrices are represented by the changes in the updated parameters. Due to these merits, iterative methods have been becoming more popular. Successful model updating applications reported on cable-supported bridges are abundantly available (Brownjohn and Xia, 2000; Zhang *et al.*, 2001; Jaishi and Ren, 2005; Song *et al.*, 2006).

An FE model updating mainly includes three aspects: the objective function with constraints, updating parameters, and optimisation algorithms.

4.4.2 Objective Functions and Constraints

The objective functions usually comprise the difference between the measured quantities and the model predictions. Selection of the measurement data type must take into account not only the sensitivity of the data to the structural parameters, but also the inherent uncertainties in the measurement. In general, as found by many researchers, frequencies can be measured with a high accuracy, but are not spatially specific and not sensitive to damages. Mode shapes have the advantage of being spatially specific. However, it is difficult to measure the high order mode shapes, which are sensitive to local damages. The advantage of using modal flexibility is that the flexibility matrix converges quickly with the first few modes only. However, measuring the modal flexibility with a sufficient spatial resolution is often experimentally impractical. So is the modal strain energy method.

In the objective functions, different types of data can be weighted differently accounting for their importance and measurement accuracy. For example, it is widely accepted that frequencies of a structure can be measured more accurately than mode shapes. Thus higher weights can be given to the frequency data. An objective function considering this can be (Xia, 2002)

$$J = \sum_{i=1}^{nm} W_{fi}^2 \left[f_i (\{p\})^A - f_i^E \right]^2 + \sum_{i=1}^{nm} W_{\phi i}^2 \sum_{j=1}^{np} \left[\phi_{ji} (\{p\})^A - \phi_{ji}^E \right]^2 \qquad (4.21)$$

where $\{f_i\}$ and $\{\phi_{ji}\}$ are the ith natural frequency and associated mode shape at the jth point, respectively; W_f and W_ϕ are the weight coefficients of the frequency and mode shape, respectively; *nm* denotes number of modes, *np* number of measurement points; subscripts "*A*" and "*E*" denote the items from the analytical model and the experiment, respectively; {*p*} includes a vector of the structural parameters to be updated. In Equation (4.21), the relative difference of frequencies can be used instead of the absolute difference (Friswell *et al.*, 1998). From the computational point of view, using eigenvalues as in Equation (4.22) instead of frequencies is preferred as calculation of the eigenvalue sensitivity is more convenient than that of the frequency sensitivity (Hao and Xia, 2002).

$$J = \sum_{i=1}^{nm} W_{\lambda i}^2 \left[\frac{\lambda_i(\{p\})^A - \lambda_i^E}{\lambda_i^E} \right]^2 + \sum_{i=1}^{nm} W_{\phi i}^2 \sum_{j=1}^{np} \left[\phi_{ji}(\{p\})^A - \phi_{ji}^E \right]^2 \tag{4.22}$$

One advantage of using Equations (4.21) and (4.22) is that the measured modal data can be compared with the analytical ones at the corresponding points and modes directly as 1) the number of identified modes is limited by the frequency range due to limitation of the excitation level and identification techniques; 2) the number of measurement locations is limited and is always less than that of the analytical model; and 3) some DOFs, such as rotational and internal ones, cannot be measured with present technologies. Therefore, model reduction or data expansion can be avoided. Special attention, however, should be paid to the following issues:

• The analyst should roughly know the quality of the measurement data. Inclusion of all data does not necessarily lead to more meaningful results because some modes may have higher uncertainty than others.

• Only corresponding modes can be compared in the above equations. Therefore mode matching is necessary and MAC is usually employed for this purpose. During the updating process, sequence of some modes may change and thus mode tracking is necessary in each iteration. The MAC value of two mode shapes {ϕ_i^E} and {ϕ_j^A} is defined by (Ewins, 2000)

$$MAC\left(\{\phi_i^E\},\{\phi_j^A\}\right) = \frac{\left|\{\phi_i^E\}^T \{\phi_j^A\}\right|^2}{\left(\{\phi_i^E\}^T \{\phi_i^E\}\right)\left(\{\phi_j^A\}^T \{\phi_j^A\}\right)} \tag{4.23}$$

• The measured mode shapes may have different scales from the analytical ones which are usually mass-normalized. In the case of ambient vibration tests or where the response of the driving point is not measured in force vibration tests, mass-normalized mode shapes are not available. The modal scale factor (MSF) can be employed so that the two pairs of mode shapes have the same phase. This quantity is defined as (Ewins, 2000)

$$MSF\left(\{\phi_i^E\},\{\phi_i^A\}\right) = \frac{\{\phi_i^E\}^T \{\phi_i^A\}}{\{\phi_i^A\}^T \{\phi_i^A\}} \tag{4.24}$$

Then the analytical mode shape is adjusted by multiplying the modal scale factor.

In model updating, the range of the updating parameters usually needs to be constrained with bounds, such that the updated parameters are in an acceptable and reasonable range. For example, many structural parameters should be larger than zero. In case of damage detection, the damaged structure is usually weaker than the intact one. Optimization with constraints is generally more difficult to solve than the unconstrained one.

4.4.3 Parameters for Updating

Selection of parameters to be updated is very critical to a successful model updating and requires engineering judgements. If incorrect model parameters are selected, the updated model either cannot reproduce the required dynamic properties accurately or does not represent the real structure. In the latter case, although the objective function may reduce to a value below the specified threshold, the updated model has no real meaning and the obtained parameters are, in effect, the compensation for the real model parameters. Therefore, the analyst should first know the type and location of the inaccuracy of the model that needs updating, based on the features of the measurement data and his knowledge of the model. Apparently this is closely associated with the modelling of the structure of interest.

The second issue is the number of updating parameters, which depends on the computational resource. For a small structure in size of system matrices, the updating parameters can be selected to the elemental level and each element may have single or multiple updating parameters. For a large-scale structure such as a suspension bridge, the element number of the model is usually more than thousands. In that case, calculation of eigensolutions and the associated eigensensitivities is very time-consuming. Updating can then be carried out at the component level or substructure level only with the present computational ability. The substructuring based model updating (Weng *et al.*, 2011) can be a promising method and merits further development.

A set of updating parameters may include various types of structural parameters whose values may be in a broad range. On the other hand, different parameters have different sensitivities. Normalising the parameters can make the changes in these updating parameters at a similar level. Consequently a poor condition during the optimisation is avoided. For example, in the widely used element-by-element model updating, the element bending rigidity is usually employed as the elemental stiffness parameter. Then the element stiffness matrix is proportional to the parameter as

$$[K_i] = \alpha_i [K^e] \tag{4.25}$$

where $[K_i]$ is the ith element stiffness matrix, α_i is the element bending rigidity, and $[K^e]$ has the identical form and size for the same element type, which is the function of geometry of the element. The global stiffness matrix is the assembly of all *ne* elements as

$$[K] = \sum_{i=1}^{ne} [K_i] = \sum_{i=1}^{ne} \alpha_i [K^e] \tag{4.26}$$

We assume the geometrical information of the element is accurate and only the bending rigidity contributes to the model inaccuracy and needs to be updated. Consequently, the global stiffness matrix of the updated structure can be similarly assembled as

$$\left[\tilde{K} \right] = \sum_{i=1}^{ne} \left[\tilde{K}_i \right] = \sum_{i=1}^{ne} \tilde{\alpha}_i \left[K^e \right] \tag{4.27}$$

In this manner, the matrix properties and the structural connectivity are preserved after updating, as described previously. One can define a stiffness reduction factor as the actual updating parameter:

$$p_i = \frac{\tilde{\alpha}_i - \alpha_i}{\alpha_i} \tag{4.28}$$

Then all the updating parameters have a value larger than -1 because the updated bending rigidity should be larger than zero.

4.4.4 Optimization Algorithms

The optimization algorithm is the technique to minimise the objective functions. According to the nature of the objective functions and constraints, optimisation methods can be classified as linear and nonlinear ones. Another classification is unconstrained and constrained optimization problems. Constraints can be classified as equality and inequality ones.

As there are quite a few different methods in solving general optimization problems, we do not plan to introduce all of these but focus on the model updating problem which can be solved in a more efficient way by utilising its characteristics. The problem of interest here such as Equations (4.21) or (4.22) is actually a nonlinear least-squares problem, which is the sum of squares of error functions. To exemplify the algorithm, only eigenvalues are considered in Equation (4.22) with unit weights for all items and the equation is rewritten in a vector form as

$$J = \sum_{i=1}^{nm} \left[\lambda_i \left(\{p\} \right) - \lambda_i^E \right]^2 = \left\{ \lambda - \lambda^E \right\}^T \left\{ \lambda - \lambda^E \right\} \tag{4.29}$$

The minimum of J is reached when the gradient is zero. By differentiating J with respect to parameter p, one has

$$\frac{\partial J}{\partial p} = 2S^T \left\{ \lambda - \lambda^E \right\} = \{0\} \tag{4.30}$$

where S is the sensitivity matrix of eigenvalues with respect to the parameters and usually is calculated using Nelson's method (Nelson, 1976). The eigenvalues can be approximated to the first-order Taylor series expansion with respect to the current point as

$$\{\lambda\} = \{\lambda\}^k + S^k \{\Delta p\}^k \tag{4.31}$$

where superscript k denotes the quantity in the kth iterate. For brevity superscript k on S is omitted hereinafter. Substituting Equation (4.31) into (4.30) leads to

$$S^T S \{\Delta p\}^k = S^T \left(\{\lambda^E\} - \{\lambda\}^k \right) = S^T \{\Delta \lambda\}^k \tag{4.32}$$

where $\Delta\lambda$ is the eigenvalue residual. At each iterate k, the eigenvalue residual can be calculated and then the parameter change can be solved. The above approach is called the Gauss-Newton method. In practice, the updated function J may not be lower than that in the previous iterate. A fraction of the increment β is employed such that $\{p\}^{k+1} = \{p\}^k + \beta\{\Delta p\}^k$. Search of such an optimal fraction is referred to as linear search, which is very commonly used in optimisation. If the optimal fraction is close to zero, implying that the increment direction is not effective, the so-called Levenberg-Marquardt algorithm can be employed and the corresponding equation is revised as

$$\left(S^T S + mI\right)\{\Delta p\}^k = S^T \{\Delta\lambda\}^k \tag{4.33}$$

where m is known as the Marquardt parameter and I is usually an identity matrix. It is noted that the increment direction is identical to that of the Gauss-Newton method when $m = 0$ and the steepest descent direction when m tends to infinity (Nocedal and Wright, 1999). An additional benefit of the Levenberg–Marquardt algorithm is that including a matrix in the left-hand side of Equation (4.33) improves the matrix condition as sensitivity matrix S might be ill-conditioned.

It is noted that the original nonlinear least-squares problem (Equation (4.29)) has been transferred into a series of linear least-squares problems at each iteration. In practice, one would not solve Equations (4.32) and (4.33) using inversion of the matrix directly, but via the Cholesky decomposition, QR decomposition, or singular value decomposition.

As mentioned previously, the updated parameters have physical meaning and are bounded, which makes the problem (Equation (4.29)) a constrained optimisation. The upper and lower bounds can then be set beforehand as

$$l \leq p \leq u \tag{4.34}$$

where l and u are respectively lower and upper bounds on parameter p. If the parameter is normalized and p is defined as Equation (4.28) in damage detection, for example, the bounds can be set as $-0.99 \leq p \leq 0$ in practice. The constrained optimisation can be solved with active set methods (Bjorck, 1996).

The active set methods divide the constraints into active and inactive sets. An active constraint refers to the parameter on the lower bound or upper bound, i.e., the equality is satisfied, and an inactive constraint implies the parameter falls inside the bound, i.e., the inequality is satisfied. Accordingly the parameters are divided into fixed variables and free variables. The former will not change during the iteration and can be removed from the problem. Consequently the problem is reduced with a set of free variables without constraints, which can be solved with the Gauss-Newton method or Levenberg-Marquardt method. The main drawback of the active set methods is that the active and inactive sets need to be updated at each iteration by adding or dropping one constraint. The gradient projection method that allows the active set to change rapidly is more efficient. Readers may refer to the work by More and Toraldo (1989) on the topic.

Among many optimization algorithms, the evolutionary algorithm, in particular, the Genetic Algorithm, is worth mention as it has been used in a number of model updating and system identification exercises in structural engineering communities (Larson and Zimmerman, 1993; Mares and Surace, 1996; Friswell *et al.*, 1998; Perry *et al.*, 2006; Hao and Xia, 2002). The Genetic Algorithm is based on the principle of the evolutionary theory such as natural selection and evolution (Holland, 1975). As compared with the traditional optimization and search algorithms, the Genetic Algorithm differs in several fundamental ways:
- It works with the coding of the decision variables, not the decision variables themselves;

- It searches from a population of points in the region of the whole solution space rather than a single point, and therefore it is a global optimisation method;
- It has the advantage of easy implementation because only an objective function is required and derivatives or other auxiliary information are not necessary; and
- It is based on probabilistic transition rules, not deterministic rules.

4.5 SHM-ORIENTED FINITE ELEMENT MODEL OF THE TSING MA BRIDGE

The Tsing Ma Suspension Bridge is used as a case study to demonstrate the establishment of an SHM-oriented FE model for long-span suspension bridges, including hybrid modelling in a commercial software package, ambient vibration measurement and modal identification, and sensitivity based model updating.

In particular, modelling and model updating of the bridge are executed using MSC/PATRAN as the model builder and MSC/NASTRAN as the FE solver. The modelling work is based on the previous model developed by Wong (2002) with the following principles: (1) model geometry should accurately represent actual geometry; (2) one analytical member should represent one real member; (3) stiffness and mass should be simulated and quantified properly; (4) boundary and continuity conditions should accurately represent the reality; and (5) the model should be detailed enough at both global and local levels to facilitate subsequent model updating and SHM applications such as buffeting-induced stress analysis (Liu *et al.* 2009; Duan *et al*, 2011). For different applications, two models have been constructed, one is the spine beam type and the other is the hybrid type. For brevity, only the latter is introduced here, and the former can be found in Xu *et al.* (1997).

4.5.1 Modelling of the Bridge

4.5.1.1 The Bridge Deck

The bridge deck is a hybrid steel structure consisting of Vierendeel cross-frames supported on two longitudinal trusses with stiffened steel plates that carry the upper and lower highways. The bridge deck is effectively modelled and assembled by a number of bridge deck modules. These bridge deck modules include (1) deck module at the main span; (2) deck module at the Ma Wan tower; (3) deck module at the Ma Wan approach span; (4) deck module at the Tsing Yi tower; and (5) deck module at the Tsing Yi approach span.

The bridge deck at the main span is a suspended deck and the structural configuration is typical for every 18 m segment. Figure 4.3(a) illustrates a typical 18 m suspended deck module consisting of mainly longitudinal trusses, cross frames, highway decks, railway tracks, and bracings.

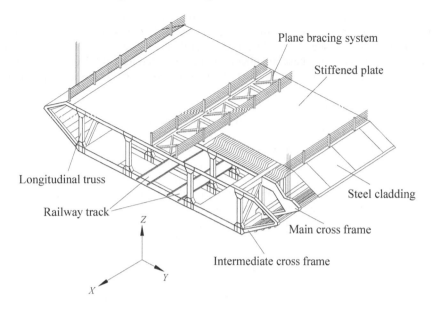

(a) An isometric view of the deck segment

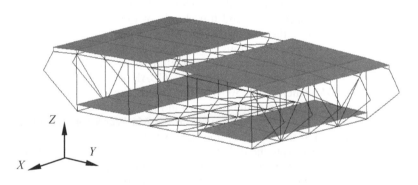

(b) The FE model of the deck segment

Figure 4.3 A typical 18-m long deck module at the main span

The upper and lower chords of the longitudinal trusses are of box section while the vertical and diagonal members of the longitudinal trusses are of I-section. They are all modelled as 12-DOF classical beam elements (CBAR) based on the principle of one element for one member. The actual section properties are computed by the program automatically. The upper and lower chords of the cross frames are of T-section predominantly except for some segments with I-section for the cross bracing systems. The inner struts, outer struts, and upper and lower inclined edge members of the cross frames all are of I-section. All the members in the cross frames are modelled as CBAR beam elements with actual section properties except for the edge members which are assigned a large elastic modulus and significantly small density to reflect the real situation where the joint is heavily stiffened for the connection with the suspender. All the members in the cross bracings are of box section while all the members in the sway bracings are of circular hollow section. These members all are modelled as CBAR beam elements with actual section properties. Each railway track is modelled as an equivalent beam by special 14-DOF beam elements (CBEAM) which are similar to CBAR but with additional properties such as variable cross-section along the beam, shear centre offset from the neutral axis, and wrap coefficient. It is feasible to have this simplification when the serviceability and the safety of running trains are not concerned. The railway tracks are meshed every 4.5 m according to the interval of the adjacent cross frames. The modulus of elasticity, the density, and Poisson's ratio for all members, except for the edge members, are taken as 2.05×10^{11} N/m^2, 8,500 kg/m^3, and 0.3, respectively. The material properties of the edge members will be determined through the model updating.

Deck plates and deck troughs comprise orthotropic decks, and the accurate modelling of stiffened deck plates is complicated. To keep the problem manageable, two-dimensional anisotropic quadrilateral flat shell elements (CQUAD4) are employed to model the stiffened deck plates. The equivalent section properties of the elements are estimated roughly by a static analysis and the material properties of steel are used first but will be updated subsequently. The deck plates are meshed by adjacent cross frames in the longitude direction. Along the cross section of the bridge deck, the top highway deck plates in each bay are further divided into two elements at the position of longitudinal trusses. The connections between the deck plates and the chords of the cross frames and the longitudinal trusses involve the use of MPC (Multi-Point Connection). Proper offsets of neutral axes for the connections between the components are considered to maintain the original configuration. In the modelling of the relevant typical 18-m deck module, a total of 130 nodes with 172 CBAR elements, 16 CBEAM elements, 24 CQUAD4 elements, and 50 MPCs are used. The skeleton view of the 3-D FE model of the 18-m deck module is shown in Figure 4.3(b).

The deck modules at the Ma Wan tower, at the Ma Wan approach span, at the Tsing Yi tower, and at the Tsing Yi approach span are constructed using the same principle while considering the differences in the shape and size of members. For convenience in integrating these bridge deck modules to form a complete bridge deck model, a global coordinate system for the whole bridge and a profile for the bridge deck have been set up before building up these deck modules. In the global coordinate system (X-Y-Z), the X-axis is along the longitudinal bridge axis (from west to east), originating from the location of the Ma Wan abutment bearings and ending at the location of the Tsing Yi abutment bearings with a total length of 2160 m; the Y-axis is along the lateral direction (perpendicular to the bridge axis) with a positive direction from the Hong Kong side (south) to the New Territories side (north); the Z-axis is along the vertical direction initiating from the Principal Datum Hong Kong and thus the z-ordinates are the same as the elevation levels used in the construction drawings. Since the bridge deck is structurally formed by 481 cross frames interconnected by the longitudinal trusses, the route profile datum line of the deck can be geometrically illustrated by the locations of these cross frames in terms of the upper freeway level.

4.5.1.2 The Bridge Towers

The bridge towers are represented by multilevel portal frames. The tower legs are modelled using CBAR elements. The tower leg from its foundation to the deck level is meshed with elements of a length of 5 m each. At the deck level, the tower leg is meshed according to the positions of the lateral bearings. Though the dimension of the cross section of the tower leg varies from its bottom to the top, the geometric properties of the beam elements are assumed to be constant along its axis with an average value based on the design drawings. The four portal beams of both towers are also modelled using CBAR elements with the real section geometric properties. The deck-level portal beam of each tower is divided at the four particular positions, which correspond to the four vertical bearings between the bridge deck and the tower. The mass density, the Poisson's ratio, and the modulus of elasticity of RC for the towers are estimated to be 2,500 kg/m^3, 0.2, and 3.4×10^{10} N/m^2, respectively. The effect of prestress in the cross beams is considered insignificant for predicting the global structural behaviour of the bridge.

4.5.1.3 The Bridge Piers

As there are no sensors installed on the piers, Piers M1, T2, and T3 are similarly modelled as a portal frame using CBAR beam elements. Pier M2 is also modelled as a portal frame using 12 CBAR elements, in which the upper portal beam is meshed according to the four vertical bearing positions. The wall panel of pier T1 is represented by an equivalent portal frame with 25 CBAR elements. The mass density, the Poisson's ratio, and the modulus of elasticity of RC for the piers are taken as 2,500 kg/m^3, 0.2, and 3.4×10^{10} N/m^2, respectively.

4.5.1.4 Cables and Suspenders

The cable system is the major system supporting the bridge deck. It consists of two main cables, 95 pairs of suspender units, and 95 pairs of cable bands. The two main cables are 36 m apart, each of which consists of 91 strands of parallel galvanized steel wires in the main span and 97 strands in the side spans. The number of wires per strand is 360 or 368 with 5.38 mm in diameter. Each cable has a cross sectional area of 0.759 m^2 consisting of 33,400 wires in the main span and 0.800 m^2 consisting of 35,224 wires in the side span. The resultant cable has an overall diameter of approximately 1.1 m after compacting. Each suspender unit is made of a pair of wire ropes of 76 mm diameter that passes over the cable bands on the main cables and is then attached to the chords of the bridge deck by steel sockets. There are four strands in each suspender unit, held apart by spacer clamps. The distance between two suspender units is 18 m along the longitudinal axis of the bridge deck. The two parallel main cables are seen as horizontal sagged cables fixed at the tower saddles and moving together with the towers and transferring the loadings into the anchorages. The Tsing Yi side span cables are seen as inclined sagged cables with the top ends fixed at the tower saddles and the lower ends fixed at the main anchorage. The Ma Wan side span cables are also inclined sagged cables with the top ends fixed at the tower saddles and the lowers fixed at the main anchorage through the saddles on pier M2. On each main cable, there are 19 suspender units within the Ma Wan approach span and 76 suspender units in the main span. In the Tsing Yi approach span, there are no suspender units.

CBEAM elements are used to model the main cables. The cable between the adjacent suspender units is modelled by one beam element of a circular cross section. The DOFs for the rotational displacements of each beam element are released at both ends because the cable is considered to be capable of resisting tensile force only. 77 beam elements are used to model each main cable in the main span while 26 and 8 elements are used to model one main cable on the Ma Wan side span and Tsing Yi side span, respectively. Each suspender unit is modelled by one CBEAM element to represent the four strands. A total of 190 elements are used to model all the suspender units. The connections between the main cables and suspenders are achieved by simply sharing their common nodes.

To model the cable system, the geometry of cable profile should be determined. The geometric modelling of the two parallel main cables follows the profiles of the cables under the design dead load at a design temperature of 23°C based on the information from the design drawings. The horizontal tension in the main cable from pier M2 to the Ma Wan anchorage is 400,013 kN and 405,838 kN in other parts of the main cable. The tension forces in the suspenders on the Ma Wan side span are taken as 2,610 kN and 4,060 kN in other suspenders. The mass densities for both cables and suspenders are taken as 8,200 kg/m^3. The cross sectional area is 0.759 m^2 for the main cables and 0.018 m^2 for the suspenders. The modulus of elasticity is greatly influenced by the tension in the main cables and suspenders, which is respectively estimated as 1.95×10^{11} N/m^2 and 1.34×10^{11} N/m^2 at design temperature of 23°C and will be updated subsequently.

4.5.1.5 Connections and Supports

The major cable fixture components of the Tsing Ma Bridge include two pairs of tower saddles, one pair of pier saddles, the Ma Wan anchorage and Tsing Yi anchorage. The tower saddles are the intermediate bridge components used to fix the main cables on the top of the bridge towers and as guiders to change curvature of the main cables between the main span and the approach spans. One pair of tower saddles is placed on the top of each tower and each saddle weighs about 5,000 kN. The pier saddles located at the column top ends of pier M2, which change the geometric profiles of the main cables in the Ma Wan approach span before the main cables are finally held by the Ma Wan anchorage. The two ends of the main cables are firmly fixed at the Ma Wan anchorage and the Tsing Yi anchorage, respectively, to transfer the cable tension forces to the anchorages. The two anchorages, which are gravity structures, are integrated with the deck abutments. The Tsing Yi anchorage is largely below the ground in a 290,000 m^3 rock excavation. In contrast, the Ma Wan anchorage is only partially buried. As the main cable is split into bundles of strands with each bundle fixed to the anchor block at different inclinations, it is very difficult to model the connections between the main cables and anchorages in an exact way in the global bridge model. Both the Ma Wan and Tsing Yi anchorages are not included in the global bridge model, whereas fixed supports are assigned at the ends of both main cables. Furthermore, the tower saddles are very stiff and their stress distributions are not considered in the global bridge model. MPCs are used to connect the main cables to the towers with proper offsets to make the geometric configuration of the cable close to the original one. In a similar way, MPCs are used to connect the main cables to the column top ends of pier M2 to represent the connection between the main cables and pier M2 via the saddles.

For the suspended deck units in the main span and in part of the Ma Wan approach span, the deck is supported by the suspenders hung from the main cables. The suspenders are connected to the main cross frames at the suspension points. In modelling the connections between the deck and suspenders, the method of sharing common nodes is also adopted. For each connection, the suspender is connected to the intersection of the two inclined edge members of the main cross frame.

At the Ma Wan tower, the bridge deck is connected to the bottom cross beam of the tower through four articulated link bearings and to the tower legs through four lateral bearings. The locations of the four bottom bearings correspond to the four vertical members (inner and outer struts) of the bearing cross frame of the bridge deck. The locations of the four lateral bearings (two on each side) correspond to the upper and lower cross beams of the bearing cross frame. The articulated link bearings allow the deck to move within the horizontal plane (x-y) but restrict the movement in the vertical direction (z). The lateral bearings are to restrain the lateral movement (y) of the deck but to allow movement within the vertical plane (y-z). Therefore, the deck is allowed to move along the longitudinal direction of the bridge. At the Tsing Yi tower, there are also four bottom bearings connecting the deck to the lowest cross beam of the tower and four lateral bearings connecting the deck to the tower legs. The only difference of the bearings at the Tsing Yi tower from those at the Ma Wan tower is that the four bottom bearings at the Tsing Yi tower are rollers rather than rockers as used at the Ma Wan tower. In the modelling, all the vertical bearings between the deck and the towers are represented by swing bar elements with pinned ends to allow free longitudinal motion of the bridge deck. All the lateral (horizontal) bearings between the deck and the towers are represented by swing bar elements with pinned ends to restrict the lateral motion of the deck only.

Piers M1, T2 and T3 are free-standing piers of similar design, which provide only bottom bearings as their connections with the bridge decks. Pier M2 provides both bottom bearings and lateral bearings to the bridge decks, and Pier T1 is part of the approach road and slip road structure on the Tsing Yi side, and also provides both bottom and lateral bearings to the bridge decks. Similar to the modelling of the connections between the deck and the towers, all the vertical bearings between the deck and the piers are modelled by the swing bar elements with pinned ends to allow free longitudinal motion of the bridge and all the horizontal bearings are represented by swing bar elements to restrict the lateral motion of the deck only.

The boundary conditions of the global bridge model are addressed using different supports. Fixed supports are used at the bottom of the foundations for all five piers (M1, M2, T3, T2 and T1) and both towers. Fixed supports are also used at the ends of main cables as mentioned before. Hinge supports are adopted at the deck end on the Ma Wan side. The hinge supports with constraints on translational directions along x-, y-, and z- directions but without constraints on rotations are adopted to replicate the effects of Ma Wan anchorages on the bridge deck. This support condition is applied to all the nodes of the lower cross beam of the bearing cross frame at the deck end on the Ma Wan side. Sliding supports are adopted at the deck end on the Tsing Yi side. The supports are modelled as rollers which allow the movement of deck along the longitudinal direction. The vertical roller supports for modelling bottom bearings are achieved by applying the boundary conditions with constraints on the y- and z- displacements to the nodes of the bottom cross beam of the bearing cross frame. The horizontal roller supports for modelling lateral bearings are achieved by applying the boundary conditions with constraint on the y- displacement to the edge nodes of the bearing cross frame at the levels of the upper and lower cross beams.

4.5.1.6 The Global Model of the Bridge

By integrating the bridge components with the proper modelling of the connections and boundary conditions, the global bridge model is established as shown in Figure 4.4. The establishment of this global bridge model involves 12,898 nodes, 21,946 elements (2,906 plate elements and 19,040 beam elements) and 4,788 MPCs.

   In the establishment of the Tsing Ma global bridge model, the coordinates of the bridge structure are taken from as-built drawings. Therefore, the configuration of the bridge model is the target one. The process for finding the target configuration of the bridge in the equilibrium state under dead loads is referred to as "shape-finding". The task of shape-finding is accordingly performed through iteration to form the final global bridge model for the subsequent dynamic analysis and model updating.

Figure 4.4 3-D finite element model of the Tsing Ma Bridge

4.5.2 Ambient Vibration Test of the Bridge

The ambient vibration test serves in obtaining the dynamic characteristics of the bridge and then for model updating. The ambient measurement of the Tsing Ma Bridge, after completion of the bridge deck, was carried out by The Hong Kong Polytechnic University (HKPU, 1996). In the measurement, for obtaining the global dynamic characteristics, the accelerometers were located such that the longitudinal, lateral, torsional, and vertical motions of the bridge deck, the main cables and the towers were measured. Because of the limitation of accelerometers available, the bridge was divided into 30 measurement sections, among which two reference sections were selected at approximately the quarter point and three-quarter point along the main span, respectively. Vibration signals in the vertical, lateral, and longitudinal directions were recorded one at a time by re-orientation of the sensors. Cross reference of all recorded signals through the measurements with sensors in the reference sections was allowed. Each of the sensors inside the bridge deck unit was located close to the centroidal axis of the deck section at a point close to the suspender, mounted onto the structural steelwork with a magnetic stand. When monitoring the main cables, the accelerometer was mounted on a magnetic stand fixed to the cable band at a selected location.

The first 18 measured natural frequencies of Tsing Ma Bridge, ranging from 0.069 Hz to 0.381 Hz, are given in Table 4.1 (Xu *et al.*, 1997).

Table 4.1 Comparison of the measured and calculated modal parameters before updating.

Mode No.	Measurement (Hz)	Calculation (Hz)	Difference (%)	MAC
1 (L1*)	0.069	0.073	5.8	0.84
2 (V1)	0.113	0.129	14.2	0.82
3 (V2)	0.139	0.156	12.2	0.90
4 (L2)	0.164	0.167	1.8	0.87
5 (V3)	0.184	0.210	14.1	0.87
6 (L3)	0.214	0.235	9.8	0.87
7 (L4)	0.226	0.244	8.0	0.87
8 (L5)	0.236	0.275	16.5	0.60
9 (L6)	0.240	0.276	15.0	0.57
10 (V4)	0.241	0.266	10.4	0.87
11 (T1)	0.267	0.253	-5.2	0.89
12 (V5)	0.284	0.300	5.6	0.84
13 (L7)	0.297	0.306	3.0	0.71
14 (T2)	0.320	0.244	-23.8	0.69
15 (V6)	0.327	0.361	10.4	0.78
16 (L8)	0.336	0.353	5.1	0.87
17 (L9)	0.352	0.346	-1.7	0.98
18 (L10)	0.381	0.370	-2.9	0.90

* L, V, T denote the lateral, vertical and torsional modes, respectively.

From the table, it can be seen that the natural frequencies of the bridge are spaced very closely. This is one characteristic of long-span suspension bridges. The first lateral, vertical, and torsional modes are illustrated in Figure 4.5, Figure 4.6, and Figure 4.7, respectively. Other modes can be found in Xu *et al.* (1997). The results, not shown here for brevity, have shown the coupling between the modes, the interaction between the main span and side spans, and the interaction between the cables, deck, and towers. In particular, the lateral vibration of the deck always appears in-phase and out-of-phase with the lateral vibration of the cables.

The table also lists the frequencies calculated from the initial FE model described in Section 4.5.1 and correlation of the calculated mode shapes with the measured ones in terms of MAC. It can be seen that the largest difference between the calculated and measured natural frequencies is 23.8% in the second torsional mode. Also relatively small MAC values appear in some modes.

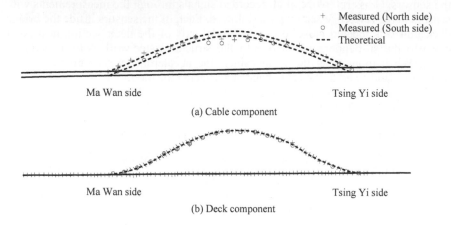

Figure 4.5 The first lateral mode in lateral plane

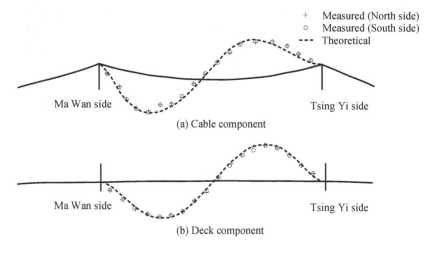

Figure 4.6 The first vertical mode in vertical plane

4.5.3 Sensitivity-based Model Updating of the Bridge

To reduce the discrepancies between the measured and computed natural frequencies, the initial FE model of the Tsing Ma Bridge is updated. From a theoretical viewpoint, both natural frequencies and mode shapes can be used in the objective function for the model updating. However, including mode shapes in the model updating costs a lot of computational resources because calculating the

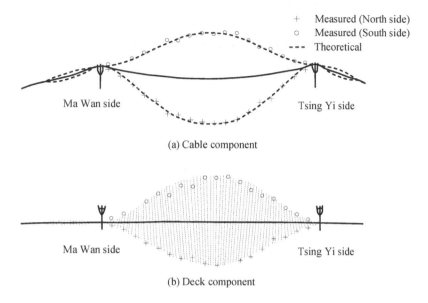

(a) Cable component

(b) Deck component

Figure 4.7 The first torsional mode in vertical plane

sensitivity of mode shapes with respect to structural parameters is very time-consuming. Moreover, the measured mode shapes are often less accurate than the measured natural frequencies in practice. This is particularly true for the Tsing Ma Bridge because the measurement points in the mode shape are 25 only. Therefore, only the first 18 natural frequencies are used in the model updating here, and the mode shapes are used for matching modes only via MAC.

Because the geometric features and supports of the bridge deck have been modelled in great detail in the initial 3-D FE model, the model updating here focuses on material properties. The modelling discrepancies in this study mainly come from four sources: (1) the simplified modelling of stiffened plates; (2) the uncertainties in pavement mass and others; (3) the uncertainties in the stiffness of bearings; and (4) the rigid connection assumption. It is assumed that the discrepancies due to the above sources can be minimized by updating material properties of the relevant components of the bridge model. Table 4.2 lists the material properties selected for updating. They include mass density M_{p1}, M_{p2}, elastic modulus E_{x1}, E_{y1}, E_{x2}, E_{y2}, and shear modulus G_{xy1}, G_{xy2} for the upper and bottom stiffened plates, respectively to mainly update the stiffened plates. By updating mass density M_d for all the beam elements of the bridge deck and mass density M_c for all the cable elements, the masses of pavement and other accessories could be included in the bridge model. Finally, elastic moduli E_h for horizontal bearings and E_v for vertical bearings are updated to overcome the uncertainties in the stiffness of bearings. Furthermore, the lower and upper bounds for the mass densities are set to avoid physically meaningless updating results and impossible updated parameter values. The variations of mass densities are set as 20% of the initial values.

Table 4.2 Selected parameters for model updating.

Component	Description	Parameter	Initial value	Updated value
Anisometric upper stiffened plates	Mass density (kg/m³)	M_{p1}	8500	9304
	Elastic modulus in x (Pa)	E_{x1}	2.05×10^{11}	1.85×10^{11}
	Elastic modulus in y (Pa)	E_{y1}	2.05×10^{11}	1.81×10^{12}
	Shear modulus (Pa)	G_{xy1}	7.885×10^{10}	1.2×10^{11}
Anisometric bottom stiffened plates	Mass density (kg/m³)	M_{p2}	8500	9107
	Elastic modulus in x (Pa)	E_{x2}	2.05×10^{11}	1.98×10^{11}
	Elastic modulus in y (Pa)	E_{y2}	2.05×10^{11}	1.807×10^{12}
	Shear modulus (Pa)	G_{xy2}	7.885×10^{10}	9.017×10^{11}
Beam elements in deck	Mass density (kg/m³)	M_d	8500	9817
Cable elements	Mass density (kg/m³)	M_c	8200	9173
Horizontal bearings	Elastic modulus (Pa)	E_h	3.4×10^{10}	3.078×10^{11}
Vertical bearings	Elastic modulus (Pa)	E_v	3.4×10^{10}	1.879×10^{11}

Table 4.3 Comparison of the measured and calculated modal parameters after updating.

Mode No.	Measurement (Hz)	Calculation (Hz)	Difference (%)	MAC
1 (L1)	0.069	0.069	0.0	0.84
2 (V1)	0.113	0.122	8.0	0.82
3 (V2)	0.139	0.147	5.8	0.90
4 (L2)	0.164	0.160	-2.4	0.89
5 (V3)	0.184	0.198	7.6	0.87
6 (L3)	0.214	0.222	3.7	0.87
7 (L4)	0.226	0.231	2.2	0.87
8 (L5)	0.236	0.233	-1.3	0.56
9 (L6)	0.240	0.243	1.3	0.57
10 (V4)	0.241	0.250	3.7	0.87
11 (T1)	0.267	0.258	-3.4	0.93
12 (V5)	0.284	0.283	-0.4	0.84
13 (L7)	0.297	0.300	1.0	0.77
14 (T2)	0.320	0.282	-11.9	0.78
15 (V6)	0.327	0.340	4.0	0.78
16 (L8)	0.336	0.336	0.0	0.85
17 (L9)	0.352	0.327	-7.1	0.98
18 (L10)	0.381	0.350	-8.1	0.90

To reflect the relative importance of the measured data, the weight coefficients are set as 10 for the first two lateral modes (Nos. 1 and 4), the first two vertical modes (Nos. 2 and 3), and the first two torsional modes (Nos. 11 and 14) and as 1 for all other modes.

The initial and updated values of the selected parameters are listed in Table 4.2. It can be seen that the elastic moduli of plate-bending elements in the *y* direction and their shear moduli have significant changes. The major reason for the change is that the initial values of elastic moduli of the plate elements are selected based on an isometric assumption but they are actually not in the actual stiffened plates. It can also be seen that in consideration of pavement mass and others, the mass densities of both cables and deck elements increase. Furthermore, the initial values for the stiffness of bearings are too small and the updated values increase significantly.

The natural frequencies and MAC values computed from the updated FE model are listed in Table 4.3. The discrepancy between the computed natural frequencies and the measured ones after model updating is reduced significantly. The largest difference in the measured and calculated natural frequency is 11.9% as compared with 23.8% in the initial FE model. The average difference in the natural frequencies is 4% for the updated model as compared with 9% in the initial model. However, the MAC values have not been improved because the mode shapes are not included in the objective function.

4.6 REFERENCES

Abdel-Ghaffer, A.M. and Scanlan, R.H., 1985, Ambient vibration studies of Golden Gate bridge: I. Suspended structure. *Journal of Engineering Mechanics, ASCE*, **111(4)**, pp. 463-82.

Andersen, P., Brincker, R. and Kirkegaard, P.H., 1996, Theory of covariance equivalent ARMAV models of civil engineering structures. In *Proceedings of the 14ᵗʰ International Modal Analysis Conference*, Dearborn, MI, edited by Wicks, A. L., (Bethel, Connecticut : Society for Experimental Mechanics), pp. 518-524.

Avitabile, P., 1999, Modal space – in our own little world. *Experimental Techniques*, **5**, pp. 17-19.

Baruch, M., 1978, Optimization procedure to correct stiffness and flexibility matrices using vibration tests. *AIAA Journal*, **16(11)**, pp. 1208-1210.

Bathe, K.J., 1996, *Finite Element Procedures*, 2ⁿᵈ ed., (New Jersey: Prentice-Hall Inc. Englewood Cliffs).

Bendat, J.S. and Piersol, A.G., 1993, *Engineering Applications of Correlation and Spectral Analysis*, 2ⁿᵈ ed., (New York: John Wiley & Sons).

Berman, A. and Nagy, E.J., 1983, Improvement of large analytical model using test data. *AIAA Journal*, **21(8)**, pp. 1168-1173.

Bishop, R.E.D. and Gladwell, G.M.L., 1963, An investigation into the theory of resonance testing. *Philosophical Transactions of the Royal Society of London A*, **255(1055)**, pp. 241-280.

Bjorck, A, 1996, *Numerical Methods for Least Squares Problems*, (Philadelphia: Society for Industrial and Applied Mathematics).

Branch, M.A., Coleman, T.F. and Li, Y., 1999, A subspace, interior, and conjugate gradient method for large-scale bound-constrained minimization problems. *SIAM Journal on Scientific Computing*, **21(1)**, pp. 1-23.

Brownjohn, J.M.W., Dumanoglu, A.A., Severn, R.T. and Blakeborough, A., 1989, Ambient vibration survey of the Bosporus suspension bridge. *Earthquake Engineering and Structural Dynamics*, **18**, pp. 263-283.

Brownjohn, J.M.W., Lee J. and Cheong, B., 1990, Dynamic performance of a curved cable-stayed bridge. *Engineering Structures*, **21(11)**, pp. 1015-1027.

Brownjohn, J.M.W., Moyo, P., Omenzetter, P. and Lu, Y., 2003, Assessment of highway bridge upgrading by dynamic testing and finite-element model updating. *Journal of Bridge Engineering*, **8**, pp. 162-172.

Brownjohn, J.M.W. and Xia, P.Q., 2000, Dynamic assessment of curved cable-stayed bridge by model updating. *Journal of Structural Engineering*, **126(2)**, pp. 252-260.

Brownjohn, J.M.W., Xia, P.Q., Hao, H. and Xia, Y., 2001, Civil structure condition assessment by FE model updating: methodology and case studies. *Finite Elements in Analysis and Design*, **37**, pp. 761-775.

Caicedo, J.M., Dyke, S.J. and Johnson, E.A., 2004, Natural excitation technique and eigensystem realization algorithm for phase I of the IASC-ASCE benchmark problem: simulated data. *Journal of Engineering Mechanics, ASCE*, **130(1)**, pp. 49-60.

Chang, C.C., Chang, T.Y.P. and Zhang, Q.W., 2001, Ambient vibration of long-span cable-stayed bridge. *Journal of Bridge Engineering*, **6(1)**, pp. 46-53.

Chung, W. and Sotelino, E.D., 2006, Three-dimensional finite element modeling of composite girder bridges. *Engineering Structures*, **28**, pp. 63-71.

Clough, R.W. and Penzien, J., 2003, *Dynamics of Structures*, 3rd ed., (Berkeley, USA: Computers and Structures).

Cunha, A., Caetano, E. and Delgado, R., 2001, Dynamic tests on large cable-stayed bridge. *Journal of Bridge Engineering, ASCE*, **6(1)**, pp. 54-62.

Doebling, S.W, Farrar, C R, Prime, M.B and Shevitz, D.W, 1996, *Damage Identification and Health Monitoring of Structural and Mechanical Systems from Changes in their Vibration Characteristics: A Literature Review*. Report LA-13070-MS, (Los Alamos: Los Alamos National Laboratory).

Duan Y.F., Xu, Y.L., Fei, Q.G., Wong, K.Y., Chan, W.Y., Ni, Y.Q. and Ng, C.L., 2011, Advanced finite element model of the Tsing Ma Bridge for structural health monitoring. *International Journal of Structural Stability and Dynamics*, **11(2)**, pp.313-344.

Ewins, D.J., 2000, *Modal Testing-Theory, Practice and Application*, (Baldock, Hertfordshire, UK: Research Studies Press Ltd).

Farhat, C. and Hemez, F.M., 1993, Updating finite element dynamic models using an element-by-element sensitivity methodology. *AIAA Journal*, **31(9)**, pp. 1702-1711.

Fei, Q.G., Xu, Y.L., Ng, C.L., Wong, K.Y., Chan, W.Y. and Man, K.L., 2007, Structural health monitoring oriented finite element model of Tsing Ma bridge tower. *International Journal of Structural Stability and Dynamics*, **7(4)**, pp. 647-668.

Friswell, M.I. and Mottershead, J.E., 1995, *Finite Element Model Updating in Structural Dynamics*, (Boston: Kluwer Academic Publishers).

Friswell, M.I., Penny, J.E.T. and Garvey, S.D., 1998, A combined genetic and eigensensitivity algorithm for the location of damage in structures. *Computers and Structures*, **69(5)**, pp. 547-556.

Ge, L. and Soong, T.T., 1998, Damage identification through regularization method I: Theory. *Journal of Engineering Mechanics, ASCE*, **124(1)**, pp. 103–108.

Goldberg, D.E., 1989, *Genetic Algorithms in Search, Optimisation and Machine Learning*, (Reading, Massachusetts: Addison-Wesley Longman).

Hao, H. and Xia, Y., 2002, Vibration-based damage detection of structures by genetic algorithm. *Journal of Computing in Civil Engineering, ASCE*, **16(3)**, pp. 222-229.

HKPU, 1996, *Lantau Fixed Crossing Vibration Monitoring of Bridge Towers-Field Measurement Report on the Tower-Cable System of Tsing Ma Bridge*. (Hong Kong: Department of Civil and Structural Engineering, The Hong Kong Polytechnic University).

Holland, J.H., 1975, *Adaptation in Natural and Artificial Systems*, (Ann Arbor: The University of Michigan Press).

Jaishi, B. and Ren, W.X., 2005, Structural finite element model updating using ambient vibration test results. *Journal of Structural Engineering*, **131(4)**, pp. 617-628.

James III, G.H., Carne, T.G. and Lauffer, J.P., 1995, The natural excitation technique (NExT) for modal parameter extraction from operating structures. *International Journal of Analytical and Experimental Modal Analysis*, **10(4)**, pp. 260-277.

Juang, J.N., 1994, *Applied System Identification*, (New Jersey: Prentice-Hall Inc. Englewood Cliffs).

Juang, J.N. and Pappa, R.S., 1985, An eigensystem realization algorithm for modal parameter identification and model reduction. *Journal of Guidance, Control and Dynamics*, **8(5)**, pp. 620-627.

Kumarasena, T., Scanlan, R.H. and Morris, G.R., 1989, Deer Isle bridge: field and computed vibrations. *Journal of Structural Engineering, ASCE*, **115**, pp. 2313-2328.

Larson, C.B. and Zimmerman, D.C., 1993, Structural model refinement using a genetic algorithm approach. In *Proceedings of the 11th International Modal Analysis Conference, Society for Experimental Mechanics*, Kissimmee, Florida, edited by DeMichele, D.J., (Bethel, Connecticut: Society for Experimental Mechanics), pp. 1095-1101.

Li, Z.X. Zhou, T.Q. Chan, T.H.T. and Yu, Y., 2007, Multi-scale numerical analysis on dynamic response and local damage in long-span bridges. *Engineering Structures*, **29**(7), pp. 1507-1524.

Link, M. and Friswell, M., 2003, Working group 1: generation of validated structural dynamic models – results of a benchmark study utilizing the GARTEUR SM-AG19 test-bed. *Mechanical Systems and Signal Processing*, **17(1)**, pp. 9-20.

Liu, T.T., Xu, Y.L., Zhang, W.S., Wong, K.Y., Zhou, H.J. and Chan, K.W.Y., 2009, Buffeting-induced stresses in a long suspension bridge: Structural health monitoring oriented stress analysis. *Wind and Structures-An International Journal*, **12(6)**, pp. 479-504.

Maia, N.M.M., Silva, J.M.M., He, J., Lieven, N.A.J., Lin, R.M., Skingle, G.W., To, W., and Urgueira, A.P.V., 1997, *Theoretical and Experimental Modal Analysis*, (England: Research Studies Press Ltd).

Mares, C. and Surace, C., 1996, An application of genetic algorithms to identify damage in elastic structures. *Journal of Sound and Vibration*, **195(2)**, pp. 195-215.

Mitchell, M., 1996, *An Introduction to Genetic Algorithms*, (Cambridge, Massachusetts: MIT Press).

More, J.J. and Sorensen, D.C., 1983, Computing a trust region step. *SIAM Journal on Scientific and Statistical Computing*, **3**, pp. 553-572.

More, J.J. and Toraldo, G., 1989, Algorithm for bound constrained quadratic programming problems. *Numerical Mathematic*, **55**, pp. 377-400.

Mottershead, J.E. and Friswell, M.I., 1993, Model updating in structural dynamics: a survey. *Journal of Sound and Vibration*, **167(2)**, pp. 347-375.

Nelson, R.B., 1976, Simplified calculation of eigenvector derivatives. *AIAA Journal*, **14**, pp. 1201-1205.

Nocedal, J. and Wright, J., 1999, *Numerical Optimization*, (New York: Springer).

Okauchi, I., Miyata, T., Tatsumi, M. and Sasaki, N., 1997, Field vibration test of a long span cable-stayed bridge using large exciters. *Journal of Structural Engineering/Earthquake Engineering*, Tokyo, Japan, **14(1)**, pp. 83-93. .

Peeters, B. 2000, *System Identification and Damage Detection in Civil Engineering*, Ph.D. thesis, Department of Civil Engineering, Catholic University of Leuven, Belgium.

Perry, M.J., Koh, C.G. and Choo, Y.S., 2006, Modified genetic algorithm strategy for structural identification. *Computers and Structures*, **84**, pp. 529-540.

Ren, W.X., Harik, I.E., Blandford, G.E., Lenett, M. and Basehearh, T., 2004, Roebling suspension bridge. II: ambient testing and live load response. *Journal of Bridge Engineering, ASCE*, **9(2)**, pp. 119-26.

Richardson, M. and Formenti, D.L., 1982, Parameter estimation from frequency response measurements using rational fraction polynomials. In *Proceedings of the 1st International Modal Analysis Conference*, Orlando, Florida, edited by DeMichele, D. J., (Bethel, Connecticut : Society for Experimental Mechanics), pp. 167-181.

Richardson, M. and Formenti, D.L., 1985, Global curve-fitting of frequency response measurements using the rational fraction polynomial method. In *Proceedings of the 3rd International Modal Analysis Conference*, Orlando, Florida, edited by DeMichele, D.J., (Bethel, Connecticut: Society for Experimental Mechanics), pp. 390-397.

Sanayei, M. and Saletnik, M.J., 1996, Parameter estimation of structures from static strain measurements, part I: formulation. *Journal of Structural Engineering, ASCE*, **122(5)**, pp. 555-562.

Song, M.K., Kim, S.H. and Choi, C.K., 2006, Enhanced finite element modelling for geometric non-linear analysis of cable-supported structures. *Structural Engineering and Mechanics*, **22(5)**, pp. 575-597.

Steihaug, T., 1983, The conjugate gradient method and trust regions in large scale optimization. *SIAM Journal on Numerical Analysis*, **20**, pp. 626-637.

Van Overschee, P. and De Moor, B., 1996, *Subspace Identification for Linear Systems: Theory-Implementation-Applications*, (Boston: Kluwer Academic Publishers).

Weng, S., Xia, Y., Xu, Y.L. and Zhu, H.P., 2011, Substructuring approach to finite element model updating. *Computers and Structures*, **89(9-10)**, pp. 772-782.

Wilson, J.C. and Liu, T., 1991, Ambient vibration measurements on a cable-stayed bridge. *Earthquake Engineering and Structural Dynamics*, **20(8)**, pp. 723-747.

Wong, K.Y., 2004, Instrumentation and health monitoring of cable-supported bridges. *Structural Control and Health Monitoring*, **11(2)**, pp. 91-124.

Xu, Y.L., Ko, J.M. and Zhang, W.S., 1997, Vibration studies of Tsing Ma suspension bridge. *Journal of Bridge Engineering*, **2(4)**, pp. 149-156.

Zhang, Q.W., Chang, T.Y.P. and Chang, C.C., 2001, Finite element model updating for the Kap Shui Mun cable-stayed bridge. *Journal of Bridge Engineering*, **6(4)**, pp. 285-293.

Zienkiewicz, O.C. and Taylor, R.L., 1994, *The Finite Element Method, Vol. 1: Basic Formulation and Linear Problems*, 4th ed., (Berkshire, England: McGraw-Hill).

Zivanovic, S., Pavic, A. and Reynolds, P., 2007, Finite element modeling and updating of a lively footbridge: the complete process. *Journal of Sound and Vibration*, **301**, pp. 126-145.

4.7 NOTATIONS

A	Effective cross sectional area.
A_{jkr}	The residue for mode r.
E_{eq}, E	Equivalent modulus of elasticity, effective modulus of elasticity.
E_h, E_v	Elastic moduli for horizontal and vertical bearings.
$f_i, \omega_i,$	The ith natural frequency (Hz), the ith circular frequency (rad/sec).
G_{xy}	Shear modulus of stiffened plates.
$[H]$	FRF matrix.
h_{jk}	Receptance defined as the response displacement at coordinate j due to a harmonic excitation force at coordinate k.
J	Objective function.
$[K^e]$	General element stiffness matrix.
$[K_e], [K_g], [K_t]$	Elastic stiffness matrix, geometric stiffness matrix, tangent stiffness matrix of the cable element.
L_c	Chord length of cable.
$\bar{l}$	Horizontal projected length.
l, u	Lower and upper bounds of the updating parameter.
$[M], [C], [K]$	Mass matrix, viscous damping matrix, stiffness matrix.
M_p, M_d, M_c	Mass density of stiffened plates, bridge deck and cable elements.
m	Marquardt parameter.
N	Number of degrees of freedom, size of system matrices.
ne	Number of elements.
nm, np	Number of modes, number of measurement points.
$\{p\}$	Structural parameters to be updated.
S	Sensitivity matrix.
s_r	The pole for mode r.
T	Mean cable tension.
W_f, W_ϕ, W_λ	Weight coefficients of the frequency, mode shape, eigenvalue.
$\{x\}, \{f\}$	Nodal displacement and force vector.

$\{X\}, \{F\}$	Fourier transformation of displacement and force vector.
α_i	Element bending rigidity.
α, β	Rayleigh damping factors.
$\rho,$	Density.
g	Gravity acceleration.
ξ_r	Damping ratio of the rth mode.
$\lambda, [\Lambda]$	Eigenvalue, eigenvalue matrix.
$\{\phi\}, [\Phi]$	Mode shape vector, mode shape matrix.

CHAPTER 5

Monitoring of Railway Loading Effects

5.1 PREVIEW

When a long-span suspension bridge carries railways, the bridge is subjected to specific forces due to moving loads of railway vehicles. These forces include vertical forces from vertical and rocking effects of railway vehicles, longitudinal forces from acceleration or deceleration of railway vehicles, lateral forces caused by irregularities at the wheel-to-rail interface, and centrifugal forces due to track curvature. Railway vehicles vary greatly with respect to weight, number of axles, and axle spacing. This variability requires a representative live load model for design that provides a safe and reliable estimation of characteristics of railway vehicles within the design life of the bridge (Unsworth, 2010). In this regard, the railway loading and its effects on the bridge must be monitored to provide information on the actual railway loading compared with the design railway loading and to ensure the functionality and safety of both the bridge and railway vehicles.

This chapter first describes sensors and sensing systems used for railway monitoring of long-span suspension bridges. The techniques for ascertaining the railway traffic condition and railway load distribution from the measurement data are introduced. The dynamic analysis of a coupled train-bridge system is then presented with wheel-rail interaction included to provide a theoretical background of the railway loading and its effect. Finally, the Tsing Ma suspension bridge is taken as an example to illustrate the monitoring of railway traffic, railway loading, and predominating railway loading effects on the bridge.

5.2 RAILWAY MONITORING SYSTEM

Railway monitoring in a long-span suspension bridge starts with the selection and arrangement of a sensor system and the associated data acquisition system. A variety of sensors are applicable for railway monitoring of long-span suspension bridges. Different sensors are designated to measure different physical quantities. The selection of the number and type of sensors is generally a compromise of many factors such as budget, application objectives, site conditions, measurement duration, and user's experience. Traditional sensors widely used for railway monitoring of a long-span suspension bridge include force (load) transducers for measuring the railway traffic and loading, displacement transducers for measuring railway-induced bridge displacement responses, accelerometers for measuring railway-induced bridge acceleration responses, and strain gauges for measuring railway-induced bridge strain responses.

A vast number of force transducer types have been developed over the years. Force transducers often use either strain gauges (foil or semiconductor) as sensing elements or piezoelectric sensing elements to achieve certain design natural frequencies, weight, and sensitivity characteristics (McConnell, 1995). The force transducer is a relatively complicated instrument since it often interacts directly with structural members of the bridge under measurements. Therefore, the selection and installation of force transducers are important in order to measure the relevant railway loading correctly. Furthermore, the calibration of a force transducer is periodically required in order to determine that the force transducer is functioning properly and has not changed with use and/or misuse. Calibration is a process where we generate a known input to a force transducer and then record its output so that the transducer's input-output relationship can be established to determine its sensitivity and linearity over a range of frequencies.

The dynamic displacement response of the bridge can be measured using displacement transducers or obtained by integrating the acceleration records twice with time. For instance, the displacement transducers are often used to measure the relative displacements between the towers and bridge deck. The GPS technique can also be used to measure railway-induced displacement responses of a long-span suspension bridge. The field equipment of GPS installed at each of the base reference station and the rover stations basically consists of a GPS antenna, a GPS receiver, and all necessary signal and power connecting cables. The GPS antennas are often permanently installed at the designated locations such as the top of bridge towers and both sides of the bridge deck, securely fixed on the bracket/stand on the bridge structure, and equipped with permanent cabling to be connected to the GPS receivers. The antenna must be positioned at a sufficiently high level to obtain a clear sky window of 15 degrees above their horizontal plane with minimum obstruction to obtain the best conditions for satellite signal reception at all times. The GPS receivers are often housed in cabinets of protection standard, securely fixed inside the interior of the bridge structure and accessible for inspection and maintenance. The GPS receiver must be capable of operating at static, rapid static, and real-time kinematic (RTK) modes. Motion simulation tables were also developed to assess static and dynamic displacement measurement accuracy of GPS in three orthogonal directions for civil engineering application (Chan *et al.*, 2006). For displacement monitoring of the vertical structure of the bridge, tiltmeters may be deployed to measure the flexural deflection along the height of the towers and the upper section of the piers. The GPS and tiltmeter equipment must be capable of properly functioning under their respective operating environment, such as temperature, humidity, vibration and exposure. Adequate protection devices must be provided to all sub-systems against lightning surge and electromagnetic interference and induction. The monitored results from tiltmeters can be integrated with the GPS monitored results in producing global displacements of the bridge.

For measuring railway-induced acceleration responses of a long-span suspension bridge, piezoelectric or servo-type tri-axial accelerometers can be used. To simultaneously measure the horizontal, vertical, and rotational acceleration responses of the bridge deck, two accelerometers can be symmetrically installed on the opposite sides of a bridge deck section with respect to the section centre. The mean value of the records from the two accelerometers in either horizontal or vertical or longitudinal direction is taken as the actual translational acceleration measurement. The difference in the records from the two accelerometers in a designated direction, after having divided by the distance between the two anemometers, represents the rotational acceleration response of the bridge with respect to the deck section centre. The selection of the number and location of the anemometers must consider not only the measurement of railway-induced acceleration responses but also the measurement of other load effects such as wind-induced acceleration responses and the identification of bridge dynamic characteristics. The accelerometers must be fixed on the bridge firmly so that the accelerometers and the bridge are in one absolute motion and there is no relative motion between them.

Strain measurements fulfil the following three functions: (1) to measure stress time histories for fatigue and strength assessments; (2) to measure stress influence lines for loading monitoring; and (3) to measure the loads in structural members for design verification. If the strain gauges are used, they must be durable, reliable and thermally stable. All strain gauges must be provided with temperature compensation over the range of temperature and humidity defined. The maximum data sampling rate must be selected to allow accurate reproduction and processing of waveforms of frequencies up to and including 10 Hz. It is therefore anticipated that a minimum sampling frequency of 100 Hz will be required for weldable foil strain gauges. Electrical checks must be performed for all installed strain gauges using a proprietary strain gauge tester. The electrical checks must demonstrate that the gauge has not been damaged during installation. All strain sensors must be secured to the structural components with adequate protection against physical impact and influx of rain, splashes and contaminants. The mounting system must enable accurate positioning of a replacement sensor in the future to maintain consistency and continuity in measurement of data and reference of monitoring.

5.3 MONITORING OF RAILWAY TRAFFIC CONDITION AND LOAD DISTRIBUTION

The railway traffic condition and railway load distribution over the deck of a long suspension bridge must be monitored and analysed so that a clear understanding of railway loading effects on the bridge can be achieved. In this regard, bogie load data must be generated from the original data recorded by the force transducers of the SHM system. Typical bogie load data include the information on the instant of a train arriving at the location of the force transducer, train moving lane, train direction, train speed, the total number of train bogies, bogie load, and axle spacing.

The bogie load data must be pre-processed to yield a high quality database. The valid data are then analysed to provide the information on the train traffic condition and railway load distribution. Traffic volume and vehicle composition of trains are two important indexes of the train traffic condition, from which the operation trend of trains with time can be inferred. Based on the bogie load records, the monthly or annual number of trains passing through the bridge can be calculated and the types of trains can also be estimated in terms of the number of bogies in a train. The information on railway load distribution includes the maximum bogie load, bogie load distribution in one train, bogie load spectrum, and gross train weight (GTW) spectrum.

5.4 DYNAMIC INTERACTION OF LONG-SPAN SUSPENSION BRIDGES WITH TRAINS

Dynamic displacement, acceleration, and stress response analyses of coupled railway vehicle-suspension bridge systems are important for both bridge safety and vehicle comfort assessment (Frýba, 1996). Nevertheless, a complex finite element model of many DOFs is often required for such analyses involving a long suspension bridge and a number of trains. The mode superposition method may have to be applied to reduce computational efforts and to make the problem manageable. A general procedure for displacement, acceleration, and stress response analyses of coupled railway vehicle-long suspension bridge systems using the mode superposition method is therefore introduced in this section (Li *et al.*, 2010; Xu *et al.*, 2010), but this does not exclude many other methods such as Diana and Cheli (1989), Yang *et al.* (1997), and Xia *et al.* (2000).

5.4.1 Vehicle Model

A train usually consists of several locomotives, passenger coaches, freight cars, or a combination of these. Each vehicle is in turn composed of a car body, bogies, wheel-sets, primary suspension systems connecting the wheel-sets to the bogies, and secondary suspension systems connecting the bogies to the car body (see Figure 5.1). The car body, bogies, and wheel-sets can be modelled by either beam elements or shell elements. The primary or secondary suspension systems can be modelled by spring elements and dashpot elements. Except for linear spring elements in the suspension units, the nonlinear part of spring elements and all dashpot elements are treated as pseudo forces in the present dynamic response analysis. The finite element method coded in a commercial software package is used for the vehicle modelling.

Based on the established finite element model, the equation of motion of a vehicle can be expressed as:

$$[M_v]\{\ddot{\delta}_v\}+[C_v]\{\dot{\delta}_v\}+[K_v]\{\delta_v\}=\{f_v\} \tag{5.1}$$

where $[M_v]$, $[C_v]$ and $[K_v]$ respectively denote the mass, structural damping and linear elastic stiffness matrices excluding nonlinear part of springs and all dashpots of the vehicle; $\{\delta_v\}$, $\{\dot{\delta}_v\}$ and $\{\ddot{\delta}_v\}$ are the displacement, velocity and acceleration vectors of the vehicle; and $\{f_v\}$ denotes the combination of the pseudo force vector $\{f_{vn}(\dot{\delta}_v,\delta_v)\}$, produced by nonlinear part of springs and all dashpots, and the wheel-rail contact force vector $\{f_{vc}\}$:

$$\{f_v\}=\{f_{vn}(\dot{\delta}_v,\delta_v)\}+\{f_{vc}\} \tag{5.2}$$

It is noted that many DOFs are involved in the finite element model of an elastic vehicle but only rigid-body motions and the first few flexible-body motions of the vehicle make significant contributions to the dynamic response of a coupled vehicle-bridge system. The mode superposition method is therefore applied to not only bridge but also vehicles to further reduce the computational effort. To apply the mode superposition method, the modal shape and the modal frequency of the vehicle must be obtained from an eigenvalue analysis as follows, using the commercial software package.

$$\left([K_v] - \omega_v^2 [M_v] \right) \{\phi_v\} = \{0\} \tag{5.3}$$

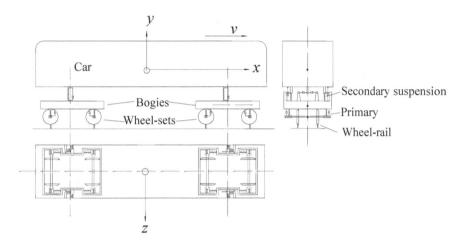

Figure 5.1 Schematic model of a railway vehicle

Let us consider the first N_v vibration modes of the vehicle. The relevant modal shape matrix normalized to the mass matrix is $[\Phi_v]$; and the diagonal modal frequency matrix is ω_v. Then, by using the mode superposition method, Equation (5.1) can be changed as

$$\{\ddot{q}_v\} + 2[\xi_v][\omega_v]\{\dot{q}_v\} + [\omega_v^2]\{q_v\} = [\Phi_v]^T \{f_v\} \tag{5.4}$$

where $\{q_v\}$ and $[\xi_v]$ are the modal coordinate vector and the modal damping matrix, respectively; and the superscript T is the transpose operation of a matrix. The pseudo forces produced by a viscous dashpot in a local coordinate can be expressed as

$$\begin{cases} f_{vn,i}^L = -C_{ij} \left(\dot{\delta}_{v,i}^L - \dot{\delta}_{v,j}^L \right) \\ f_{vn,j}^L = -f_{vn,i}^L \end{cases} \tag{5.5}$$

where $f_{vn,i}^L$ and $f_{vn,j}^L$ are the pseudo forces acting on nodes i and j, respectively, along the $\bar{x}$-axis of the local coordinates; $\dot{\delta}_{v,i}^L$ and $\dot{\delta}_{v,j}^L$ are the velocities of nodes i and j, respectively, along the $\bar{x}$-axis of the local coordinates; and C_{ij} is the viscous damping coefficient. If the dashpot has a saturation property, the corresponding pseudo forces can be expressed as

$$\begin{cases} f_{vn,i}^L = \begin{cases} -F_{maxij}(\dot{\delta}_{v,i}^L - \dot{\delta}_{v,j}^L)/v_{0ij} & \left|\dot{\delta}_{v,i}^L - \dot{\delta}_{v,j}^L\right| < v_{0ij} \\ -F_{maxij}\,sign(\dot{\delta}_{v,i}^L - \dot{\delta}_{v,j}^L) & \left|\dot{\delta}_{v,i}^L - \dot{\delta}_{v,j}^L\right| \geq v_{0ij} \end{cases} \\ f_{vn,j}^L = -f_{vn,i}^L \end{cases} \tag{5.6}$$

where F_{maxij} is the saturated damping force when the dashpot velocity exceeds v_{0ij}. Equation (5.6) can be used to simulate dry friction when v_{0ij} becomes a very small value (Frýba, 1996). Other types of nonlinear springs or dashpots can be treated in a similar way.

5.4.2 Bridge Model

When a bridge carries a railway, there is a track structure laid on the bridge deck. The forces from the wheels of a train are then transmitted to the bridge deck through the track structure (Gimsing, 1997). Two types of track structures are commonly used for a railway: ballasted track structure and non-ballasted track structure (Zhai and Cai, 2003). There are many types of bridges, which can be modelled by the finite element method using a series of beam elements, plate elements, shell elements, and others. In this study, the finite element method coded in the same commercial software package as used for the vehicle modelling is utilized to establish the equations of motion of the bridge subsystem including the track structure.

$$[M_b]\{\ddot{\delta}_b\} + [C_b]\{\dot{\delta}_b\} + [K_b]\{\delta_b\} = \{f_b\} \tag{5.7}$$

where $[M_b]$, $[C_b]$ and $[K_b]$ respectively denote the mass, structural damping and stiffness matrices of the bridge subsystem including the track structure and the geometric stiffness in the case of long span cable-supported bridges; $\{\delta_b\}$, $\{\dot{\delta}_b\}$ and $\{\ddot{\delta}_b\}$ are the displacement, velocity and acceleration vectors of the bridge subsystem; and $\{f_b\}$ is the force vector consisting of two parts:

$$\{f_b\} = \{f_{bn}(\dot{\delta}_b, \delta_b)\} + \{f_{bc}\} \tag{5.8}$$

where $\{f_{bn}(\dot{\delta}_b, \delta_b)\}$ denotes the pseudo forces produced by the nonlinear part of springs and all dashpots, which may be used to model bridge bearings, vibration control devices installed in the bridge subsystem, and the connections used in the track structures; and $\{f_{bc}\}$ denote the wheel-rail contact forces acting on the bridge, which will be discussed together with those acting on the vehicle. By using the mode superposition method, Equation (5.7) can be changed to

$$\{\ddot{q}_b\} + 2[\xi_b][\omega_b]\{\dot{q}_b\} + [\omega_b^2]\{q_b\} = [\Phi_b]^T\{f_b\} \tag{5.9}$$

where $[\xi_b]$, $[\omega_b]$ and $[\Phi_b]$ are respectively the modal damping, modal frequency, and normalized modal shape matrices with respect to the considered N_b vibration modes of the bridge subsystem; and $\{q_b\}$ denotes the modal coordinate vector of the bridge subsystem.

5.4.3 Wheel and Rail Interaction

Wheel and rail interaction is an essential element that couples the vehicle subsystem and the bridge subsystem. The interaction between a wheel and a rail involves two basic issues: the contact geometry and the contact forces. In this study, the contact forces on the two subsystems are regarded as pseudo forces. They are mathematically arranged at the right hand sides of the equations for both vehicle and bridge subsystems so that the mode superposition method can be applied to the left hand sides of the equations.

5.4.3.1 Contact Geometry

Wheel-rail contact geometry computation is the basis to determine contact forces. Consequently the contact geometry parameters should be computed first as functions of the relative lateral displacement and yawing angle of the wheel-set to the rails. Two hypotheses are adopted for this computation: 1) the wheel and rail are regarded as rigid bodies; and 2) the wheel jumping away from the rail is not allowed. The contact geometry parameters include, but are not limited to, the position of the contact point between the wheel and rail, the contact angle at contact point between the wheel and rail, the relative rolling angle of the wheel-set to the rails corresponding to the contact point concerned, and the radius of curvature at contact point for either wheel or rail.

5.4.3.2 Contact Forces

Based on the Kalker linear theory (Kalker, 1990), the creeping forces (see Figure 5.2a) between the jth wheel ($j = 1$ stands for the left and $j = 2$ for the right) and the rail can be calculated as

$$T_{xj} = -f_{11,j}\xi_{1,j}; \quad T_{zj} = -f_{22,j}\xi_{2,j} - f_{23,j}\xi_{3,j}; \quad \bar{M}_j = f_{23,j}\xi_{2,j} - f_{33,j}\xi_{3,j} \quad (5.10)$$

where T_{xj}, T_{zj} and $\bar{M}_j$ are the longitudinal creeping force, lateral creeping force and spin creeping moment, respectively; $f_{11,j}$, $f_{22,j}$, $f_{33,j}$ are, respectively, the longitudinal, lateral, lateral/spin and spin creepage coefficients, which are determined by the normal contact forces between the wheel and rail and the corresponding contact geometry parameters; $\xi_{1,j}$, $\xi_{2,j}$, and $\xi_{3,j}$ are the creepage ratios in the longitudinal, lateral and spin directions, respectively, and they can be expressed as follows:

$$\xi_{1,j} = \frac{v_{wx,j} - v_{rx,j}}{v}; \quad \xi_{2,j} = \frac{v_{wz,j} - v_{rz,j}}{v}; \quad \xi_{3,j} = \frac{\Omega_{w,j} - \Omega_{r,j}}{v} \quad (5.11)$$

where v is the nominal travelling speed of the wheel-set; $v_{wx,j}$ and $v_{wz,j}$ are the velocities of the wheel at the contact point in the longitudinal and lateral directions, respectively; $v_{rx,j}$ and $v_{rz,j}$ are the velocities of the rail at the contact point in the longitudinal and lateral directions, respectively; $\Omega_{w,j}$ and $\Omega_{r,j}$ are the rotational velocities of spin motion of the wheel and rail at the contact point, respectively. For a large creepage ratio, the Kalker's linear creepage theory may cause errors in calculating creeping forces. The Shen-Hedrick-Elkins theory (Shen *et al.*, 1983) is used to modify the Kalker linear creepage theory in this study.

The vehicle subsystem and the bridge subsystem are coupled through contacts between the wheels and rails. As shown in Figure 5.2b, the contact forces transmitted from the wheels to the left and right rails in the vertical and lateral directions can be expressed as

$$\begin{cases} F_{y1} = N_1 \cos(\delta_1 + \theta_{w1}) - T_{z1} \sin(\delta_1 + \theta_{w1}) \\ F_{z1} = N_1 \sin(\delta_1 + \theta_{w1}) + T_{z1} \cos(\delta_1 + \theta_{w1}) \\ F_{y2} = N_2 \cos(\delta_2 - \theta_{w2}) + T_{z2} \sin(\delta_2 - \theta_{w2}) \\ F_{z2} = -N_2 \sin(\delta_2 - \theta_{w2}) + T_{z2} \cos(\delta_2 - \theta_{w2}) \end{cases} \tag{5.12}$$

where subscripts "1" and "2" stand for the left and right wheels, respectively; δ_1 and δ_2 are the contact angles with respect to the left and right wheels, respectively; θ_{w1} and θ_{w2} are the rolling angles; F_{y1} and F_{y2} are the contact forces in the vertical direction; F_{z1} and F_{z2} denote the contact forces in the lateral direction; N_1 and N_2 are the normal contact forces; and T_{z1} and T_{z2} are the lateral creeping forces.

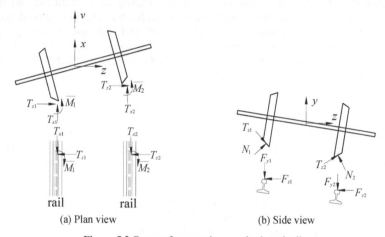

(a) Plan view   (b) Side view

Figure 5.2 Contact forces acting on wheels and rails

There are four equations in Equation (5.12) but there are eight unknown contact forces required to be determined. Two more equations can be found from Equation (5.10) because the lateral creep forces T_{z1} and T_{z2} are related to the normal contact forces N_1 and N_2, according to the Kalker creepage theory. The other two additional equations can be found by considering the equilibrium conditions of the wheel-set as shown in Figure 5.3.

$$\begin{cases} F_{y1} + F_{y2} - M_{ws}\ddot{y}_{ws} - F_{s1} - F_{s2} - 2W = 0 \\ (F_{y1} - F_{y2})b - J_{ws}\ddot{\theta}_{ws} - (F_{s1} - F_{s2})b_1 - (F_{z1} + F_{z2})r_0 = 0 \end{cases} \tag{5.13}$$

where M_{ws} and J_{ws} are the mass and rolling mass moment of inertia of the wheel-set; F_{s1} and F_{s2} are the vertical forces acting at the two sides of the wheel-set due to the primary suspension; W is the static wheel load; b is half width of the nominal rail gauge; b_1 is half width of the transverse distance between the two springs of the primary suspension unit; r_0 is the nominal wheel radius; and y_{ws} and

θ_{ws} are the bouncing and rolling motions of the wheel-set, which can be expressed as

$$y_{ws} = (y_{w1} + y_{w2})/2, \quad \theta_{ws} = (y_{w1} - y_{w2})/2/b \tag{5.14}$$

where y_{w1} and y_{w2} are the vertical motions of the left and right wheels, which can be determined based on the wheel non-jump assumption

$$y_{w1} = y_{r1}, \quad y_{w2} = y_{r2} \tag{5.15}$$

where y_{r1} and y_{r2} are the vertical motions of the left and right rails including the bridge motion and track irregularity.

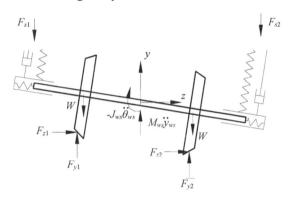

Figure 5.3 Dynamic equilibrium of a wheel-set in vertical plane

The above wheel non-jump model assumes that the vertical displacement of a wheel is dependent on the rail under the wheel. There are no vertical DOFs assigned to the wheels of a vehicle. Therefore, the vertical forces produced by the motion of the wheel-set on the bogie should be found so that the motions of the vehicle subsystem (excluding the vertical motion of the wheels) can be determined. The vertical force F_{byj} acting on the bogie due to the motion of the wheel-set can be expressed as (see Figure 5.4)

$$F_{byj} = (y_{ws} + (-1)^{j+1}\theta_{ws}b_1)k_{1y} + (\dot{y}_{ws} + (-1)^{j+1}\dot{\theta}_{ws}b_1)c_{1y} \tag{5.16}$$

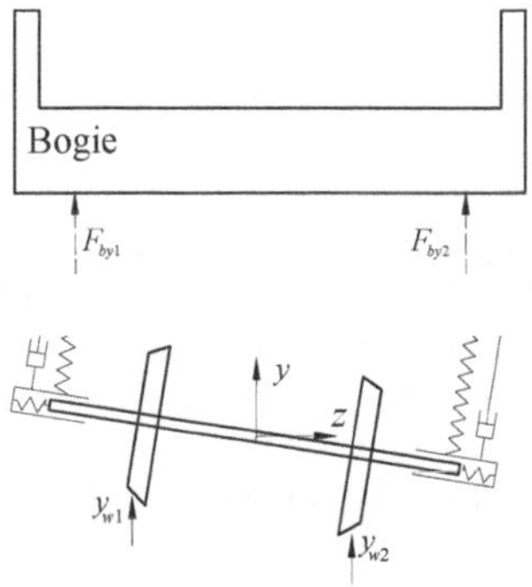

Figure 5.4 Vertical forces acting on a bogie due to the motion of the wheel-set

where k_{1y} and c_{1y} are the vertical stiffness and damping coefficient of the primary suspension unit. It should be noted that Equation (5.16) is only applicable when the primary suspension unit is composed of the linear elastic stiffness and viscous damping in the vertical direction.

5.4.4 Numerical Solution

For this study, the coupled vehicle-bridge system is divided into two subsystems, one is the vehicle subsystem and the other is the bridge subsystem. Equations of motion of the two subsystems are expressed by Equations (5.4) and (5.7). The terms on the right hand side of Equations (5.4) and (5.7) represents the pseudo forces, which are functions of the responses of both the vehicle subsystem and bridge subsystem. Equations (5.4) and (5.7) are solved iteratively using the explicit integration method. This iterative scheme can provide a higher convergence rate compared with those using the Wilson-θ method and the Newmark-β method (Li *et al.*, 2010).

5.4.5 Stress Analysis

Once the generalized coordinates of the bridge have been computed, the element stress vector of the jth element without considering initial strains and stresses, $\{\sigma_{b,j}^e\}$ can be obtained as follows:

$$\{\sigma_{b,j}^e\} = [\Gamma_{b,j}]\{q_b\} \tag{5.17}$$

where $[\Gamma_{b,j}]$ denotes the modal stress matrix which can be determined as

$$[\Gamma_{b,j}] = [D_{b,j}][L_{b,j}][N_{b,j}][\Phi_{b,j}] \tag{5.18}$$

where $[\Phi_{b,j}]$ is the modal shape matrix with respect to nodes of the jth element; $[N_{b,j}]$ is the shape function transforming node displacements to the element displacement field; $[L_{b,j}]$ is the differential operator transforming the element displacement field to the element strain field; and $[D_{b,j}]$ denotes the elastic matrix representing stress-strain relationship.

5.5 MONITORING OF RAILWAY LOADING EFFECTS ON THE TSING MA BRIDGE

The Tsing Ma Bridge in Hong Kong is the longest suspension bridge in the world carrying both highway and railway. It is the key structure of the most important transportation network in Hong Kong that links the Hong Kong International Airport to the commercial centres of Hong Kong Island and Kowloon. The operation condition of a railway in a long-span suspension bridge and the railway loading effect on the bridge should be monitored closely to ensure the functionality and safety of both trains and the bridge (Wong *et al.*, 2001a; HKPU, 2007). This is particularly important for the railway in the Tsing Ma Bridge. When the hourly mean wind speed recorded on site is in excess of 165 km/h, both the upper and lower decks must be closed for all road vehicles and the railway becomes the only transportation tool connecting Lantau Island including the Hong Kong International Airport to the rest of Hong Kong.

The railway connecting Lantau Island including the Hong Kong International Airport to the existing commercial centres of Hong Kong Island and Kowloon via the Tsing Ma Bridge is managed by the MTR Corporation Hong Kong Limited. The Tsing Ma Bridge was opened to the public in July 1997 and the railway operation on the Tsing Ma Bridge began in June 1998. Most of the trains running on the Tsing Ma Bridge were 7-car trains before 2003. With the increasing demand of train passengers, most of the trains became 8-car trains at the end of 2005. Between 2003 and 2005, there was a transition period with 7-car and 8-car trains in concurrent operation. The total length of a 7-car and 8-car train is approximately 158 m and 182 m, respectively. Each coach (car) in a train has two identical bogies and each bogie is supported by two identical wheel sets. The full length of one coach is about 22.5 m. Figure 5.5 shows the composition of a 7-car train. Seven coaches in the train have different functions, leading to different bogie loads or a bogie load distribution.

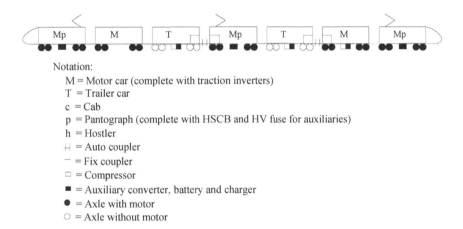

Notation:

    M = Motor car (complete with traction inverters)

    T = Trailer car

    c = Cab

    p = Pantograph (complete with HSCB and HV fuse for auxiliaries)

    h = Hostler

    ⊢ = Auto coupler

    ⁻ = Fix coupler

    □ = Compressor

    ■ = Auxiliary converter, battery and charger

    ● = Axle with motor

    ○ = Axle without motor

Figure 5.5 Configuration of a 7-car train

Railway tracks supported on the bottom chords of the cross frames of the bridge deck are constituted of track plates, rail waybeams and tee diaphragms. The rail waybeams are the main longitudinal members which exist in pairs in each railway track. The rail waybeams are stiffened by the T-shaped diaphragms interconnecting the two parallel rail waybeams in the orthogonal direction. On top of the stiffened rail waybeams, 20 mm thick steel track plates are laid.

To monitor the train traffic flow and identify the bogie load distribution, a set of strain gauges were installed on the inner waybeam of each pair of waybeams under the two rail tracks. Through a proper calibration, the signals from the strain gauges can be converted to the bogie load data, by which the requested information on the train traffic flow and bogie load distribution can be obtained. This special measurement system was put into operation in 2000. The bogie load data include the instant of the train arriving at the location of the strain gauges, train direction (lane), train speed, the total number of train bogies, bogie load, and axle spacing. The bogie load data recorded by the strain gauges must be pre-processed. The primary function of data pre-processing is to remove the unreasonable data while maintaining the useful data for the subsequent statistical data analysis.

5.5.1 Monitoring of Railway Traffic Condition

Traffic volume and traffic composition of trains are two important indexes of train (railway) traffic condition, from which the operation trend of trains with time can be inferred. Based on the bogie load records, the annual number of trains passing through the Tsing Ma Bridge is calculated in terms of the number of bogies and shown in Figure 5.6 for six years from 2000 to 2005. The monthly number of trains running through the Tsing Ma Bridge in 2005 is also calculated in terms of the number of bogies and shown in Figure 5.7. Major observations can be summarised as follows:

- There is a significant increase in the annual train count in 2003 and afterwards the annual train count becomes stable around 150,000.
- Before 2003, 14-bogie trains (7-car trains) are dominant, at about 96% of all the trains passing through the bridge annually.
- Since 2003, the number of 16-bogie trains (8-car trains) has exceeded that of 14-bogie trains.
- After the opening of the Hong Kong Disneyland in September 2005, 16-bogie trains become dominant. In November 2005, the percentage of the 16-bogie trains is already about 90%.
- November and December of 2005 can be considered as the months representing the current train traffic condition accurately. The monthly train count is about 12,000 and more than 90% of the trains are 16-bogie trains.

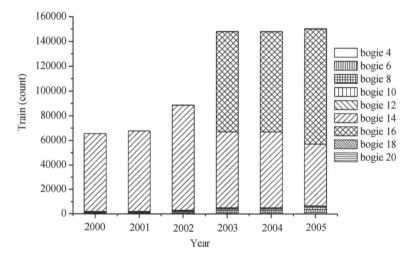

Figure 5.6 Annual train count (2000-2005)

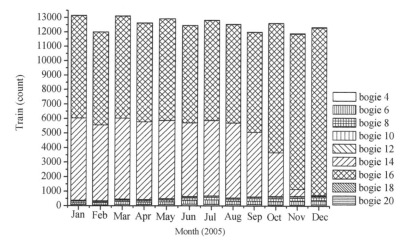

Figure 5.7 Monthly train count in 2005

5.5.2 Monitoring of Railway Loading

The bogie load data recorded in November 2005 are chosen as a database to investigate the bogie load distribution. Because about 90% of the trains, that is, a total of 10,705 trains, passing through the Tsing Ma Bridge in November 2005 are 16-bogie trains, the investigation of the bogie load distribution focuses on 16-bogie trains. The bogie load distribution in a 16-bogie train is actually the distribution of 16 bogie loads along the train. For each 16-bogie train passing through the Tsing Ma Bridge in either the Airport bound or the Kowloon bound direction, one bogie load distribution can be obtained.

The bogie load distribution with the maximum bogie load and the minimum bogie load in November 2005 is shown in Figure 5.8 for the Airport bound trains. The mean of the bogie load distribution is also plotted in the figure for the Airport bound trains together with the design bogie load distribution provided by the MTR Corporation Hong Kong Limited. It can be seen that that the pattern of the measured bogie load distributions follows the design pattern. The bogie load distribution with the maximum bogie load is below the design bogie load distribution. The same observations are made for the Kowloon bound trains.

A bogie load spectrum is defined as either the probability density function or the probability distribution of bogie loads, by which the railway loads on the Tsing Ma Bridge can be described statistically. In November 2005, a total of about 10,705 eight-car trains passed through the Tsing Ma Bridge. By taking the 13th bogie in an 8-car train as an example, the bogie load spectrum is calculated to see the probability distribution of bogie loads from the 13th bogie of all the 8-car trains passing through the Tsing Ma Bridge in November 2005. The result for the Airport bound trains is shown in Figure 5.9. It can be seen that the most frequent load from the 13th bogie of all the 8-car trains in November 2005 is about 25.5 tonnes for the

Airport bound. The bogie load from the 13th bogie with the exceedance probability of 1% is 28.5 tonnes.

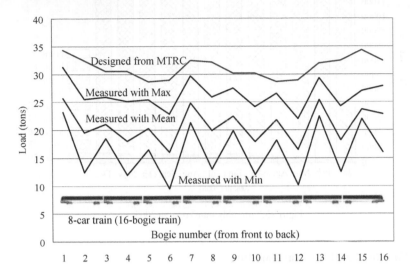

Figure 5.8 Comparison of the measured and designed bogie load distributions in November 2005 (Airport bound)

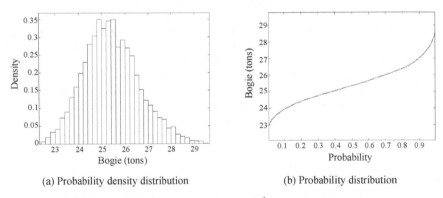

(a) Probability density distribution   (b) Probability distribution

Figure 5.9 Bogie load spectrum of the 13th bogie (Airport bound)

When considering all the bogie loads without reference to a particular bogie, a database of about 171,280 bogie loads is used to work out the bogie load spectrum. The results are shown in Figure 5.10 for the Airport bound trains. It can be seen that the most frequent bogie load among all the bogie loads in November 2005 is about 21.5 tonnes for the Airport bound, which are smaller than those from the 13th bogie only. The bogie load with the exceedance probability of 1% is 28.0 tonnes for the Airport bound.

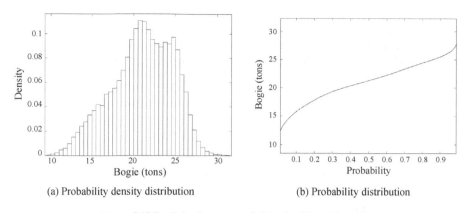

(a) Probability density distribution   (b) Probability distribution

Figure 5.10 Bogie load spectrum of all bogies (Airport bound)

A gross train weight (GTW) spectrum is defined as either the probability density function or the probability distribution of gross train weights. In November 2005, a total of about 10,705 eight-car trains passed through the Tsing Ma Bridge. Figure 5.11 shows the gross train weight spectrum for the Airport bound from all the 8-car trains passing through the Tsing Ma Bridge in November 2005. It can be seen that the most frequent GTW among all the 8-car trains in November 2005 is about 350 tonnes for the Airport bound. The GTW with the exceedance probability of 1% is 405 tonnes for the Airport bound, which is below the design maximum load of 498.4 tonnes.

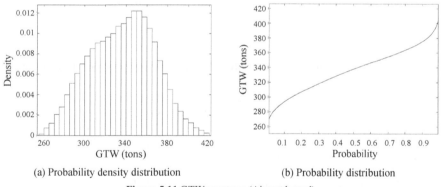

(a) Probability density distribution   (b) Probability distribution

Figure 5.11 GTW spectrum (Airport bound)

5.5.3 Monitoring of Railway Loading Effects on Bridge Displacements

5.5.3.1 Background

The displacement response of a bridge reflects the safety and integrity of the bridge. The displacement responses of the Tsing Ma Bridge are monitored by three types of sensory system: displacement transducers, lever sensing stations and GPS (Wong *et al.*, 2001b). In consideration of the nature of railway loads and the features of railway loading effects on the bridge, only the displacement responses of the bridge deck in the vertical direction recoded by the GPS stations will be analysed to ascertain railway loading effects in this study. There are in total 14 GPS stations installed in the Tsing Ma Bridge: eight on the bridge deck; four on the top of the bridge towers; and two on the main cables. Their locations are shown in Figure 5.12.

It can be seen that the four sections of the bridge deck are mounted with eight GPS stations: each section has one on the south and the other on the north. The mid-span of the Ma Wan side span is noted with one pair of GPS stations. The other three pairs of GPS stations are located at one quarter, one half and three quarters of the main span of the bridge deck with reference from the Ma Wan tower. The displacement response of the bridge deck measured by GPS is a positive moment in the corresponding three directions if the bridge component moves toward the Tsing Yi Island, sways to the north side and goes upward, respectively. The GPS data are collected at a sampling rate of 10 Hz. The GPS stations were implemented in the Tsing Ma Bridge in 2000 with on-going updating of the GPS stations until September 2002.

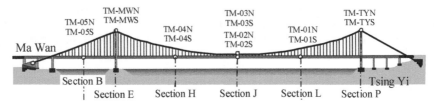

Figure 5.12 Locations of GPS stations in the Tsing Ma Bridge

The Tsing Ma Bridge is not only subjected to highway and railway traffic loads but also subjected to wind loads and temperature loads. As the bridge is continuously under different combinations of the four primary bridge loadings, it is difficult to evaluate the bridge response induced by railway loading and excluding the effects from the others. Furthermore, railway loading effects on the bridge displacement response depend on the number, speed and direction of trains. Therefore, the proper methods should be proposed to extract the train-induced displacement response from the measured total bridge displacement response, to standardize train-induced displacement response time histories so as to obtain train-induced displacement influence line, and to apply the influence line for the case of trains on two lines.

5.5.3.2 Data Collection and Pre-processing

In order to study railway loading effects, the GPS data should be carefully selected to keep all other loading effects as low as possible so that the railway loading effect can be identified. The data collected in November 2005 is considered as a desirable dataset for this purpose because both the railway and highway traffic in this month were steady and no strong winds were reported. The average hourly mean wind speed for this month was 2.8 m/s only. Furthermore, the data from this month represent the current traffic condition of Tsing Ma Bridge as the condition has been changed after the opening of the Hong Kong Disneyland in September 2005.

Pre-processing of the GPS data is required prior to the railway loading effect analysis. The main steps of the pre-processing of GPS data include: (1) the conversion of GPS data from the HK80 geographic coordinate to the Universal Transverse Mercator grid coordinate; (2) to compute the bridge displacement response with reference to the mean coordinate which was set for each GPS when it was installed; (3) to remove unreasonable and undesirable data in terms of the number of satellites, the dilution of precision, and spike while maintaining the quality data for the subsequent statistical data analysis; and (4) to calculate the bridge displacement responses in the global longitudinal, lateral and vertical directions of the bridge.

5.5.3.3 Railway Loading Effects on Bridge Displacements

After completion of the data pre-processing, the displacement induced by railway loadings is extracted and analysed by the following procedures: (1) the extraction of the vertical bridge-deck displacement due to railway loading effects by selecting a proper frequency range and reconstructing the new displacement time history through the wavelet packet transform; (2) the standardization of the displacement time history and establishment of the database for displacement influence line of a train on the bridge; (3) the establishment of three statistical displacement lines from the database of displacement influence lines: the mean line, the upper limit line, and the low limit line; and (4) the reconstruction of the displacement time history of two trains on the bridge, and subsequently estimating the most unfavourable situations.

The GPS displacement data in the hour from 01:00 to 02:00 on November 2, 2005 collected by the GPS receivers located at Section J (middle of the main span of the deck) in the vertical direction are plotted in Figure 5.13(a) and taken as an example to demonstrate how to extract the train-induced displacement response. The hourly mean wind speed in the hour period is 2.48 m/s only. It is observed together with the information from strain gauges that the vertical displacement of the deck is suddenly increased when a train is running on the bridge. Figure 5.13(a) also shows that there is a significant fluctuating displacement even though there is no train running on the bridge. This implies that the vertical displacement is influenced not only by the railway loading but also by the effects from other loadings. The power spectrum of the relevant one-hour vertical displacement of the bridge deck at the middle of the main span shows that the dominant bridge vibration energy occurs in the low frequency range less than 0.08 Hz.

To exclude the effects from other loadings, the displacement response due to the railway loading effect would be extracted from the raw data by the wavelet transform. Wavelet transform is a mathematical tool that can decompose a temporal signal into a summation of time-domain functions of various resolutions. The wavelet packet transform is used to decompose the original GPS displacement signal to wavelet packet component signals. After j-level decomposition, the original signal $f(t)$ can be expressed as:

$$f(t) = \sum_{i=1}^{2^j} f_j^i(t) \tag{5.19}$$

The signal energy E_f is defined as:

$$E_f = \sum_{i=1}^{2^j} E_{f_j^i} \tag{5.20}$$

The wavelet packet component energy, $E_{f_j^i}$, can be considered to be the energy stored in the component signal $f_j^i(t)$ during the time length of data T:

$$E_{f_j^i} = \int_0^T f_j^i(t)^2 dt \tag{5.21}$$

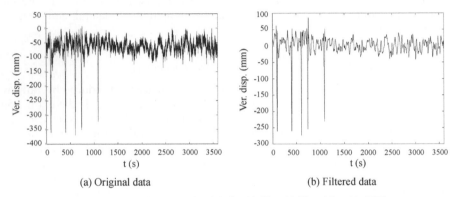

(a) Original data   (b) Filtered data

Figure 5.13 GPS data at Section J during 01:00 to 02:00 on Nov. 02, 2005

In order to extract the data due to railway loading effects from the GPS displacement data, the main frequency range of railway loading should first be determined. The signal within the frequency range of 0 ~ 0.0025 Hz can be excluded after considering the span-length of bridge and the speed of train. The wavelet packet component energy of the relevant frequency range is calculated based on both 9- and 11-levels decomposition, and comparisons are made between the case of with train and without train running on the bridge. It was found that the railway loading takes a predominating effect in the frequency range of 0.0025 ~ 0.08 Hz. The displacement response within the frequency range of 0.0025 ~ 0.08 Hz can be considered to be dominated by railway loading effects. The processed time history with a frequency range between 0.0025 to 0.08 Hz is plotted in Figure 5.13(b).

In order to compare the processes of trains with different speeds and heading-directions, it is necessary to convert the time coordinate to the space coordinate and to convert the two heading-directions to a standard direction as well. The database of train-induced displacement lines can be established subsequently after performing the conversion. The conversion of the time coordinate to the space coordinate can be completed if the train speed is known. The calculation of the train speed is based on the assumption that the vertical displacement reaches its minimum when the middle of a train is at the location of a sensor and the train speed is constant. Two sensors located at Section J (middle of the main span) and Section B (middle of the Ma Wan side span) respectively are chosen to determine the time difference of minimum values. As the distance between Section B and Section J is known, the train speed can be calculated by the following equation:

$$T_{speed} = \frac{L_B - L_J}{t_B - t_J}$$ (5.22)

In Equation (5.22), t_J and t_B are the instants of minimum of displacement response in the time period of a train running on the bridge collected by GPS sensors located at Sections J and B; L_J and L_B are the space coordinates of Sections J and B, and T_{speed} is the mean speed of the train running on the bridge. The conversion of heading direction can be determined from the vertical displacement data collected by two GPS receivers on the south and the north track. The track with train is denoted as 'Mainway', and the other track is denoted as 'Sideway'. The vertical displacement of the track on which the train is running is relatively larger than that of the track without the train. The time periods of 01:00~02:00 and 05:00~06:00 every day in November 2005 are chosen to establish the database due to one train on the bridge because during these time periods only one train runs on the bridge and road vehicle traffic is relatively small. 180 displacement influence lines for different sections are then obtained after conversion and they are plotted in Figure 5.14.

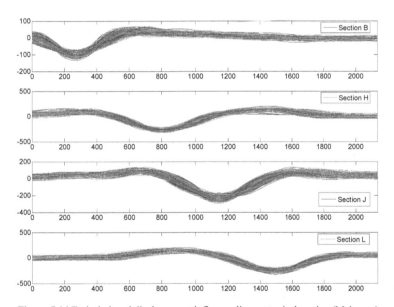

Figure 5.14 Train-induced displacement influence lines at typical section (Mainway)

The vertical displacement influence lines due to the railway loading are analysed statistically in terms of mean value, lower limit, and upper limit in consideration of different gross train weights. The mean displacement influence lines of four sections for 'Mainway' and 'Sideway' are plotted in Figure 5.15. It is found that the largest values of mean displacement line at the quarter main span (Sections H and L) are greater than that at the middle of main span (Section J) while relatively small mean displacement is observed at the side span of the bridge (Section B). The mean displacement at the "Mainway" is slightly greater than that at the "Sideway". It can also be seen that in general, each mean influence line has one minimum value (downward), one maximum value (upward) on the left and one maximum value on the right with reference to the minimum value. The three extreme values and their locations of each mean influence line of train-induced displacement are summarized in Table 5.1. The locations of the extreme values in the table refer to the Ma Wan anchorage.

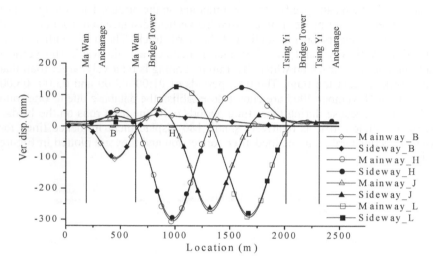

Figure 5.15 Mean influence lines of train-induced displacement at typical sections

Table 5.1 Extreme values and locations of mean influence lines.

Section	Track	Max. on left		Min. in middle		Max. on right	
		Location (m)	Value (mm)	Location (m)	Value (mm)	Location (m)	Value (mm)
B	Mainway	-	-	270	-105.8	727	37.0
	Sideway	-	-	270	-102.1	727	37.0
H	Mainway	-	-	801	-304.3	1443	124.0
	Sideway	-	-	801	-296.1	1443	123.6
J	Mainway	665	56.5	1145	-273.5	1625	40.8
	Sideway	665	56.2	1145	-264.9	1625	40.7
L	Mainway	853	127.7	1498	-291.7	-	-
	Sideway	853	126.5	1498	-284.8	-	-

Apart from the mean displacement influence lines, the statistical upper and lower limit lines of the train-induced bridge vertical displacements are also computed to reflect the effects of gross train weight. The statistical upper and lower limit lines are determined based on the exceedance probabilities of 1% and 99% respectively. There are 180 samples for each location in a set of 180 train-induced displacement influence lines for a particular section. The displacements of exceedance probabilities of 1% and 99% are chosen from the 180 samples to have the upper and low limit values, respectively. Repeating this procedure for a series of locations, the statistical upper and lower limit lines can be obtained. Figure 5.16 shows the probability density function and the probability distribution of the displacements at the location of 1498 m using the set of 180 train-induced displacement influence lines for Section L as shown in Figure 5.14. Based on the probability distribution, the upper and lower limit values can be easily obtained. In this way, the statistical displacement lines (mean line, lower limit line and upper limit line) due to one train load at Sections B, H, J and L are calculated and plotted in Figure 5.17. It can be seen that in general, each influence line has one minimum value (downward), one maximum value (upward) on the left and one maximum value on the right with reference to the minimum value.

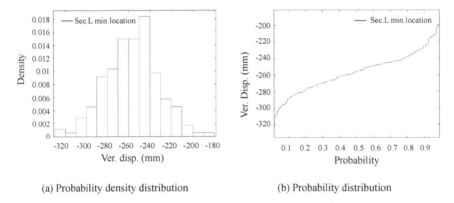

(a) Probability density distribution   (b) Probability distribution

Figure 5.16 Distribution of displacements for Section L at location of minimum value

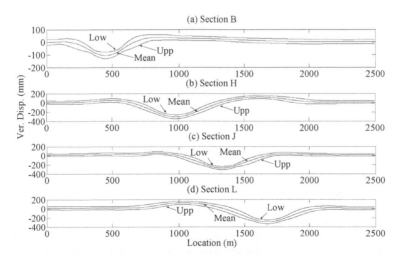

Figure 5.17 Statistical one train-induced displacement influence lines (Mainway)

The three extreme values and their locations of each set of train-induced displacement influence lines are summarized in Table 5.2. Based on the results listed in Table 5.2, the maximum and minimum displacements caused by one 8-car train at each of the four sections with the GPS stations can be worked out at the designated probability level. The results are shown in Figure 5.18, and one train-induced displacement range is 209.2 mm, 536.3 mm, 427.0 mm and 522.9 mm for Sections B, H, J, and L respectively.

Table 5.2 Extreme values and locations of statistical influence lines (Mainway).

Sections		Max. on left		Min. in middle		Max. on right	
		Location (m)	Value (mm)	Location (m)	Value (mm)	Location (m)	Value (mm)
B	Upper (99%)	-	-	270	-71.5	727	69.9
	Mean (50%)				-105.8		37.0
	Low (1%)				-139.3		11.8
H	Upper (99%)	-	-	801	-247.3	1443	171.3
	Mean (50%)				-304.3		124.0
	Low (1%)				-365.0		73.6
J	Upper (99%)	665	100.7	1145	-226.4	1625	88.2
	Mean (50%)		56.5		-273.5		40.8
	Low (1%)		16.1		-326.3		-12.6
L	Upper (99%)	853	170.3	1498	-220.0	-	-
	Mean (50%)		127.7		-291.7		
	Low (1%)		76.7		-352.6		

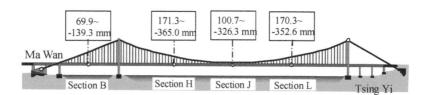

Figure 5.18 Displacement ranges at measured sections due to one train

The results obtained above could be used to estimate the displacement of the bridge with two trains if the displacement of the bridge is assumed to vary linearly with train load and the train is assumed to run at a constant speed. Based on these two assumptions, the principle of superposition can be applied and the statistical displacement influence line due to one train can be used to construct the displacement time histories of two trains running on the bridge and to anticipate the most unfavourable situation occurring on the bridge with two trains. The procedure of reconstructing the time history of two trains running on the bridge should include: (1) determining the train speed and heading direction, and subsequently, the relative locations of two trains on the bridge; and (2) determining the vertical displacements induced by the train on 'Mainway' and the train on 'Sideway', and adding them together to obtain a total displacement induced by the two trains.

To check the validity of the method, a case with two trains running on the bridge recorded within 11:00~12:00 on 2 Nov. 2005 is taken as an example. These two trains arrive at Section J in opposite directions at the time instant of 11:49:12 and 11:49:37 respectively. Comparison between the collected GPS displacement data and the reconstructed time history of three statistical lines – mean line, upper limit line, and lower limit line – is shown in Figure 5.19 for the four instrumented sections. It can be seen that the GPS data coincide with the mean line quite well and the GPS data are within the lower and upper limit lines.

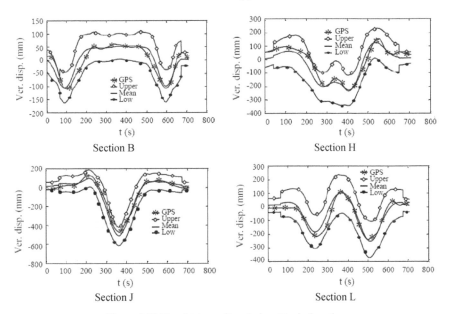

Figure 5.19 Time history of two trains at typical sections

The most unfavourable situation with two trains meeting at the location of the GPS station is investigated. The results from the cases with two trains meeting at Sections B, H, J and L are plotted separately in Figure 5.20. It is found that the vertical displacement of the bridge deck does reach the minimum value when the two trains meet at the locations of GPS stations. The extreme displacement values due to two trains meeting at the measured sections are listed in Table 5.3. Based on the results listed in Table 5.3, the maximum and minimum displacements caused by two 8-car trains meeting at each of the four sections with GPS stations can be worked out at the designated probability level. The results are shown in Figure

5.21, and the two-train induced displacement range is 381.1 mm, 973.1 mm, 839.2 mm and 923.1 mm for Sections B, H, J, and L respectively.

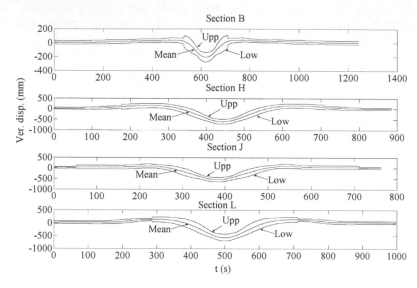

Figure 5.20 The most unfavourable case of the displacement time history due to two trains

Table 5.3 Extreme displacements due to two trains meeting at the measured sections.

	Section	Minimum (mm)	Maximum (mm)
	Upper (99%)	-138.5	105.6
B	Mean (50%)	-207.6	33.5
	Low (1%)	-277.5	7.6
	Upper (99%)	-485.1	254.2
H	Mean (50%)	-600.3	157.4
	Low (1%)	-718.9	50.6
	Upper (99%)	-439.5	189.8
J	Mean (50%)	-538.4	97.1
	Low (1%)	-649.4	4.7
	Upper (99%)	-434.3	224.1
L	Mean (50%)	-576.4	137.7
	Low (1%)	-699.0	40.6

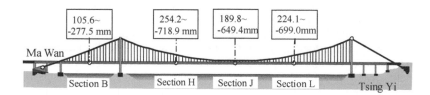

Figure 5.21 Worst displacement ranges at the measured sections due to two trains

5.5.4 Monitoring of Railway Loading Effects on Bridge Accelerations

5.5.4.1 Background

The acceleration response of a combined highway and railway bridge is closely related to the serviceability and functionality of both vehicles and bridge. The acceleration responses of the Tsing Ma Bridge are measured by accelerometers installed in the bridge deck and main cables. This section investigates mainly railway loading effects and traffic (consider both railway and highway) loading effects on the acceleration response of the bridge deck in the vertical direction. There are totally 24 channels' accelerometers installed in the deck, main cables, and Ma Wan anchorage of the Tsing Ma Bridge. The acceleration responses of the bridge deck are measured by a total of 12 channels' accelerometers installed in the four sections (Sections B, F, I and J) as shown in Figure 5.22(a). At each section, there are one bi-axial accelerometer (labelled as "AB") in the south measuring accelerations in the vertical and lateral directions and one uni-axial accelerometer (labelled as "AS") in the north measuring acceleration in the vertical direction, as shown in Figure 5.22(b). Hourly acceleration response time histories of the bridge deck from 12-channels were recorded on tapes with the sampling frequency of 25.6 Hz before 2001 and 51.2 Hz from 2002 onward. The measurements from accelerometers are directly in unit of millimetre per second squares (mm/s^2) and in positive value if the deck sways to the north side in lateral direction and goes upward in the vertical direction.

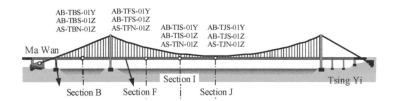

(a) Location of accelerometers in the bridge deck

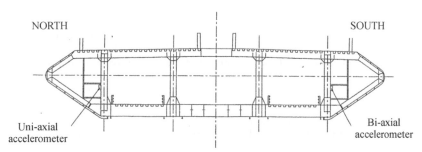

(b) Position of accelerometers at cross sections of the bridge deck

Figure 5.22 Location of accelerometers in Tsing Ma Bridge

The bridge acceleration responses are induced by a combination of four major types of load effects, referring to wind, temperature, highway, and railway loads. It is not an easy task to distinguish one loading effect from other loading effects. In consideration of the nature of traffic (both highway and railway) load and the features of traffic loading effects on the bridge, only the acceleration responses of the bridge deck in the vertical direction recorded by the accelerometers will be analyzed to ascertain traffic loading effects on bridge acceleration responses in this study. No attempt is made to distinguish highway loading- and railway loading-induced bridge accelerations.

5.5.4.2 Data Collection and Pre-processing

In order to study traffic loading effects, the acceleration data should be carefully selected to keep all other loading effects as low as possible so that the traffic loading effects can be identified. The temperature load does not make a significant contribution to bridge acceleration response compared with wind and traffic loads. Therefore, the acceleration data collected in November 2005 is considered as a desirable dataset for this purpose due to the reason described above. A few hourly records with hourly mean wind speed more than 6 m/s are excluded from the database. The remaining acceleration responses are mainly attributed to the effects of highway and railway loads.

Pre-processing of the acceleration response data is required prior to the traffic loading effect analysis. The main steps of the pre-processing of acceleration data include: (1) to obtain zero-mean acceleration response time history by removing non-stationary trend from the original time history; (2) to remove impulsive spikes by using a median filter according to the difference between the filtered signal and the original signal; and (3) to calculate the acceleration response of the bridge deck in the vertical direction by averaging the signals from the two vertical accelerometers. The pre-processing of all the acceleration response data is done automatically with computer programs using MATLAB as a platform.

5.5.4.3 Railway Loading Effects on Bridge Accelerations

As compared with the displacement response extraction, it is more difficult to identify the frequency range of acceleration responses governed by the railway loading. Furthermore, the approximate instant of a train running on the bridge is hard to identify from the measured acceleration response time history. Let us consider two data samples collected at 01:00~02:00 and 11:00~12:00 on Nov. 2, 2005, which represent the cases of low and heavy highway traffic conditions, respectively. The original hourly acceleration response time histories recorded from the four instrumented sections at 01:00~02:00 on Nov. 2, 2005 are plotted in Figure 5.23. Clearly, the instant of the train running on the bridge is difficult to identify from the original acceleration time histories.

Similar to the method used in processing the train-induced displacement responses, the original acceleration time histories are decomposed using the wavelet analysis and only the components within the frequency range from 0.0125 to 0.1 Hz are plotted in Figure 5.24 for the four instrumented sections. From the figure, the approximate time of a train running on the bridge and the effect of the running train on the bridge acceleration response can be estimated. It is observed that the train-induced acceleration response is much smaller than the total acceleration response. As the railway loading is within a low frequency range, the contribution to the total acceleration is attributed to higher frequency components especially the highway loading. Consequently one may conclude that the railway-induced bridge acceleration response is very small as compared with the highway-induced bridge acceleration response.

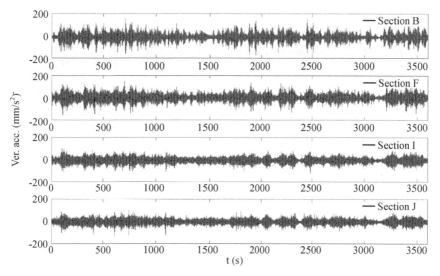

Figure 5.23 Original acceleration data recorded at 01:00~ 02:00 on Nov. 2, 2005

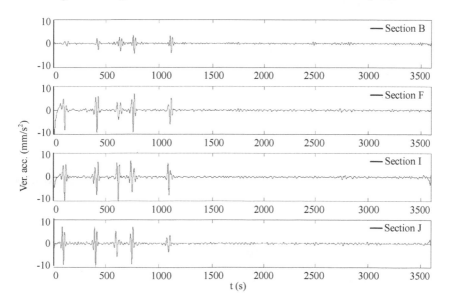

Figure 5.24 Filtered acceleration data recorded at 01:00~ 02:00 on Nov. 2, 2005 (0.0125~0.1Hz)

To further confirm this observation, the original acceleration response time histories collected at 11:00~12:00 on November 2, 2005 from the accelerometers located at the middle of the main span (Section J) are also analysed. The original and filtered data within the frequency range between 0.0125 and 0.1 Hz are plotted in Figure 5.25. The figure also clearly shows that the railway loading effect on bridge acceleration response is small as compared with the highway loading effect. This is particularly true for the heavy highway traffic condition.

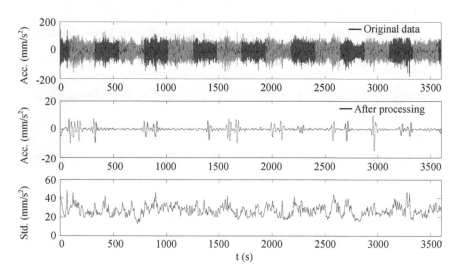

Figure 5.25 Original and filtered acceleration data at Section J during 11:00~12:00 on Nov. 2, 2005

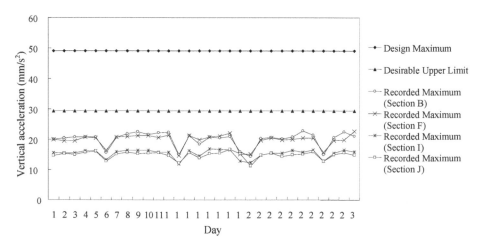

Figure 5.26 Peak accelerations of the bridge deck in November 2005

Based on the above observations and from a practical point of view, it is advisable to consider traffic (both highway and railway) loading effects on bridge acceleration response other than only railway or highway loading effects.

The peak acceleration is an important index for the evaluation of human comfort. The design maximum operating acceleration is 490 mm/sec². The desirable upper limit of operating acceleration on the bridge deck for LAR is 294 mm/sec². The daily maximum peak acceleration of the bridge deck in the vertical direction recorded by the accelerometer is plotted in Figure 5.26 for each of the four instrumented sections during November 2005. It can be seen that all the measured daily peak accelerations are below the desirable upper limit of operating acceleration and far below the design maximum operating acceleration. The daily peak accelerations of the bridge deck at Sections B and F are higher than those at Sections I and J. Daily peak accelerations exhibit approximate periodic wave: relatively small peak acceleration occurs every Sunday.

5.5.4.4 Case Study

The dynamic interaction of a long-span suspension bridge with running trains was investigated numerically by taking the Tsing Ma Bridge as an example (Xia *et al.*, 2000). A three-dimensional finite element model was used to represent the Tsing Ma Bridge. The computed natural frequencies and mode shapes were verified through the comparison with the measured results and the details can be found in Xu *et al.* (1997). The first twenty natural frequencies up to 0.38 Hz and the associated mode shapes were finally used in this case study since these natural frequencies and mode shapes were validated by the measurement data. The damping ratios in the lateral and vertical vibration modes were taken as 1% while the damping ratios in the torsional modes were 0.5% with reference to the measurement data. The train concerned in the case study consisted of eight passenger coaches. Each coach has two identical bogies and each bogie is supported by two identical wheel-sets. Each 4-axle coach was modelled by a 27-degrees-of-freedom dynamic system. The eight passenger coaches were assumed to be identical in this case study. The principal vibration mode frequency of the coach was about 1.04 Hz in the vertical direction and 0.68 Hz in the lateral direction. The dynamic interaction between the bridge and train was achieved through the contact forces between the wheels and track. The track vertical, lateral and torsional irregularities were taken into consideration by using the measured data from one of the main railways in China because no measurement data were available for the Tsing Ma Bridge. By applying a mode superposition technique to the bridge and taking the measured track irregularities as known quantities, the number of degrees of freedom of the bridge-train system was significantly reduced and the coupled equations of motion were efficiently solved. The dynamic response of the bridge-train system and the derail and offload factors related to the running safety of the train were computed. The results showed that the formulation presented here could well predict dynamic behaviours of both the bridge and train with reasonable computation efforts. It was also found that the dynamic responses of the long suspension bridge under running train are relatively small and the effects of the bridge motion on the runability of railway vehicles are insignificant. Further details on this work can be found in the literature (Xia *et al.*, 2000).

5.6 SUMMARY

The major sensors and sensing systems used for railway monitoring of long-span suspension bridges have been described in this chapter. The techniques for ascertaining the railway traffic condition and railway load distribution from the measurement data have been introduced. The finite element method-based analytical backgrounds of dynamic interaction of long-span suspension bridges with running trains have been discussed. The Tsing Ma suspension bridge has been taken as an example to illustrate the monitoring of the railway traffic condition and railway loading effects on the bridge. The railway traffic condition of the Tsing Ma Bridge has been established in terms of the train traffic volume, bogie load distribution, bogie load spectrum and gross train weight spectrum. A practical method has been proposed to investigate the railway loading effects on the vertical displacement response of the bridge deck based on the GPS measured displacement data. The railway loading effects on the vertical acceleration response of the bridge deck have been compared with the traffic (both highway and railway) loading effects on the same quantity. The measured traffic-induced peak accelerations have been also compared with the design values. The railway loading effects on bridge acceleration response have been found to be small as compared with the highway loading effects. This is particularly true for the heavy highway traffic condition. From a practical point of view, traffic (both highway and railway) loading effects on bridge acceleration response other than only railway or highway loading effects should be considered.

5.7 REFERENCES

Chan, W.S., Xu, Y.L., Ding, X.L., Xiong, Y.L. and Dai, W.J., 2006, Assessment of dynamic measurement accuracy of GPS in three directions. *Journal of Surveying Engineering*, ASCE, **132(3)**, pp. 108-117.

Diana, G. and Cheli, F. 1989, Dynamic interaction of railway systems with large bridges. *Vehicle System Dynamics*, **18**, pp. 71-106.

Frýba, L. 1996, *Dynamics of Railway Bridges*, (London: Thomas Telford House).

Gimsing, N.J., 1997, *Cable Supported Bridges*, (New York: John Wiley & Sons).

HKPU, 2007, *Establishment of Bridge Rating System for Tsing Ma Bridge: Statistical Modelling and Simulation of Predominating Railway Loading Effects.* Report No. 6, (Hong Kong: Department of Civil and Structural Engineering, The Hong Kong Polytechnic University).

Kalker, J.J., 1990, *Three-Dimensional Elastic Bodies in Rolling Contact*, (Dordrecht, Netherlands: Kluwer Academic Publishers).

Li, Q., Xu, Y.L., Wu, D.J. and Chen, Z.W., 2010, Computer-aided nonlinear vehicle-bridge interaction analysis. *Journal of Vibration and Control*, **16(12)**, pp. 1791-1816.

McConnell, K.G., 1995, *Vibration Testing*, (New York: John Wiley & Sons).

Shen, Z.Y., Hedrick, J.K., and Elkins, J.A., 1983, A comparison of alternative creep force models for rail vehicle dynamic analysis. In *Proceedings of the 8th IAVSD Symposium*, Cambridge, MA, edited by Hedrick, J.K. (London: Swets & Zeitlinger Publishers), pp. 591-605.

Unsworth, J.F., 2010, *Design of Modern Steel Railway Bridges*, (Boca Raton, Florida: CRC Press).

Xia, H., Xu, Y.L., and Chan, T.H.T., 2000, Dynamic interaction of long suspension bridges with running trains. *Journal of Sound and Vibration,* **237(2)**, pp. 263-280.

Xu, Y.L., Ko, J.M., and Zhang, W.S., 1997, Vibration studies of Tsing Ma long suspension bridge. *Journal of Bridge Engineering, ASCE,* **2(4)**, pp. 149-156.

Xu, Y.L., Li, Q., Wu, D.J. and Chen, Z.W., 2010, Stress and acceleration analysis of coupled vehicle and long-span bridge systems using mode superposition method. *Engineering Structures,* **32 (5)**, pp. 1356-1368.

Yang, Y.B., Yau, J.D. and Hsu, L.C., 1997, Vibration of simple beams due to trains moving at high speeds. *Engineering Structures,* **19(11)**, pp. 936-944.

Wong, K.Y, Man, K.L and Chan, W.Y.K, 2001a, Monitoring of wind load and response for cable-supported bridges. In *Proceedings of SPIE 6th International Symposium on NDE for Health Monitoring and Diagnostics, Health Monitoring and Management of Civil Infrastructure Systems*, Newport Beach, California, edited by Chase and Aktan, **4337**, (Bellingham, Washington: SPIE), pp. 292-303.

Wong, K.Y., Man, K.L and Chan, W.Y.K, 2001b, Application of global positioning system to structural health monitoring of cable-supported bridges. In *Proceedings of SPIE 6th International Symposium on NDE for Health Monitoring and Diagnostics, Health Monitoring and Management of Civil Infrastructure Systems*, Newport Beach, California, edited by Chase and Aktan, **4337**, (Bellingham, Washington: SPIE), pp. 390-401.

Zhai, W.M. and Cai, C.B., 2003, Train/track/bridge dynamic interactions: simulation and applications. *Vehicle System Dynamics,* **37(supplement)**, pp. 653-665.

5.8 NOTATIONS

b	Half width of the nominal rail gauge.
b_1	Half width of the transverse distance between the two springs of the primary suspension unit.
c_{1y}	Vertical damping coefficient of the primary suspension unit.
$[D_{b,j}]$	Elastic matrix representing stress-strain relationship.
E_f	Signal energy.
$f(t)$	Original signal before decomposition.
$f_{11,j}, f_{22,j}, f_{33,j}$	Longitudinal, lateral, lateral/spin and spin creepage coefficients.
$\{f_{bc}\}$	Wheel-rail contact forces acting on the bridge.
$\{f_{vc}\}$	Wheel-rail contact force vector.
$\left\{f_{vn}^L\right\}$	Pseudo force vector.
F_{byj}	Vertical force acting on the bogie.
F_{maxij}	Saturated damping force.
F_{s1}, F_{s2}	Vertical forces acting at the two sides of the wheel-set.
$F_y, F_z,$	Contact forces in the vertical and lateral direction.

J_{ws}	Rolling mass moment of inertia of the wheel-set.
k_{1y},	Vertical stiffness of the primary suspension unit.
$[L_{b,j}]$	Differential operator transforming the element displacement field to the element strain field.
M_{ws}	Mass of the wheel-set.
$[M_b]$, $[C_b]$, $[K_b]$	Mass, damping and stiffness matrices of the bridge subsystem.
$[M_v]$, $[C_v]$, $[K_v]$	Mass, damping and stiffness matrices of the vehicle.
N_1, N_2	Normal contact forces.
$[N_{b,j}]$	Shape function transforming node displacements to element displacement field.
$\{q_b\}$, N_b	Modal coordinate and the quantity of modes of the bridge subsystem.
$\{q_v\}$, N_v	Modal coordinate and the quantity of modes of the vehicle.
r_0	Nominal wheel radius.
T_{speed}	Mean speed of the train running on the bridge.
T_{xj}, T_{zj}, $\bar{M}_j$	Longitudinal creeping force, lateral creeping force and spin creeping moment.
T_{z1}, T_{z2}	Lateral creeping forces.
v	Nominal travelling speed of the wheel-set.
W	Static wheel load.
y_w, y_r	Vertical motions of the wheels and rails.
y_{ws}, θ_{ws}	Bouncing and rolling motions of the wheel-set.
$\{\delta_b\}$, $\{\dot{\delta}_b\}$, $\{\ddot{\delta}_b\}$	Displacement, velocity and acceleration of the bridge subsystem.
$\{\delta_v\}$, $\{\dot{\delta}_v\}$, $\{\ddot{\delta}_v\}$	Displacement, velocity and acceleration of the vehicle.
$[\xi_b]$, $[\omega_b]$, $[\Phi_b]$	Modal damping, modal frequency, and normalized mode shape matrices of the bridge subsystem.
$[\xi_v]$, $[\omega_v]$, $[\Phi_v]$	Modal damping, modal frequency, and normalized modal shape matrices of the vehicle.
δ, θ_w	Contact angles, the rolling angles.
$\xi_{1,j}$, $\xi_{2,j}$, $\xi_{3,j}$	Creepage ratios in the longitudinal, lateral and spin directions.
$\Omega_{w,j}$, $\Omega_{r,j}$	Rotational velocities of spin motion of the wheel and rail at the contact point.
$[\Gamma_{b,j}]$	Modal stress matrix.

CHAPTER 6

Monitoring of Highway Loading Effects

6.1 PREVIEW

Modern highway traffic has evolved over a period of several hundred years. As a consequence, many types of road vehicles have been developed and the number and speed of road vehicles running on highways have increased significantly (Taly, 1998). When a long-span suspension bridge carries a highway, the bridge is therefore subjected to a variety of non-stationary loads, such as those due to motorcycles, cars, buses, trucks, and heavy goods vehicles. Highway loadings are rather complex. At any given time, a bridge deck may be loaded randomly with a multitude of road vehicles. The effect of highway loadings on a suspension bridge is a function of several parameters, such as the axle loads, axle configuration, and gross vehicle weight (GVW), the number of vehicles on the bridge, the speed of the vehicles, the stiffness characteristics of the bridge, and the bridge configuration. Together, these parameters introduce analytical complexity and affect the load distribution in the bridge structure and its components. In this regard, the highway loading and its effects on the bridge must be monitored to provide information on the actual highway loading compared with the design highway loading and to ensure the functionality and safety of both the bridge and highway vehicles (Mufti, 2001; Balageas, 2006; Karbhari and Ansari, 2009)

This chapter first describes sensors and sensing systems used for highway monitoring of long-span suspension bridges. The techniques for ascertaining the highway traffic condition and highway load distribution from the measurement data are introduced. The dynamic analysis of coupled road vehicle-bridge systems is then presented with a wheel-road surface interaction included to provide a theoretical background of the highway loading and its effect. Finally, the Tsing Ma suspension bridge is taken as an example to illustrate the monitoring of highway traffic, highway loading, and predominating highway loading effects on the bridge. The highway load spectrum, including both the axle load spectrum and GVW spectrum, are calculated using the measurement data from the weigh-in-motion (WIM) systems. Stress data collected from strain gauges installed in the key members of the bridge are analysed. The fatigue life of the relevant bridge components due to the highway loading only, railway loading only, and a combination of railway and highway loadings is estimated respectively.

6.2 HIGHWAY MONITORING SYSTEM

Similar to railway monitoring, highway monitoring of a long-span suspension bridge also starts with the selection and arrangement of a sensor system and the associated data acquisition system. A variety of sensors are applicable for highway monitoring of long-span suspension bridges (Mufti *et al.*, 2008). Traditional sensors widely used for highway monitoring of a long-span suspension bridge include dynamic WIM stations for measuring the highway traffic and loading, displacement transducers for bridge displacements, accelerometers for acceleration responses, and strain gauges for strain responses. In some cases, additional sensors to those used in the railway monitoring system may be needed and their locations may be determined according to the requirement of highway monitoring.

WIM stations are designed to record traffic conditions and measure vehicle axle and gross weights and speed statistics as they drive over the station. Unlike older static weigh stations, current WIM stations do not require the subject vehicles to stop, making them much more efficient. Gross vehicle and axle weight monitoring is useful in an array of applications including the pavement design and monitoring, bridge design and monitoring, vehicle size and weight enforcement, legislation and regulation, and administration and planning. A dynamic WIM station mainly consists of a metal housing with lightning protection, bending plate sensors and sensor processor board, induction loop detection and loop processor board, central processing unit, and power supply. The metal housing must have sufficient room to house an inner cabinet of suitable size where the control electronics with power supply and maintenance-free backup batteries can be located. Proprietary plug-in circuit boards are designed with individual lightning protection for the bending sensors monitoring, induction loops monitoring, data storage RAM and additional serial interfaces. One dynamic WIM station, which is installed across the whole width of one traffic lane, comprises of four bending plate sensors and four induction loops. Four bending plate sensors are installed in line with two rows and with induction loop detectors (wheel detectors) in between and low power recording electronics to obtain wheel load data. The standard distance between the two rows of sensors must not be less than 3 metres with other distance programmable. In fixing the bending plate, the plate must be bevelled on both sides at the longitudinal borders and supported for two bevelled strips in a foundation frame, which must be tied firmly in the road surface by a special installation procedure. Each bending plate needs to be installed separately in a single frame. The loop processor board accepts inputs from four induction loop detectors. All signals must be operated or processed in real-time and not multiplexed to achieve high quality of data. After the installation of WIM stations, they need to be tested by applying loads to the sensors and checking the outputs with appropriate units.

Similar to railway monitoring, the dynamic displacement responses of the bridge to highway loadings can be measured using displacement transducers or GPS or obtained by integrating the acceleration records twice with time. The dynamic acceleration responses of the bridge to highway loadings can be measured using piezoelectric or servo-type accelerometers. The dynamic stress responses of the bridge to highway loadings can be measured using strain gauges or optical fiber sensors. Nevertheless, attention must be paid to the effective frequency range of the sensors because the frequency range of dynamic responses of the bridge to highway loadings is quite different from that to railway loadings. It is also noted that one of the major dynamic effects of traffic loadings (consider both highway and railway) on a long-span steel suspension bridge is the fatigue problem. Stress data collected from strain gauges installed in key members of the bridge must therefore be analysed to estimate the fatigue life of the bridge due to traffic loadings.

6.3 MONITORING OF HIGHWAY TRAFFIC CONDITION AND LOAD DISTRIBUTION

The highway traffic condition and highway load distribution over the deck of a long-span suspension bridge must be monitored and analysed so that a clear understanding of highway loading effects on the bridge can be achieved. In this regard, the M-class vehicle classification system can be used to classify types of vehicles running on the bridge according to their main features such as the number of axles, axle distance and gross weight (HKTD, 2008). Typical vehicle load data recorded by WIM stations include vehicle arrival date and time, lane number, vehicle speed, vehicle class according to M-class vehicle classification system, number of axles, axle weight, and axle spacing. From the WIM recorded raw data, different statistical data can be extracted.

The vehicle load data recorded by WIM stations must be pre-processed to yield a high quality database. The valid data can then be analysed to provide the information on the highway traffic condition and highway load distribution. Vehicle amount and composition of different vehicle types are two important indexes to represent the vehicle traffic condition, from which the operation trend of road vehicles with time can be inferred. The information on the vehicle load distribution includes the axle load distribution and gross vehicle weight distribution. From a structural design point of view for traffic loading determination, the derived numbers of axles and vehicles falling into their corresponding individual load ranges must be presented in ascending series in a probabilistic way to form an axle load spectrum or a gross vehicle weight (GVW) spectrum. These spectra can then be used as a basis for deriving the equivalent number of passages of a standard fatigue vehicle for estimating fatigue life of the bridge. The axle load spectrum actually displays the axle load distribution in terms of the exceedance probability of the axle load greater than a given value. The GVW spectrum can be developed in a similar way based on the recorded GVW distribution.

6.4 DYNAMIC INTERACTION OF LONG-SPAN SUSPENSION BRIDGES WITH ROAD VEHICLES

Dynamic road vehicle-bridge interaction problems have been studied by many investigators since the middle of the 19th century (Olsson, 1985; Yang and Lin, 1995; Cheung *et al.*, 1999). Because of the limitation of computation capacity in the early stage, only simplified models of road vehicle-bridge systems could be considered. For instance, a moving vehicle was modelled as a moving load without considering the effect of the inertia force (Timoshenko *et al.*, 1974), and later a moving-mass model was used instead of a moving load to include inertia force effects (Blejwas *et al.*, 1979). Nowadays, the volume of traffic and the speed of road vehicles have increased considerably, and the configurations of road vehicles have also changed dramatically. More sophisticated and rational models and computerised approaches are thus needed.

This section introduces a fully computerised approach for assembling equations of motion of any types of coupled road vehicle-bridge systems (Guo and Xu, 2001). Heavy road vehicles are idealised as a combination of a number of rigid bodies connected by a series of springs and dampers while the bridge is modelled using the conventional finite element method. The mass matrix, stiffness matrix, damping matrix, and force vector of the coupled road vehicle-bridge systems are automatically assembled using the fully computerised approach and taking account of the road surface roughness. The required input data about road vehicles to the computer program are only the dynamic properties and positions of the rigid bodies, springs, dampers, and constraint conditions. The derived equations of motion of the coupled road vehicle-bridge system are then solved by the direct integration method. It is noted that there are many other methods available for dynamic analysis of coupled vehicle-bridge systems (Wang and Huang, 1992; Au *et al.*, 2001; Yang and Wu, 2001).

6.4.1 Modelling of Road Vehicles

There are a variety of configurations of vehicles in reality, including the tractor with trailers having different axle spacings. In this study, a vehicle is modelled as a combination of several rigid bodies connected by a series of springs, damping devices, and pivots (see Figure 6.1). The coordinate system for the vehicle is the same as the coordinate system for the bridge. The X-axis, Y-axis, and Z-axis are set in the longitudinal, lateral, and vertical directions of the bridge, respectively, following the right hand rule.

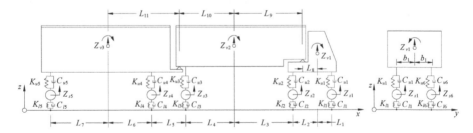

Figure 6.1 Schematic model of a tractor-trailer

The rigid bodies can be used to represent the vehicle bodies, axles, wheels, or others. The centre of gravity of each rigid body is taken as a node. In a most general case, a node has six DOFs: three translational degrees and three rotational degrees. The corresponding displacements and rotations are assumed to remain small throughout the analysis so that the sines of the angles of rotation may be taken as equal to angles themselves and the cosines of the angles of rotation may be taken as unity. The mass and/or the mass moments of inertia of each rigid body are calculated from the weight distribution and dimension of the body with respect to the local coordinate originated at its node.

The contact between the bridge deck and the moving tyre of the vehicle is assumed to be a point contact. With the assumption that the road surface profile is not too rough to make the vehicle jump or leave the surface, the tyres of the vehicle remain in contact with the bridge deck at all times. No matter whether the mass of the tyre is lumped at the contact point or not, the vertical displacement of the tyre is not an independent DOF, and actually it can be expressed in terms of the relevant DOFs of the bridge and the road surface profile.

The springs can be used to model the suspension system, the flexibility of a tyre, or others. Each spring is assumed to be massless. Besides the stiffness coefficient of each spring, the positions of the two ends of the spring connecting two rigid bodies or connecting one rigid body and one contact point are required as input data. The energy dissipation capacity of the suspension system, the tyre, or others can be modelled by damping devices. If the damping device is of the viscous type, the damping coefficient can be used as a sole parameter of the device. The positions of the two ends of the damping device connecting two rigid bodies or one rigid body and one contact point should also be identified as input data. The pivots may be used to connect the trailer to the tractor, for which the constraint equations will be correspondingly developed.

In summary, the independent DOFs of a vehicle in the coupled road vehicle-bridge system are the sum of the DOFs of all the rigid bodies, excluding the masses at the contact points in the vertical direction, minus the number of the constraint equations. The required input data about vehicles to the computer program at a given time are the dynamic properties and positions of all the rigid bodies and springs and damping devices, positions of all the contact points, and the constraint conditions for all the pivots. Based on these input data, the mass, damping, and stiffness matrices of the vehicle itself and the coupled mass, damping, and stiffness matrices and the contact force vectors of the vehicle with the bridge can be automatically assembled using the fully computerised approach.

6.4.2 Modelling of Bridge

A long-span suspension bridge model can be established using different types of finite elements such as beam elements, cable elements, plate elements, and solid elements (Xu *et al.*, 1997). The stiffness matrix, mass matrix and force vector of the bridge can be obtained using the conventional finite element method. The structural damping of the bridge is assumed to be Rayleigh damping. Consequently the damping matrix can be expressed as the function of the mass and stiffness matrices

$$[C_b] = \alpha_b[M_b] + \beta_b[K_b] \tag{6.1}$$

where $[M_b]$, $[C_b]$ and $[K_b]$ are the mass, damping and stiffness matrices of the bridge, respectively; and α_b and β_b are the Rayleigh damping factors which can be evaluated if the two structural damping ratios and the associated frequencies are known. The equations of motion of the bridge alone without moving vehicles can be expressed as

$$[M_b]\{\ddot{v}_b\} + [C_b]\{\dot{v}_b\} + [K_b]\{v_b\} = \{P_{be}\} \tag{6.2}$$

where $\{v_b\}$, $\{\dot{v}_b\}$ and $\{\ddot{v}_b\}$ are the nodal dynamic displacement, velocity, and acceleration vectors of the bridge, respectively; and $\{P_{be}\}$ is the external force vector of the bridge.

6.4.3 Modelling of Road Surface Roughness

Many investigations have shown that the roughness of the bridge surface is an important factor that affects the dynamic responses of both the bridge and vehicle (Wang and Huang, 1992). The road surface roughness may be described as a realisation of a random process that can be described by a power spectral density (PSD) function. The following PSD functions were proposed by Dodds and Robson (1973) for highway road surface roughness:

$$S(\bar{\phi}) = A_r (\frac{\bar{\phi}}{\bar{\phi}_0})^{-w_1}, \text{ when } \bar{\phi} \le \bar{\phi}_0 \tag{6.3}$$

$$S(\bar{\phi}) = A_r (\frac{\bar{\phi}}{\bar{\phi}_0})^{-w_2}, \text{ when } \bar{\phi} > \bar{\phi}_0 \tag{6.4}$$

where $S(\bar{\phi})$ is the PSD function (m^3/cycle) for the road surface elevation; $\bar{\phi}$ is the spatial frequency (cycle/m); $\bar{\phi}_0$ is the discontinuity frequency of 0.5π (cycle/m); and A_r is the roughness coefficient (m^3/cycle) and its value depends on the road condition. Power exponents w_1 and w_2 vary from 1.36 to 2.28 (Dodds, 1972). To simplify the description of the road surface roughness, Huang and Wang (1992) suggested the following PSD function:

$$S(\bar{\phi}) = A_r (\frac{\bar{\phi}}{\bar{\phi}_0})^{-2} \tag{6.5}$$

The road surface roughness is assumed to be a zero-mean stationary Gaussian random process. Therefore, it can be generated through an inverse Fourier transform:

$$r(x) = \sum_{k=1}^{N} \sqrt{2S(\bar{\phi}_k)\Delta\bar{\phi}} \cos(2\pi\bar{\phi}_k x + \theta_k) \tag{6.6}$$

where θ_k is the random phase angle uniformly distributed from 0 to 2π. Figure 6.2 shows a typical vertical road surface profile averaged from five simulations for the good road.

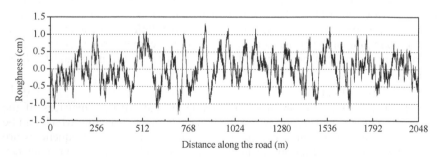

Figure 6.2 Vertical good road surface profile

6.4.4 Principle of Computerized Approach

In most of the previous studies, the equations of motion of a vehicle are derived by hand using either D'Alembert's principle or the principle of virtual work or Hamilton's principle. The derived equations of motion of the vehicle are then combined with those of the bridge. However, for a long-span suspension bridge with a group of moving vehicles of various configurations, it may not be easy to use this way to assemble the equations of motion of the coupled road vehicle-bridge systems. In this section, a computerised approach is suggested to expand the equations of motion of the bridge, obtained by the finite element method, to those of the coupled road vehicle-bridge system. The computerised approach uses the principle of virtual work to fulfil such a task.

Assume that the coupled road vehicle-bridge system has N_{vb} independent DOFs denoted by the displacement vector $\{v\}$, in which the vehicle displacement vector $\{v_v\}$ has DOFs of N_v and the bridge displacement vector $\{v_b\}$ has DOFs of $N_b=N_{vb}-N_v$. The virtual displacement vectors can then be expressed as $\{\delta v\}$, $\{\delta v_v\}$ and $\{\delta v_b\}$ for the coupled system, vehicle, and bridge, respectively; and the external force vectors as $\{P\}$, $\{P_v\}$, and $\{P_b\}$ accordingly. In the computerised approach, the following steps are taken to expand the equations of motion of the bridge alone to the equations of motion of the coupled road vehicle-bridge system.

1. Express the relative displacement of each spring in the vehicle as the function of displacement vector $\{v\}$. Express the relative velocity of each viscous damper as the function of velocity vector $\{\dot{v}\}$. For the spring connecting two rigid bodies of the vehicle, this can be done by relating the given positions of the two ends of the spring to the displacements of the two rigid bodies, which may need the constraint conditions to be considered. For the spring connecting one rigid body and one contact point, the road surface profile and the relevant displacements of the bridge are involved. A similar way is used to handle the viscous dampers.

2. Compute the inertia forces, elastic forces, and damping forces acting on the vehicle based on the given spring stiffness and damper damping coefficients, the computed mass and mass moment of inertia, and the computed relative displacements and velocities. This will lead to many terms and each term falls into one of the four categories: $C_m\ddot{v}_j$, $C_c\dot{v}_j$, $C_k v_j$ and P_r, where P_r is the force due to the road surface roughness and $1 \le i \le N_{vb}$.

3. Compute the virtual work done by all the forces on each virtual displacement. This will again produce many terms and each term falls into one of the four categories: $\delta v_i C_m \ddot{v}_j$, $\delta v_i C_c \dot{v}_j$, $\delta v_i C_k v_j$ and $\delta v_i P_r$, where $1 \le i \le N_{vb}$.

4. Identify the mass or mass moment coefficients, stiffness coefficients, damping coefficients, and force components and place them into proper positions of the system matrices and force vector. For instance, $\delta v_i C_m \ddot{v}_j$ indicates that C_m should be added to the mass coefficient m_{ij} at the ith row and j^{th} column of the system mass matrix. In a similar way, C_c in $\delta v_i C_c \dot{v}_j$ should be added to the damping coefficient C_{ij} at the ith row and j^{th} column of the system damping matrix, and C_k in $\delta v_i C_k v_i$ added to the stiffness coefficient k_{ij} at the ith row and j^{th} column of the system stiffness matrix. $\delta v_i P_r$ indicates that force P_r should be added to the ith force coefficient P_i of the system force vector.

It should be pointed out that since the virtual work is scalar quantity and the total virtual work of the vehicle done by all forces can be determined by the algebraic summation of the virtual work done by each force, the coefficient in each system matrix can be assembled algebraically. This approach is particularly suitable for the expansion of equations of motion of the bridge to those of the coupled road vehicle-bridge system.

6.4.5 Equations of Motion of Coupled Vehicle-bridge System

The use of the fully computerised approach can easily lead to the following equations of motion of the coupled vehicle-bridge system, which is established from the static equilibrium position of the system.

$$
\begin{bmatrix} M_b + M_{bbv} & 0 \\ 0 & M_v \end{bmatrix} \begin{Bmatrix} \ddot{v}_b \\ \ddot{v}_v \end{Bmatrix} + \begin{bmatrix} C_b + C_{bbv} & C_{bv} \\ C_{vb} & C_{v1} + C_{v2} \end{bmatrix} \begin{Bmatrix} \dot{v}_b \\ \dot{v}_v \end{Bmatrix} + \begin{bmatrix} K_b + K_{bbv} & K_{bv} \\ K_{vb} & K_{v1} + K_{v2} \end{bmatrix} \begin{Bmatrix} v_b \\ v_v \end{Bmatrix}
$$
$$
= \begin{Bmatrix} P_{bvg} + P_{bvr1} + P_{bvr2} + P_{bvr3} \\ P_{vvr2} + P_{vvr3} \end{Bmatrix} \tag{6.7}
$$

In Equation (6.7), $[M_b]$, $[C_b]$, and $[K_b]$ are the mass, damping, and stiffness matrix of the bridge alone, respectively, obtained by the conventional finite element method. Matrix $[M_{bbv}]$ is related to the inertia forces of all the masses of the vehicle at the contact points due to the bridge accelerations while matrix $[M_v]$ corresponds to the inertia forces of all the rigid bodies of the vehicle, excluding the masses at the contact points. The inertia forces of all the masses of the vehicle at the contact points due to the road surface roughness constitute the force vector $\{P_{bvr1}\}$. For the dampers of which the relative velocities are the function of the DOFs of the vehicle only, the damper forces lead to matrix $[C_{v1}]$. For the dampers connected to the contact points, their relative velocities depend on not only the DOFs of the vehicle but also DOFs of the bridge and the road surface roughness. As a result, coupled damping matrices $[C_{bv}]$ and $[C_{vb}]$, additional damping matrix $[C_{bbv}]$ to bridge damping matrix $[C_b]$, additional damping matrix $[C_{v2}]$ to vehicle damping matrix $[C_{v1}]$, additional force vector on the bridge $\{P_{bv2}\}$ due to the road surface roughness, and additional force vector on the vehicle $\{P_{vvr2}\}$ due to the road surface roughness are generated. Similarly, for the springs of which the relative displacements are the function of DOFs of the vehicle only, the spring forces lead to matrix $[K_{v1}]$. From the springs connected to the contact points, stiffness matrices $[K_{bv}]$, $[K_{vb}]$, $[K_{bbv}]$, $[K_{v2}]$ and additional force vectors due to road surface roughness $\{P_{bvr3}\}$ and $\{P_{vvr3}\}$ are constituted. The last but not least is the external forces on the bridge due to the gravity forces of the vehicle denoted by force vector $\{P_{bvg}\}$.

6.4.6 An Example

Heavy tractor-trailer vehicles that frequently run on long-span suspension bridges are taken as example vehicles in this study. The vehicle model comprises 13 rigid bodies: one for tractor, two for 2 trailers, and ten for 5 axle sets including weights of tyres and brakes (see Figure 6.1). The suspension system at each axle set, connecting the axle to either the tractor or trailer, is represented by two identical units across the width of the vehicle. The two units are parallel combinations of a linear elastic spring of stiffness K_{ui} and a viscous damper of damping coefficient C_{ui}. The parallel combination of a linear elastic spring of stiffness K_{li} and a viscous damper of damping coefficient C_{li} is used to represent the dynamic characteristics of the tyre. There are also two identical units connecting each axle to the bridge deck. The horizontal distance between the two units is $2b_1$. If only the vertical vibration of the vehicle is considered in the analysis, the rigid bodies representing either the tractor or trailers are each assigned three DOFs: the vertical displacement (Z_{vi}), the rotation about the transverse axis (pitch θ_{vi}), and the rotation about the longitudinal axis (roll ϕ_{vi}). The rigid body representing the axle set is assigned only one DOF in the vertical direction (Z_{sli} or Z_{sri}). Since the tractor and trailer are connected by a frictionless pivot and the two trailers are also connected by a frictionless pivot, the two constraint equations should be introduced. The total number of DOFs of the vehicle is 17 rather than 19.

By using the conventional finite element approach, mass matrix $[M_b]$, damping matrix $[C_b]$, and stiffness matrix $[K_b]$ of the bridge alone can be easily assembled. Then, displacement vector of the bridge $\{v_b\}$ is expanded to include the displacement vector of the vehicle $\{v_v\}$:

$$\{v_v\}=\{Z_{v1}\ \theta_{v1}\ \phi_{v1}\ Z_{v2}\ \phi_{v2}\ Z_{v3}\ \phi_{v3}\ Z_{sl1}\ Z_{sl2}\ Z_{sl3}\ Z_{sl4}\ Z_{sl5}\ Z_{sr1}\ Z_{sr2}\ Z_{sr3}\ Z_{sr4}\ Z_{sr5}\} \qquad (6.8)$$

where Z_{v1}, Z_{v2} and Z_{v3} are the vertical displacements of the tractor and two trailers, respectively, at their own centroids measured from the position of static equilibrium; θ_{v1} is the rotation about the transverse axis of the tractor; ϕ_{v1}, ϕ_{v2}, and ϕ_{v3} denote the rotations about the longitudinal axis of the tractor and two trailers, respectively; Z_{sli} and Z_{sri} ($i = 1, 2, \cdots, 5$) are the left and right vertical displacements of the five axle sets, respectively, at their own centroids measured from the position of static equilibrium. Clearly, the above 17 displacements are independent.

To expand the mass, damping, stiffness matrices and force vector of the bridge to those of the coupled road bridge-vehicle system, the inertial forces, damping forces and elastic forces of the vehicle should be computed. Since there is no mass at the contact point, it is very easy to determine the inertial forces acting on the rigid bodies. To determine the elastic forces and damping forces, the relative displacement of each spring and the relative velocity of each damper should be computed based on the given geometric information and sign convention. For instance, the deformations of the upper springs can be expressed as

$$\Delta_{ul1} = Z_{v1} - L_1\theta_{v1} - b_1\phi_{v1} - Z_{sl1} \qquad (6.9)$$

$$\Delta_{ur1} = Z_{v1} - L_1\theta_{v1} + b_1\phi_{v1} - Z_{sr1} \qquad (6.10)$$

$$\Delta_{ul2} = Z_{v1} + L_2\theta_{v1} - b_1\phi_{v1} - Z_{sl2} \qquad (6.11)$$

$$\Delta_{ur2} = Z_{v1} + L_2\theta_{v1} + b_1\phi_{v1} - Z_{sr2} \qquad (6.12)$$

$$\Delta_{ul3} = Z_{v2} + L_4\theta_{v2} - b_1\phi_{v2} - Z_{sl3} \qquad (6.13)$$

$$\Delta_{ur3} = Z_{v2} + L_4\theta_{v2} + b_1\phi_{v2} - Z_{sr3} \qquad (6.14)$$

$$\Delta_{ul4} = Z_{v3} - L_6\theta_{v3} - b_1\phi_{v3} - Z_{sl4} \qquad (6.15)$$

$$\Delta_{ur4} = Z_{v3} - L_6\theta_{v3} + b_1\phi_{v3} - Z_{sr4} \qquad (6.16)$$

$$\Delta_{ul5} = Z_{v3} + L_7\theta_{v3} - b_1\phi_{v3} - Z_{sl5} \tag{6.17}$$

$$\Delta_{ur5} = Z_{v3} + L_7\theta_{v3} + b_1\phi_{v3} - Z_{sr5} \tag{6.18}$$

where Δ_{uli} and Δ_{uri} ($i = 1, 2, \cdots, 5$) are the deformations of the upper left and right springs at the five axle sets, respectively. The two constraint conditions can be expressed as

$$Z_{v1} + L_8\theta_{v1} = Z_{v2} - L_9\theta_{v2} \tag{6.19}$$

$$Z_{v2} + L_{10}\theta_{v2} = Z_{v3} - L_{11}\theta_{v3} \tag{6.20}$$

Thus, θ_{v2} and θ_{v3} can be expressed in terms of the independent DOFs of the vehicle as:

$$\theta_{v2} = \frac{Z_{v2} - Z_{v1} - L_8\theta_{v1}}{L_9} \tag{6.21}$$

$$\theta_{v3} = \frac{L_9 Z_{v3} - (L_9 + L_{10})Z_{v2} + L_{10}Z_{v1} + L_8 L_{10}\theta_{v1}}{L_9 L_{11}} \tag{6.22}$$

Accordingly, the deformations of all the upper springs (see Equations (6.9) ~ (6.18)) can be expressed in terms of the independent DOFs of the vehicle only. The deformations of the lower springs are

$$\Delta_{lli} = Z_{sli} - Z_{li} \quad (i = 1, 2, \cdots, 5) \tag{6.23}$$

$$\Delta_{lri} = Z_{sri} - Z_{ri} \quad (i = 1, 2, \cdots, 5) \tag{6.24}$$

where Δ_{lli} and Δ_{lri} ($i = 1, 2, \cdots, 5$) are the deformations of the lower left and right springs at the five axle sets, respectively; and Z_{li} and Z_{ri} denote the vertical displacements of the left contact point and right contact point, respectively. If the ith left and right tyres of the vehicle run on the bridge, the vertical displacement of each contact point can be expressed in terms of the relevant DOFs of the bridge and the road surface roughness as

$$Z_{li} = N_{li}(x)\{\delta\}_{bi}^e + r_{li}(x) \tag{6.25}$$

$$Z_{ri} = N_{ri}(x)\{\delta\}_{bi}^e + r_{ri}(x) \tag{6.26}$$

where r_{li} and r_{ri} are the road surface roughness under the ith left and right contact points, respectively; $\{\delta\}_{bi}^e$ is the displacement vector of the bridge element over which the ith contact point runs and it is also the subset of the displacement vector of the bridge $\{v_b\}$; and N_{li} is the transfer function from the node displacements of the bridge element to the displacement of the ith left contact point. The velocity and acceleration of the ith left contact point are computed by

$$\dot{Z}_{li} = N_{li}(x)\{\dot{\delta}\}_{bi}^e + u\frac{\partial N_{li}(x)}{\partial x}\{\delta\}_{bi}^e + \frac{\partial r_{li}(x)}{\partial x}u \tag{6.27}$$

$$\ddot{Z}_{li} = N_{li}(x)\{\ddot{\delta}\}_{bi}^e + 2u\frac{\partial N_{li}(x)}{\partial x}\{\dot{\delta}\}_{bi}^e + u^2\frac{\partial^2 N_{li}(x)}{\partial x^2}\{\delta\}_{bi}^e + a\frac{\partial N_{li}(x)}{\partial x}\{\delta\}_{bi}^e$$

$$+ \frac{\partial r_{li}(x)}{\partial x}a + \frac{\partial^2 r_{li}(x)}{\partial x^2}u^2 \tag{6.28}$$

where u and a are the travelling velocity and acceleration of the vehicle, respectively. Similarly, $\dot{Z}_{li}$ and $\ddot{Z}_{li}$ can be computed in terms of the relevant DOFs of the bridge and the road surface roughness.

If the ith left and right tyres of the vehicle have not yet arrived at the bridge or they have already left the bridge, these tyres are actually running on the ground. In these cases, the vertical displacement, velocity, and acceleration of the ith left contact point, for example, should be determined by the following equations rather than Equations (6.25)~(6.28).

$$Z_{li} = r_{li}(x) \tag{6.29}$$

$$\dot{Z}_{li} = \frac{\partial r_{li}(x)}{\partial x} u \tag{6.30}$$

$$\ddot{Z}_{li} = \frac{\partial r_{li}(x)}{\partial x} a + \frac{\partial^2 r_{li}(x)}{\partial x^2} u^2 \tag{6.31}$$

The vertical displacement, velocity, and acceleration of the ith right contact point on the ground should be also computed in a similar way without a direct interaction with the bridge. Now, assume that all the displacements and rotations remain small throughout the analysis. The virtual work done by all inertial forces, damping forces, and elastic forces acting on the vehicle at a given time can be expressed as

$$\delta W_{VI} = \sum_{i=1}^{3} \left(\delta Z_{vi} M_{vi} \ddot{Z}_{vi} + \delta \theta_{vi} J_{yvi} \ddot{\theta}_{vi} + \delta \phi_{vi} J_{xvi} \ddot{\phi}_{vi} \right) + \sum_{i=1}^{5} \left(\delta Z_{sli} M_{sli} \ddot{Z}_{sli} + \delta Z_{sri} M_{sri} \ddot{Z}_{sri} \right) \tag{6.32}$$

$$\delta W_{VD} = \sum_{i=1}^{5} \left(\delta \Delta_{uli} C_{ui} \dot{\Delta}_{uli} + \delta \Delta_{uri} C_{ui} \dot{\Delta}_{uri} + \delta \Delta_{lli} C_{li} \dot{\Delta}_{lli} + \delta \Delta_{lri} C_{li} \dot{\Delta}_{lri} \right) \tag{6.33}$$

$$\delta W_{VE} = \sum_{i=1}^{5} \left(\delta \Delta_{uli} K_{ui} \Delta_{uli} + \delta \Delta_{uri} K_{ui} \Delta_{uri} + \delta \Delta_{lli} K_{li} \Delta_{lli} + \delta \Delta_{lri} K_{li} \Delta_{lri} \right) \tag{6.34}$$

where M_{v1}, M_{v2} and M_{v3} are the lump mass of the tractor and two trailers, respectively; M_{sli} and M_{sri} ($i = 1, 2, \cdots, 5$) are the left and right lump mass of the five axle sets, respectively; J_{yv1}, J_{yv2} and J_{yv3} are the pitching moments of inertia of the tractor and two trailers, respectively; and J_{xv1}, J_{xv2}, and J_{xv3} are the rolling moments of inertia of the tractor and two trailers, respectively.

Since there are no masses at the contact points, matrix $[M_{bbv}]$ and force vector $\{P_{bvr1}\}$ in Equation (6.7) are zero and the expansion of Equation (6.32) leads to matrix $[M_v]$. The expansion of the first two terms in Equation (6.34) generates stiffness matrix $[K_{v1}]$ while the expansion of the last two terms in Equation (6.34) leads to stiffness matrices $[K_{v2}]$, $[K_{v2}]$, $[K_{vb}]$, $[K_{bbv}]$, additional force vector on the vehicle $\{P_{vvr3}\}$, and additional force vector on the bridge $\{P_{bvr3}\}$. Similarly, the expansion of Equation (6.33) produces damping matrices $[C_{v1}]$, $[C_{v2}]$, $[C_{bbv}]$, $[C_{bv}]$, $[C_{vb}]$ and additional force vectors $\{P_{vvr2}\}$ and $\{P_{bvr2}\}$. For the vehicle-bridge system concerned, the ten external forces on the bridge (or on the ground) due to the gravity forces of the vehicle should be considered. The virtual work done by all external forces at a given time can be computed by

$$\delta W_{bp} = \sum_{i=1}^{5} \left(\delta Z_{li} P_{li} + \delta Z_{ri} P_{ri} \right) \tag{6.35}$$

where P_{li} and P_{ri} ($i = 1, 2, \cdots, 5$) are ten external forces on the bridge (or on the ground) at the contact points on the left side and right side respectively due to the gravity of the vehicle. The expansion of Equation (6.35) will lead to the force vector $\{P_{bvg}\}$ on the bridge (or on the ground). Finally, the computer program will provide the equations of motion of the coupled bridge-vehicle system at a given time as

$$\begin{bmatrix} M_b & 0 \\ 0 & M_v \end{bmatrix}\begin{Bmatrix} \ddot{v}_b \\ \ddot{v}_v \end{Bmatrix} + \begin{bmatrix} C_b + C_{bbv} & C_{bv} \\ C_{vb} & C_{v1} + C_{v2} \end{bmatrix}\begin{Bmatrix} \dot{v}_b \\ \dot{v}_v \end{Bmatrix} + \begin{bmatrix} K_b + K_{bbv} & K_{bv} \\ K_{vb} & K_{v1} + K_{v2} \end{bmatrix}\begin{Bmatrix} v_b \\ v_v \end{Bmatrix}$$

$$= \begin{Bmatrix} P_{bvg} + P_{bvr2} + P_{bvr3} \\ P_{vvr2} + P_{vvr3} \end{Bmatrix} \tag{6.36}$$

By using the fully computerised approach, it is very convenient to manage the case in which some or all tyres of a vehicle are not running on the bridge. For instance, if the ith left and right tyres of the investigated vehicle are running on the ground rather than on the bridge deck at a given time, their vertical displacements, velocities, and accelerations will be automatically computed according to Equations (6.29) to (6.31) rather than Equations (6.25) to (6.28). This adjustment results in changes in calculating the virtual work done by damping forces, elastic forces, and external forces in terms of Z_{li}, Z_{ri}, Δ_{lli}, Δ_{lri}, Δ_{uli}, and Δ_{uri} in Equations (6.33) to (6.35). The virtual work done by inertia forces has no change in this case study because there are no masses at the contact points. Consequently, the expansion of the virtual work done by the damping forces, elastic forces, and external forces leads to changes in sub-damping matrices $[C_{bbv}]$, $[C_{bv}]$, $[C_{vb}]$, sub-stiffness matrices $[K_{bbv}]$, $[K_{bv}]$, $[K_{vb}]$, and loading vector $\{P_{bvg}\}$ in Equation (6.36). All these will be done by the computer automatically.

The equations of motion achieved are a set of coupled second order differential equations with time varying coefficients. The Wilson-θ method or other numerical methods can be used for determining dynamic responses of both vehicles and bridge in the time domain.

6.5 MONITORING OF HIGHWAY LOADING EFFECTS ON THE TSING MA BRIDGE

The road traffic condition and highway loading on the Tsing Ma Bridge are monitored by seven dynamic WIM stations which were installed in the respective seven lanes of the two carriageways (the Airport bound way and the Kowloon bound way) at the approach to Lantau Toll Plaza, near the Lantau Administration Building as shown in Figure 6.3 (Wong *et al.*, 2001; HKPU, 2007).

6.5.1 Data Processing

The M-class vehicle classification system (HKTD, 2008) is used to classify types of vehicles running on the bridge into eight categories (see Table 6.1) according to their main features such as the number of axles, axle distance and gross weight. To simplify the classification, sometimes eight categories are grouped into only four categories based on the vehicle gross weight, as shown in Table 6.2.

Figure 6.3 Dynamic weigh-in-motion stations at Lantau Toll Plaza

Table 6.1 M-class vehicle classification system.

Class (1-8)	Vehicle category	Short string	No. of axles	Magnetic vehicle length	Axle distance (1 to 2)	Gross weight
1	Motor cycles	MC	2	< 2 m plus magnetic contour analysis		< 0.6 T
2	Cars, vans or taxis	C/V	2, 3, 4	< 6.3 m		< 3 T
3	Public service vehicles	PSV		Magnetic contour analysis	< 5.5 m	< 4 T
4	Light goods vehicles	LGV	2, 3, 4	≥ 6.3 m, < 10 m	< 3.5 m	< 5.5 T
5	Medium goods vehicles	MGV	2,3,4,5,6,7,8		≥ 3.5 m, < 7.8 m	< 24 T
6	Rigid heavy goods vehicles	RHGV	2,3,4,5,6,7,8		≥ 3.5 m, < 7.8 m	< 38 T
7	Articulated heavy goods vehicles	AHGV	2,3,4,5,6,7,8		≥ 3.5 m, < 7.8 m	< 44 T
8	Buses and coaches	BUS	2,3,4	Magnetic contour analysis	≥ 7.8 m	< 24 T

For each vehicle passing through a WIM station, a line of raw data is recorded. A sample of WIM raw data is shown in Table 6.3. The information from a WIM station includes the vehicle arrival date and time, lane number, vehicle speed, vehicle class according to the M-class vehicle classification system, number of axles, axle weight, and axle spacing. From the WIM recorded raw data, different statistical data can be extracted.

The WIM stations have recorded the traffic information since August 1998. Owing to the need to conduct major updating for the entire system, both the Airport bound and Kowloon bound stations were in suspension from July 2004 to

March 2005. It has been shown that the daily average vehicle counts increase significantly since April 2005. The derived averaged daily vehicle count (two-way) on the Tsing Ma Bridge during the period from April 2005 to December 2005 was 51,901. This number is equivalent to an average value of passenger car units (pcu) of 67,472.

Table 6.2 Simplified M-class vehicle classification system.

Class (I-IV)	Class (1-8)	Vehicle category	Short string
I	2	Cars, vans or taxis	C/V
II	3, 4	Public service vehicles, light goods vehicles	LVS
III	5, 8	Medium goods vehicles, buses and coaches	MVS
IV	6, 7	Rigid heavy goods vehicles, articulated heavy goods vehicles	HGV

Table 6.3 A sample of WIM raw data.

Bound	Date & Time	Lane	Speed	Class	No. of axle	1st axle weight	2nd axle weight	1st axle distance	3rd axle weight	2nd axle distance
1	11/2/2005 1:16:58	3	91	2	2	450	450	249		
2	11/2/2005 1:17:00	3	86	2	2	1030	750	289		
1	11/2/2005 1:17:01	2	78	2	2	780	760	281		
2	11/2/2005 1:17:12	2	75	2	2	450	450	249		
1	11/2/2005 1:17:11	2	67	4	2	3710	4600	498		
2	11/2/2005 1:17:12	1	71	7	3	4460	5540	558	4450	150
1	11/2/2005 1:17:13	2	65	3	2	1730	1190	398		

Bound: 1 Kowloon bound, 2 Airport bound   Lane: 1 slow lane, 2 middle lane, 3 fast lane
Speed: in km/hour   Class: 8 classes of vehicle category
Axle weight: in Kg   Axle distance: in cm

The detailed traffic information on each lane was not available until April 2005. Therefore, the WIM data for the whole year of 2006 are collected and analysed in this section, which best reflects the current traffic condition and provides detailed traffic lane information. The monthly variation of daily average vehicle counts on the Tsing Ma Bridge in 2006 is depicted in Figure 6.4.

The first step of data processing is the merging of traffic lanes. There is a total of seven lanes in the two carriageways (three lanes in the Airport bound way and four lanes in the Kowloon bound way) at the Lantau Toll Plaza where WIM data are collected, but there are only six lanes on the Tsing Ma Bridge: three lanes in the Airport bound way and three lanes in the Kowloon bound way. Therefore, for the subsequent analysis, the WIM data recorded on the two middle lanes at the

Lantau Toll Plaza are merged and assigned to the middle lane of Tsing Ma Bridge for the Kowloon bound way.

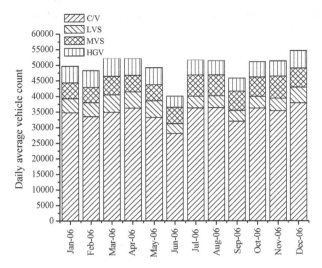

Figure 6.4 Monthly variation of daily average vehicle counts on Tsing Ma Bridge in 2006

The WIM data recorded during the system interregnum must be eliminated. One case is associated with a high wind management system for the Tsing Ma Bridge. The operation guidelines adopted for the bridge during the high wind period specify that both the upper and lower decks must be closed for all road vehicles except trains when the hourly mean wind speed recorded on site is in excess of 165 km/h. Therefore, there will be some cases in which no vehicle is recorded by the WIM stations due to the system interregnum. Correspondingly, the criterion to eliminate this kind of abnormal data is set in terms of the number of vehicles recorded in the relevant hour for the bridge.

$$\sum_{i=1}^{7}\sum_{j=1}^{8}V_{ij} = 0 \tag{6.37}$$

where V_{ij} is the number of vehicles of the jth vehicle category (j = 1, 2, …, 8) running on the ith lane (i = 1, 2, …, 7) in an hourly record.

The second kind of abnormal WIM data is mainly due to the malfunction of the WIM stations, which may yield unreasonable data such as axle loads and/or gross vehicle weights exceeding the specified maximum values considerably. This kind of abnormal data should be deleted from the database to ensure the accuracy of data analysis. The Hong Kong road traffic regulations (Chapter 374A Schedule 2) specify the maximum weight of vehicles according to vehicle category as shown in Table 6.4. In reality, overloaded vehicles running on the Tsing Ma Bridge do happen from time to time. The maximum GVW values specified in the Hong Kong road traffic regulations are increased by a factor of 1.25, and the resulting upper limits of the maximum GVW are listed in Table 6.4. The GVW data recorded by the WIM stations, which exceed the corresponding upper limits, are then eliminated from the database. Furthermore, the maximum axle load is stipulated as

10 tonnes in the Hong Kong road traffic regulations for all types of vehicle. In reality, the GVW of vehicles of class 1 to 4 is less than 6 tonnes and each vehicle has at least 2 axles. Therefore, the maximum axle load of vehicles of classes 1 to 4 must be smaller than 6 tons. In this study, a reasonable value of the maximum axle load is therefore assigned to the vehicles of classes 1 to 8, as listed in Table 6.5. The upper limit of the maximum axle load is then increased by a factor of 1.25 to consider overloaded vehicles. The recorded axle loads exceeding the upper limit of the maximum axle load should be eliminated from the database.

Table 6.4 Upper limits of maximum GVW of vehicles.

Class (1-8)	Vehicle category	Maximum GVW in Hong Kong road traffic regulations (ton)	Upper limit of maximum GVW (ton)
1	Motor cycles	0.6	0.75
2	Cars, vans or taxis	3	4
3	Public service vehicles	4	5
4	Light goods vehicles	5.5	7
5	Medium goods vehicles	24	30
6	Rigid heavy goods vehicles	38	48
7	Articulated heavy goods vehicles	44	55
8	Buses and coaches	24	30

Table 6.5 Upper limits of maximum axle load of vehicles.

Class (1-8)	Vehicle category	Maximum axle load (ton)	Upper limit of maximum axle load (ton)
1	Motor cycles	0.4	0.5
2	Cars, vans or taxis	2	2.5
3	Public service vehicles	3	3.75
4	Light goods vehicles	4	5
5	Medium goods vehicles	10	12.5
6	Rigid heavy goods vehicles	10	12.5
7	Articulated heavy goods vehicles	10	12.5
8	Buses and coaches	10	12.5

6.5.2 Monitoring of Highway Traffic Condition

Vehicle number and composition of different vehicle types are two important indexes to represent the vehicle traffic condition. The study on vehicle number and vehicle composition of the Tsing Ma Bridge is based on the valid statistical WIM data in 2006. The results in each lane are shown in Figure 6.5 for 8-class vehicle classification and in Figure 6.6 for 4-class vehicle classification. From Figure 6.5 and Figure 6.6, some observations can be made as follows:

1. In 2006, a total of 8.5 million vehicles run through the Tsing Ma Bridge for the Airport bound way and 8.8 million vehicles for the Kowloon bound way.
2. The cars, vans and taxis take up the biggest percentage among all types of vehicles: about 69.8% for the Airport bound way and 69.0% for the Kowloon bound way.

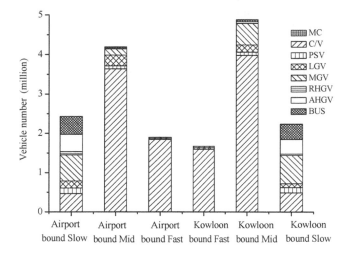

Figure 6.5 Vehicle count by 8-class vehicle classification (2006)

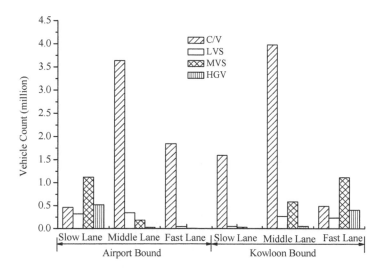

Figure 6.6 Vehicle count by 4-class vehicle classification (2006)

3. The percentage of heavy goods vehicles, including rigid heavy goods vehicles and articulated heavy goods vehicles, is about 5% to 6% of the total vehicles.
4. More vehicles run on the middle lane than other lanes for both ways, which account for 49.1% and 55.5% respectively of the total vehicles.
5. Most heavy goods vehicles use the slow lane for both ways, accounting for 95.1% and 89.1% respectively of the total heavy goods vehicles.
6. Most medium goods vehicles, including medium goods vehicles and buses, use the slow lane for both ways, accounting for 85.7% and 64.4% respectively of the total medium goods vehicles.

6.5.3 Monitoring of Highway Loading Distribution

Axle load is an important parameter for bridge pavement and bridge structures. The percentage of axle load in different loading ranges with 1 tonne interval can be determined according to the aforementioned vehicle classification system. Based on the statistical WIM data in 2006, the distributions of axle load are displayed in Figure 6.7 in terms of vehicle category, and in Figure 6.8 and Figure 6.9 in terms of bridge lane. Major observations can be summarised as follows:

1. The axle number decreases with increasing axle load. Only 12.5% of the total vehicle axles have an axle load more than 5 tonnes;
2. Most vehicle axles have an axle load less than 1 tonne, which account for 54.1% and 48.6% of the total axles respectively for the Airport bound way and the Kowloon bound way;
3. The second most frequent axle number corresponds to the axle load between 1 to 2 tons;

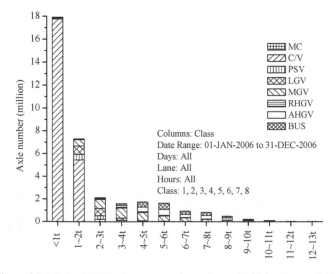

Figure 6.7 Axle load distribution versus axle number and vehicle category (2006)

4. More axles of vehicles run on the middle lane than the slow lane and fast lane, which account for about 47.2% and 55.1% of the total axles respectively for the Airport bound way and the Kowloon bound way; and

5. Most of the axles with axle load more than 5 tonnes run on the slow lane, accounting for about 95.5% for the Airport bound way and 84.2% for the Kowloon bound way.

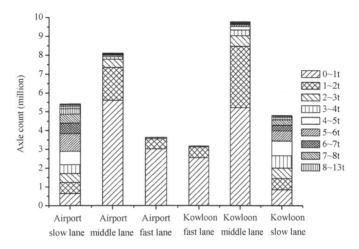

Figure 6.8 Axle load distribution versus axle number and bridge lane

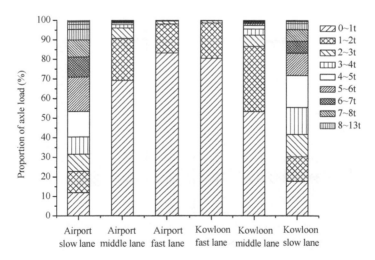

Figure 6.9 Proportion of axle loads for each lane (2006)

GVW is another important parameter for bridge pavement and bridge structures. The percentage of vehicles in different GVW ranges with 4 tonne interval can be determined according to the aforementioned vehicle classification system. Based on the WIM data of year 2006, the distributions of GVW are displayed in Figure 6.10 in terms of vehicle category, and in Figure 6.11 and Figure 6.12 in terms of bridge lanes. Major observations can be summarised as follows:

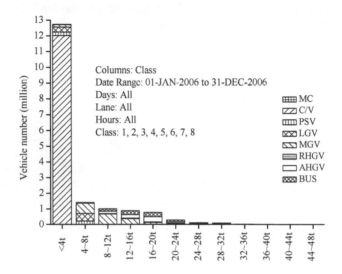

Figure 6.10 GVW distribution versus vehicle number and category (2006)

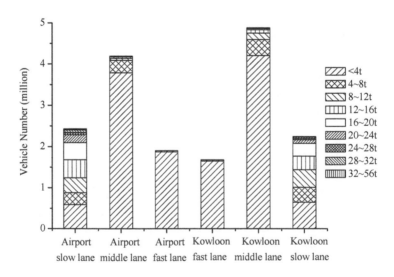

Figure 6.11 GVW distribution versus vehicle number and bridge lane (2006)

1. The vehicle number decreases with increasing GVW. Only 7.5% of the total vehicles have a GVW more than 16 tonnes.

2. Most of the vehicles have GVW less than 4 tonnes, which account for 73.2% and 73.9% of the total vehicles for the Airport bound way and the Kowloon bound way, respectively.
3. More vehicles run on the middle lane than the slow and fast lanes, accounting for about 49.1% of the total vehicles for the Airport bound way and 55.5% for the Kowloon bound way.
4. Most of the vehicles of more than 16 tonnes GVW use the slow lane, accounting for 97.6% for the Airport bound way and 90.7% for the Kowloon bound way.

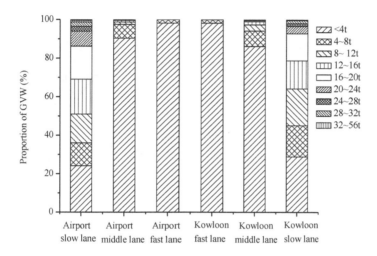

Figure 6.12 Proportion of GVW for each lane (2006)

6.5.4 Highway Loading Spectrum

From a structural design point of view for traffic loading determination, the derived numbers of axles and vehicles falling into their corresponding individual load ranges are presented in ascending series in a probabilistic way to form an axle load spectrum or a GVW spectrum. These spectra can then be used as a basis for deriving the equivalent number of passage of a standard fatigue vehicle for estimating fatigue life of the bridge. The axle load spectrum and GVW spectrum obtained based on the statistical WIM data collected in 2006 are presented and discussed in this section.

The amount and percentage of axles in each axle load range have been known from the axle load distribution obtained from the WIM data. The axle load spectrum which actually displays the axle load distribution in terms of exceeding probability of axle load greater than a given value can then be calculated. Two other axle load spectra, the load spectrum provided in BS5400 Part 10 (1980) and the design load spectrum for the Tsing Ma Bridge, are also provided for comparison in this section. The use of BS5400 Part 10 other than the latest standards is because it was taken as a reference at the time of the bridge design. A typical example of the axle load spectrum from the recorded WIM data in 2006 for the slow lane of the Airport bound way is presented in Figure 6.13 together with the two reference load spectra.

In deriving the axle load spectrum plotted in Figure 6.13, some parameters should be preset. The lower bound threshold of axle load is preset as 1 ton, and accordingly axle loads less than 1 tonne are not included in the calculation. This is because the axle loads less than 1 tonne can be neglected in the estimation of fatigue damage to the bridge. The bin size of axle load is preset to 1 tonne to be consistent with that in the statistical WIM data.

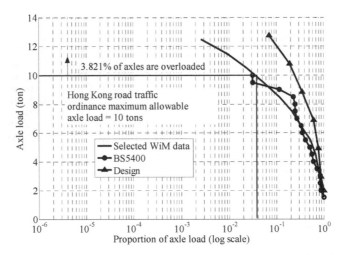

Figure 6.13 Axle load spectrum for Tsing Ma Bridge Airport bound slow lane (2006)

To have a more effective comparison between the measured axle load spectrum and the reference axle load spectrum based on BS5400, an effective axle load is defined based on axle loads in all the axle load ranges and their corresponding proportions. The formula of calculating the effective axle load is given as:

$$L = \left(\frac{1}{N} \sum_i n_i L_i^m \right)^{\frac{1}{m}}$$
(6.38)

where n_i is the repetition of the ith axle load L_i; N is the total number of repetitions $N = \Sigma n_i$; and $m = 3.0$ according to BS5400 Part 10 (1980). The effective axle loads calculated based on the above definition are shown in Figure 6.13. In Figure 6.13, some typical points in the axle load spectrum are also specified. For instance, according to the Hong Kong road traffic regulations and considering the overloaded axles, the upper limit of the allowable maximum axle load is set as 12.5 tonnes. The percentage of overloaded axles based on the measurement data in 2006 is therefore given in Figure 6.13. In addition, some important exceeding possibilities are also provided for the sake of clarification.

Some observations can be made based on the results depicted in Figure 6.13. The axle load spectrum from the measured WIM data is quite close to that based on BS5400, and the former is safer than the latter as the former effective axle load is smaller than the latter. Nevertheless, both are on the safe side as compared with the design spectrum. There is a significant proportion of overloaded axles (axle loads larger than 10 tons) on the slow lane with the percentage of 3.8% of all the vehicle axles concerned.

GVW spectrum can be developed in a similar way based on the recorded GVW data. A typical example of the GVW spectrum from the recorded WIM data in 2006 for the slow lane of the Airport bound way is presented in Figure 6.14 together with the reference GVW spectrum based on BS5400. In the figure, the lower bound threshold of GVW is set to 4 tonnes and the bin size is 4 tonnes to be consistent with that in the statistical WIM data. The allowable maximum GVW is 44 tonnes according to the Hong Kong road traffic regulations. The effective GVW is also defined by a similar equation as Equation (6.38), in which L_i is denoted as the amplitude of GVW and L is the effective GVW. Some observations can be made from Figure 6.14. The measured GVW spectrum is on the safe side compared with the reference GVW spectrum based on BS5400 because the former effective GVW is smaller than the latter one. There is a small proportion (0.157%) of overloaded vehicles on the Airport bound slow lane.

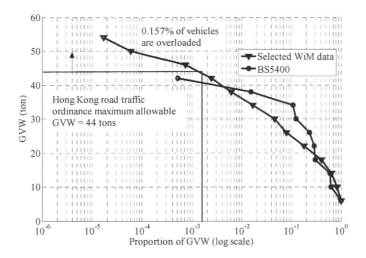

Figure 6.14 GVW spectrum for Tsing Ma Bridge Airport bound slow lane (2006)

6.5.5 Strain Analysis and Fatigue Evaluation

6.5.5.1 Background

The fatigue evaluation in this study is based on the strain data collected from a total of 110 strain gauges installed on the three typical bridge deck sections of the Tsing Ma Bridge as shown in Figure 6.15. The three typical bridge deck sections include the rail track section and deck section at around CH 24662.50, the side span deck section at CH 23488.0, and the deck at the Man Wan tower at CH

23623.00. Most of these strain gauges are attached to fatigue-susceptible members identified during the design of the health monitoring system.

Strain time histories recorded by strain gauges in the Tsing Ma Bridge are attributed to not only traffic loading, but also temperature loading and wind loading. Nevertheless, attention is paid to dynamic stresses for fatigue evaluation. As the temperature effect on dynamic stresses is insignificant, the temperature effect can be disregarded in this study. To keep wind loading effect as low as possible, the strain data for fatigue analysis should be carefully selected. The strain data collected in November 2005 are considered a desirable dataset for this study because both the railway and highway traffic in this month were steady and no strong winds were reported. The average hourly mean wind speed in this month was 2.8 m/s only. Furthermore, the data from this month represent the current traffic condition of the Tsing Ma Bridge as the condition has been changed after the opening of the Hong Kong Disneyland in September 2005.

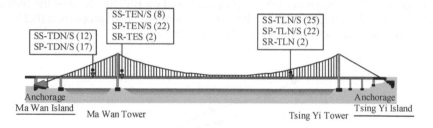

Figure 6.15 Distribution of strain gauges in Tsing Ma Bridge

The fatigue damage and fatigue life of the instrumented components of the Tsing Ma Bridge can be estimated using the method suggested in BS5400 Part 10 (1980). The effect of highway loading and railway loading on fatigue damage will be evaluated separately and then the combined effects.

Before the stress analysis and fatigue evaluation, the original data are pre-processed to obtain a reliable database. The mean value of the original hourly strain is first removed to eliminate the temperature effect and facilitate fatigue analysis. Only strain ranges are involved in the fatigue damage evaluation. The second step is to remove the abnormal strain data due to system interregnum or strain gauge malfunction. When system interregnum turns up or strain gauge loses function, the strain data will be automatically assigned a typical value exceeding the normal strain range, which is between -1000 $\mu\varepsilon$ and 1000 $\mu\varepsilon$. Therefore, the absolute strain value $|\varepsilon_i| > 1000$ $\mu\varepsilon$ is considered to be abnormal. Furthermore, one hourly strain record is not suitable for use if more than 5% of strain data are abnormal in this record. The strain gauge is also considered to be abnormal if more than 20% of the total hours in the month are abnormal. These abnormal data, hourly record, and strain sensors should not be used as they may lead to the computed fatigue damage being much larger than the actual one.

6.5.5.2 Procedure for Fatigue Evaluation

In order to distinguish the effects of highway loading from railway loading on fatigue damage and fatigue life, the hourly strain time history from each strain gauge must be properly decomposed. The time period during which a train is running on the bridge needs to be ascertained first. The original hourly time history can then be decomposed into two parts: with and without trains running on the bridge. The fatigue damage to the relevant bridge components due to the railway loading and highway loading can be finally evaluated separately using Miner's law as suggested in BS 5400.

The strain gauges which are very sensitive to the railway loading are selected to determine the time period of a train on the bridge and the train speed and direction. These strain gauges are installed at the inner waybeam of each pair of waybeams under the two rail tracks at CH 24662.5. The tag numbers of these strain gauges are SS-TLS-11, SS-TLS-12, SS-TLS-13, SS-TLS-14, SP-TLS-14, SP-TLS-15, SP-TLS-16, and SP-TLS-17. Strain gauges SS-TLS-12 and SS-TLS-14 are chosen here.

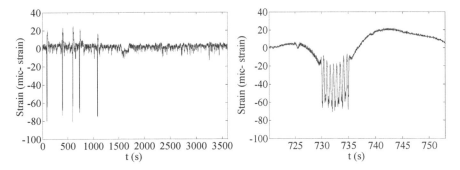

Figure 6.16 Data collected by strain gauge SS-TLS-14

Figure 6.16 (a) shows an hourly strain data recorded from SS-TLS-14 between 01:00 and 02:00 on November 2, 2005. Several sharp peaks are observed when trains run over the location of the strain gauge. Without a train, the minimum strain is about $-12\ \mu\varepsilon$ but with a train the minimum strain is smaller than $-50\ \mu\varepsilon$. A close-up of a segment with a train is shown in Figure 6.16 (b). It can be seen that the train's effect on the strain includes two parts: a fluctuation of the strain as the train is running through the location of the strain gauge and a gradual change as the train runs in and out of the bridge. Nine local peaks, which are induced by eight carriages of the train, are observed. By studying the nine local peaks, the time period of the train passing through the sensor location can be determined and the train speed can be estimated.

The train heading direction can be judged through the comparison of the strain data collected by two strain gauges (SS-TLS-12 and SS-TLS-14), which are located at one bridge deck section but different tracks. Figure 6.17 shows one same segment of the strain time histories from the two strain gauges. Since the peak strain magnitudes at SS-TLS-14 are larger than those at SS-TLS-12, the train is judged to run on the railway track installed with SS-TLS-14. After the train speed and heading direction are known, the moment when the train runs in and out of the bridge can be estimated from the first and last local strain peaks.

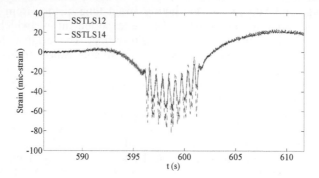

Figure 6.17 Data comparison for determining train direction

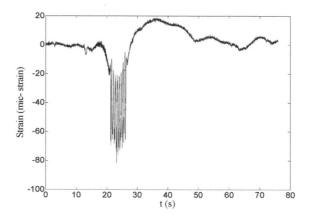

Figure 6.18 Extracted strain time history with one train running on the bridge

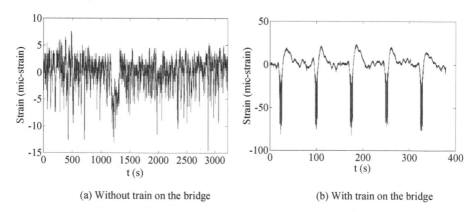

(a) Without train on the bridge   (b) With train on the bridge

Figure 6.19 Strain collected by strain gauge SS-TLS-14 (01:00~02:00 on Nov. 2, 2005)

Figure 6.18 displays the extracted strain time history with one train running on the bridge. By repeating the same procedure all the trains passing through the bridge within the hour concerned, the hourly strain time history can be divided into two parts: one without the train effect, and the other with both train and road vehicle effects. Figure 6.19 displays the two parts of one hourly strain collected by SS-TLS-14.

With the time information of the running trains, the hourly strain time history collected by other gauges can be decomposed into two parts as well. Figure 6.20 depicts the corresponding two parts of the hourly strain data collected by SS-TLS-02 at the main span cross frame. It can be observed that strain ranges in the presence of trains are much larger than those without a train on the bridge.

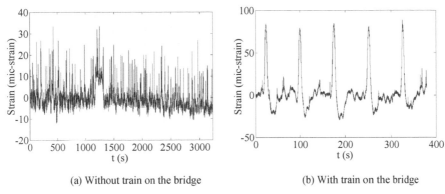

(a) Without train on the bridge   (b) With train on the bridge

Figure 6.20 Strain collected by strain gauge SS-TLS-02 (01:00~02:00 on Nov. 2, 2005)

From the two parts of hourly strain time history, the fatigue damage due to either highway loading or railway loading can be estimated using the procedure shown in Figure 6.21, in which the two parts of strain data are denoted as Time history I and Time history II, respectively, and their time lengths are denoted as T_1 and T_2 correspondingly. In particular, the fatigue damage induced by the highway loading during T_2 is calculated from Time history II. Based on the assumption that fatigue damage induced by the highway loading is uncorrelated to that induced by the railway loading, the fatigue damage induced by highway loading within one hour (= T_1+T_2) can be obtained by multiplying a factor $(T_1+T_2)/T_2$ by the fatigue damage calculated from Time history II. On the other hand, the fatigue damage induced by both railway loading and highway loading during T_1 can be calculated from Time history I. The fatigue damage induced by highway loading only within T_1 can be calculated by multiplying the factor T_1/T_2 by the fatigue damage calculated from Time history II. The fatigue damage induced by the railway loading only within one hour can be finally estimated by subtracting the fatigue damaged induced by the highway loading only within T_1 from the fatigue damage calculated from Time history I. The fatigue damage is calculated based on Miner's law as suggested in BS5400 and the remaining fatigue life of the relevant bridge components is estimated afterwards, as will be introduced in the next section.

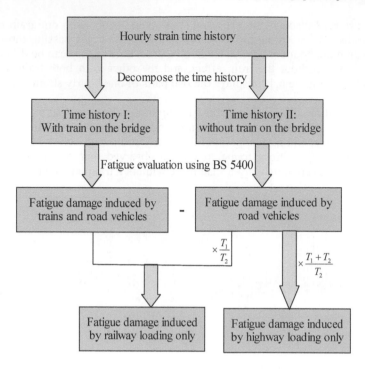

Figure 6.21 Procedure for estimating fatigue damage induced by trains and road vehicles

6.5.5.3 Fatigue Damage Evaluation

Fatigue is caused by repeated applications of a stress which is insufficient to induce failure by a single application and damage by gradual cracking of parts of a steel structure. Stress cycle is a pattern of variation of stress at a point defined by the cycle counting method and consisting of a change in stress between defined minimum (trough) and maximum (peak) values and back again. The cycle counting method counts the number of stress cycles of different magnitudes which occur in a service stress history. The particular counting method recommended in BS5400 Part 10 is the rainflow or reservoir method.

A σ_r–N curve displays the quantitative relationship between the fatigue strength σ_r and the number of cycles N corresponding to a specific probability of failure for a certain detail class. The σ_r–N curve is derived from test data. A detail class is the rating given to a particular structural detail to indicate which fatigue strength curve should be used in the fatigue assessment. The class is denoted by one of the following letters: A, B, C, D, E, F, F2, G, S, T, W or X in BS5400. This categorization takes consideration of the local stress concentration at the detail, size, and shape of the maximum acceptable discontinuity, stress direction, metallurgical effects, residual stresses, fatigue crack shape, and in some cases the welding process and a post-weld improvement method. Miner's linear damage rule is used in BS5400 and given as follows:

$$\sum_{i=1}^{n} \frac{n_i}{N_i} = \left(\frac{n_1}{N_1} + \frac{n_2}{N_2} + \cdots\cdots + \frac{n_n}{N_n} \right) \tag{6.39}$$

where n_i ($i = 1, 2, \ldots, n$) is the number of repetitions of the ith stress range σ_i, and N_i is the number of cycles to failure at σ_i. In the present study, n_i and σ_i are obtained by applying the rainflow counting method to the strain measurement data. N_i is determined using the σ_r–N curves plotted in Figure 6.22, which can be expressed by either Equation (6.40) or (6.41).

$$N \times \sigma_r^m = K_2 \tag{6.40}$$

$$Log_{10}N = Log_{10}K_2 - mLog_{10}\sigma_r \tag{6.41}$$

where K_2 and m have the values given in BS5400 for different detail classes listed in Table 6.6; and N is the number of repetitions of stress range σ_r.

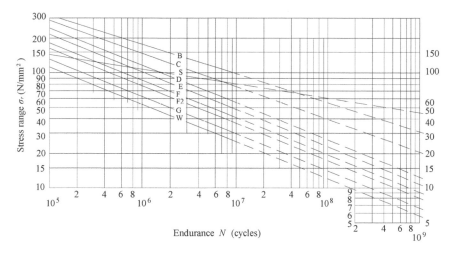

Figure 6.22 σ_r–N curves used in this study

Table 6.6 Major parameters in σ_r – N curves.

Detail class	m	K_2	σ_0 (N/mm^2)
W	3.0	0.16×10^{12}	25
G	3.0	0.25×10^{12}	29
F2	3.0	0.43×10^{12}	35
F	3.0	0.63×10^{12}	40
E	3.0	1.04×10^{12}	47
D	3.0	1.52×10^{12}	53
C	3.5	4.23×10^{12}	78
B	4.0	1.01×10^{15}	100
S	8.0	2.08×10^{22}	82

In Table 6.6, σ_0 is a limit value of the stress range and the corresponding number of repetitions N_0 is equal to 10^7. When the stress range σ_i is less than σ_0, the number of repetitions N_i should be reduced in the proportion of $(\sigma_i/\sigma_0)^2$.

Therefore, the calculation of the design number of repetitions N_i to failure of each stress range σ_i is conducted using the following equations:

$$\frac{N_i}{N_0} = \frac{N_i \sigma_i^m}{K_2} = \frac{N_i}{10^7} \left(\frac{\sigma_i}{\sigma_0}\right)^m \quad (\sigma_i \geq \sigma_0) \tag{6.42}$$

$$\frac{N_i}{N_0} = \frac{N_i \sigma_i^{m+2}}{K_2 \sigma_0^2} = \frac{N_i}{10^7} \left(\frac{\sigma_i}{\sigma_0}\right)^{(m+2)} \quad (\sigma_i < \sigma_0) \tag{6.43}$$

The major steps used to evaluate fatigue damage and fatigue life of the relevant bridge components of the Tsing Ma Bridge can be summarized as follows:

1. Determine the class of the connection detail near the relevant strain gauge according to BS7608 (1993).
2. Count the number of cycles n_i and the corresponding stress range σ_i by applying the rainflow method to Time history I or Time history II which are obtained from the recorded strain time history.
3. Determine the number of repetitions to failure N_i for each stress range in accordance with Equation (6.42), and if necessary modify N_i in the case of low stress cycles in accordance with Equation (6.43).
4. Use Equation (6.39) to evaluate the fatigue damage. In this study the fatigue damage is projected to 120 years based on the assumption that the fatigue damage estimated based on the strain data measured in November 2005 could be extended within 120 years.
5. Calculate the fatigue life in number of years as $120/(\Sigma(n_i/N_i))$.

According to the detail classification described in BS7608, the connections of steel components in the Tsing Ma Bridge can be classified as 'F' or 'F2'. Their fatigue damage and fatigue life are estimated based on the daily strain data recorded on 2 November 2005 and the monthly strain data recorded during the whole month so that the effects of data length can be investigated. By taking SSTEN03 as an example, Table 6.7 lists the number of cycles and the corresponding stress ranges identified from the strain time history measured on 2 November 2005 without differentiating the highway loading and railway loading. The identified number of cycles is projected to 120 years and the design number of cycles is calculated using Equation (6.43) with the connection detail class "F". The fatigue damage index DI is calculated as 0.097 for 120 years and the fatigue life is estimated to be 1,237 years.

With the above understanding, extensive computations are performed to evaluate fatigue life of the welding connections near all the relevant valid strain gauges in the bridge for connection classes "F" and "F2" and for the daily and monthly data length, respectively. Some results are shown in Figure 6.23, in which the number in brackets is the fatigue life due to railway loading only. Some conclusions can be drawn from the figure as follows:

1. The fatigue life estimated based on the daily data length is almost the same as that based on the monthly data length, which indicate the traffic condition throughout November 2005 is steady.
2. The fatigue life estimated based on the connection class "F" is much longer than that based on the connection class "F2".

3. The fatigue damage to the steel connections of the Tsing Ma Bridge is mainly induced by moving trains. The contribution of road vehicles to fatigue damage is rather small.

4. The fatigue life of the Tsing Ma Bridge is very long for some connections. All the instrumented components are in good integral condition and will survive for the duration of their design life.

Table 6.7 An example of fatigue life evaluation.

Stress range (MPa)	Design no. of cycles (N_i)	Measured no. of cycles (n_i) projected to 120 years	n_i/N_i
2.05 - 4.10	3.67E+12	2.01E+08	0.00005
4.10 - 6.15	2.85E+11	2.06E+07	0.00007
6.15 - 8.20	5.30E+10	6.46E+06	0.00012
8.20 - 10.25	1.51E+10	2.93E+06	0.00019
10.25 - 12.30	5.53E+09	2.06E+06	0.00037
12.30 - 14.35	2.40E+09	2.15E+06	0.00090
14.35 - 16.40	1.17E+09	2.50E+06	0.00214
16.40 - 18.45	6.27E+08	2.12E+06	0.00338
18.45 - 20.50	3.60E+08	2.50E+06	0.00694
20.50 - 22.55	2.18E+08	2.76E+06	0.01266
22.55 - 24.60	1.38E+08	2.19E+06	0.01587
24.60 - 26.65	9.12E+07	1.36E+06	0.01491
26.65 - 28.70	6.21E+07	5.69E+05	0.00916
28.70 - 30.75	4.34E+07	3.72E+05	0.00857
30.75 - 32.80	3.11E+07	2.19E+05	0.00704
32.80 - 34.85	2.28E+07	1.10E+05	0.00482
34.85 - 36.90	1.70E+07	1.10E+05	0.00647
36.90 - 38.95	1.28E+07	4.38E+04	0.00342
38.95 - 41.00	9.87E+06	0.00E+00	0.00000

Fatigue damage index $DI = \Sigma(n_i/N_i) = 0.097$
Fatigue life = 120 / DI = 1237 years

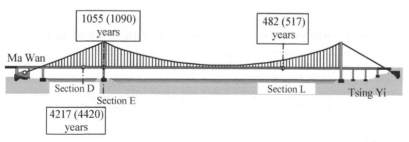

(a) Connection class F

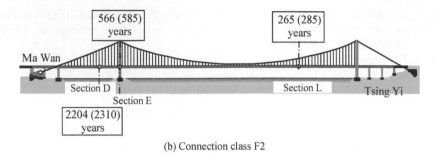

(b) Connection class F2

Figure 6.23 Shortest fatigue life estimated for each instrumented section

6.6 SUMMARY

The major sensors and sensing systems used for highway monitoring of long-span suspension bridges have been described in this chapter. The techniques for ascertaining the highway traffic condition and highway load distribution from the measurement data have been introduced. The finite element method based analytical backgrounds of dynamic interaction between the bridge and moving road vehicles have been discussed. The Tsing Ma suspension bridge has been taken as an example to illustrate the monitoring of highway traffic condition and highway loading effects on the bridge. The highway traffic condition was analysed in terms of vehicle traffic volume, vehicle traffic composition, axle load spectrum and GVW spectrum. The statistical data of the axle load distribution in 2006 manifested that the axle number decreased with increasing axle load, and only 12.5% of the total vehicle axles had an axle load more than 5 tonnes. The statistical data of the GVW distribution in 2006 also showed that the vehicle number decreased with increasing GVW, and only 7.5% of the total vehicles had a GVW more than 16 tonnes. The measured axle load spectrum and GVW spectrum have been compared with the design spectrum and the spectrum specified in BS5400. The axle load spectra from the measured WIM data in 2006 and based on BS5400 both were on the safe side compared with the design spectrum. The measured GVW spectrum was also on the safe side compared with the reference GVW spectrum based on BS5400. The strain data collected by 110 strain gauges installed at three sections of the bridge during November 2005 have been processed and analysed. A practical method in conjunction with the method proposed in BS5400 has been used to estimate the fatigue life of the bridge due to highway loading only, railway loading only, and a combination of railway and highway loading. The results of fatigue life evaluation demonstrated that the fatigue life estimated based on the daily data length was almost the same as that based on the monthly data length, indicating the traffic condition in the month is steady. The fatigue life estimated based on the connection class "F" was much longer than that based on the connection class "F2". The fatigue damage to the steel connections of the Tsing Ma Bridge was mainly induced by moving trains and the contribution of road vehicles was rather small. The fatigue life of the bridge is very long for some connections, and all the instrumented components are in good integral condition and will survive for the duration of their design life.

6.7 REFERENCES

Au, F.T.K., Cheng, Y.S. and Cheung, Y.K., 2001, Effects of random road surface roughness and long-term deflection of prestressed concrete girder and cable-stayed bridges on impact due to moving vehicles. *Computers & Structures*, **79(8)**, pp. 853-872.

Balageas, D., Fritzen, C. and Guemes, A., 2006, *Structural Health Monitoring*, (London: ISTE Ltd.).

Blejwas, T.E., Feng, C.C., and Ayre, R.S., 1979, Dynamic interaction of moving vehicles and structures. *Journal of Sound and Vibration*, **67**, pp. 513-521.

BS, 1980, *BS5400: Part 10, Code of Practice for Fatigue*, (London: British Standards Institution).

BS, 1993, *BS7608: Code of Practice for Fatigue Design and Assessment of Steel Structures*, (London: British Standards Institution).

Cheung, Y.K., Au, F.T.K., Zheng, D.Y. and Cheng, Y.S., 1999, Vibration of multi-span non-uniform bridges under moving vehicles and trains by using modified beam vibration functions. *Journal of Sound and Vibration*, **228**, pp. 611-628.

Dodds, C.J., 1972, *International Organization for Standardisation ISO/TC/108/WG9, Document No. 5*. BSI proposals for generalised terrain dynamic inputs to vehicles, (London: British Standards Institution).

Dodds, C.J. and Robson, J.D., 1973, The description of road surface roughness. *Journal of Sound and Vibration*, **31**, pp. 175-183.

Guo, W.H. and Xu, Y.L., 2001, Fully computerized approach to study cable stayed bridge-vehicle interaction. *Journal of Sound and Vibration*, **248(4)**, pp. 745-761.

HKPU, 2007, *Establishment of Bridge Rating System for Tsing Ma Bridge: Statistical Modelling and Simulation of Predominating Highway Loading Effects.* Report No. 5, (Hong Kong: Department of Civil and Structural Engineering, The Hong Kong Polytechnic University).

HKTD, 2008, *Transport Planning & Design Manual*, (Hong Kong: Hong Kong Transport Department)

Huang, D.Z. and Wang, T.L., 1992, Impact analysis of cable-stayed bridges. *Computers & Structures*, **43**, pp. 897-908.

Karbhari, V.M. and Ansari, F., 2009, *Structural Health Monitoring of Civil Infrastructure Systems*, (Cambridge, England: Woodhead Pub.).

Mufti, A.A., 2001, *Guidelines for Structural Health Monitoring*, (Winnipeg, Canada: Intelligent Sensing for Innovative Structures).

Mufti, A.A., Bakht, B. and Jaeger, L.G., 2008, *Recent Advances in Bridge Engineering*, (Winnipeg, Canada: JMBT Structures Research Inc.)

Olsson, M., 1985, Finite element modal co-ordinate analysis of structures subjected to moving loads. *Journal of Sound and Vibration*, **99**, pp. 1-12.

Taly, N., 1998, *Design of Modern Highway Bridges*, (Berkshire, England: McGraw-Hill).

Timoshenko, S, Young, D.H. and Weaver, W., 1974, *Vibration Problems in Engineering*, 4th ed., (New York: John Wiley & Sons).

Wang, T.L. and Huang, D.Z., 1992, Cable-stayed bridge vibration due to road surface roughness. *Journal of Structural Engineering*, ASCE, **118**, pp. 1354-1374.

Wong, K.Y, Man, K.L and Chan, W.Y.K, 2001, Monitoring of wind load and response for cable-supported bridges. In *Proceedings of SPIE 6th International Symposium on NDE for Health Monitoring and Diagnostics, Health Monitoring and Management of Civil Infrastructure Systems*, Newport Beach, California, edited by Kundu, T., (Bellingham, Washington: SPIE), pp. 292-303.

Xu, Y.L., Ko, J.M. and Zhang, W.S., 1997, Vibration studies of Tsing Ma suspension bridge. *Journal of Bridge Engineering, ASCE,* **2**, pp. 149-156.

Yang, Y.B. and Lin, H.B, 1995, Vehicle-bridge interaction analysis by dynamic condensation method. *Journal of Structural Engineering, ASCE,* **121**, pp. 1636-1643.

Yang, Y.B. and Wu, Y.S., 2001, A versatile element for analyzing vehicle-bridge interaction response. *Engineering Structures*, **23(5)**, pp. 452-469.

6.8 NOTATIONS

A_r	Roughness coefficient.
a	Travelling acceleration of the vehicle.
b_1	Horizontal distance between the two units.
$[C_{vl}]$	The damper force matrix of the dampers that the relative velocities are the function of the DOFs of the vehicle only.
$[C_{v1}], [C_{v2}]$	Vehicle damping matrix, additional damping matrix to vehicle damping matrix $[C_{v1}]$.
$[C_{bv}], [C_{vb}]$	Coupled damping matrices.
$[C_{bbv}]$	Additional damping matrix to bridge damping matrix.
C_m, C_k, C_c	Mass or mass moment coefficients, stiffness coefficients, damping coefficients.
C_{li}, C_{ui}	Damping coefficient of viscous damper ($i=1,2$).
DI	Fatigue damage index.
J_{yvi}, J_{xvi}	Pitching moments and rolling moments of inertia of the tractor and two trailers ($i=1,2,...,3$).
$[K_{v1}], [K_{v2}]$	Vehicle stiffness matrix, additional stiffness matrix to vehicle stiffness matrix $[K_{v1}]$.
$[K_{bv}], [K_{vb}],$	Coupled stiffness matrices.
$[K_{bbv}]$	Additional stiffness matrix to bridge stiffness matrix.
K_{li}, K_{ui}	Stiffness of linear elastic spring ($i=1,2$).
K_2	Parameter in σ_r–N curves.
L	Effective axle load.
$[M_b], [C_b], [K_b]$	Mass, damping and stiffness matrices of the bridge.
$[M_v], [C_v], [K_v]$	Mass, damping and stiffness matrices of the vehicle.
$[M_{bbv}]$	The inertia forces of the vehicle at the contact points due to the bridge accelerations.
$[M_v]$	Inertia forces of all the rigid bodies of the vehicle.
M_{sli}, M_{sri}	Left and right lump mass of the axle sets ($i=1,2,...,5$).

M_{v1}, M_{v2}, M_{v3}	Lump mass of the tractor and two trailers.
m	Parameter in $\sigma_r - N$ curves.
N_b, N_v, N_{vb}	DOFs of the bridge, the vehicle, and the coupled road vehicle-bridge system.
N_{li}	Transfer function from the node displacements of the bridge element to the displacement of the ith left contact point ($i = 1$, $2, ..., 5$).
N_i	Number of circles to failure at σ_i.
N_0	Number of repetitions to the limit value of the stress range.
n_i, N	Repetition of the ith axle load, total number of repetitions.
$\{P\}, \{P_v\}, \{P_b\}$	External force vector of coupled system, vehicle, and bridge.
$\{P_{be}\}$	External force vector of the bridge.
$\{P_{bvg}\}$	External forces on the bridge due to the gravity forces of the vehicle.
$\{P_{bvri}\}$,	Additional force vector on the bridge due to the road surface roughness ($i = 1, ... 3$).
$\{P_{bvrl}\}$	The inertia forces of the vehicle at the contact points due to the road surface roughness.
$\{P_{li}\}, \{P_{ri}\}$	External forces at the contact points due to the gravity of the vehicle ($i = 1, 2, ..., 5$).
P_i	Force coefficients.
$\{P_r\}$	Force due to the road surface roughness.
$\{P_{vvr2}\}, \{P_{vvr3}\}$	Additional force vector on the vehicle due to the road surface roughness.
r_{li}, r_{ri}	Road surface roughness under the ith left and right contact points ($i = 1, 2, ..., 5$).
$S(\bar{\phi})$	PSD function for the road surface elevation.
T_1, T_2	Time length.
u	Travelling velocity of the vehicle.
V_{ij}	The number of vehicles of the jth vehicle category running on the ith lane in an hourly record ($i = 1, 2, ..., 7; j = 1, 2, ..., 8$).
$\{v_v\}, \{\dot{v}_v\}, \{\ddot{v}_v\}$	Nodal dynamic displacement, velocity, and acceleration vectors of the vehicle.
$\{v_b\}, \{\dot{v}_b\}, \{\ddot{v}_b\}$	Nodal dynamic displacement, velocity, and acceleration vectors of the bridge.
$\{\delta v\}, \{\delta v_v\}, \{\delta v_b\}$	Virtual displacement of coupled system, vehicle, and bridge.
$\{v\}, \{\dot{v}\}, \{\ddot{v}\}$	Displacement, velocity, acceleration vectors of the coupled road vehicle-bridge system.
$\delta W_{VI}, \delta W_{VD}, \delta W_{VE}$	Virtual work done by all inertial forces, damping forces, and elastic forces acting on the vehicle at a given time.
δW_{bp}	Virtual work done by all external forces at a given time.
$Z_{vi}, \theta_{vi}, \phi_{vi}$	Vertical displacement, rotation about the transverse axis, and rotation about the longitudinal axis of the tractor and two trailers ($i = 1, 2, ..., 3$).

Z_{li}, Z_{ri}	Vertical displacements of the left contact point and right contact point (i = 1, 2, ..., 5).
Z_{sli}, Z_{sri}	Vertical displacement of the axle sets (i = 1, 2, ..., 5).
α_b, β_b	Rayleigh damping factors.
Δ_{lli}, Δ_{lri}	Deformations of the lower left and right springs at the axle sets (i = 1, 2, ..., 5).
Δ_{uli}, Δ_{uri}	Deformations of the upper left and right springs at the axle sets (i = 1, 2, ..., 5).
σ_r	Fatigue strength.
σ_0	Limit value of the stress range.
$\bar{\phi}$, $\bar{\phi}_0$	Spatial frequency, discontinuity frequency of 0.5π (cycle/m).
θ_k	Random phase angle.
$\{\delta\}_{bi}^e$	Displacement vector of the bridge element over which the ith contact point runs (i = 1, 2, ..., 5).

CHAPTER 7

Monitoring of Temperature Effects

7.1 PREVIEW

Bridges are subjected to daily and seasonal environmental thermal effects induced by the solar radiation and ambient air temperature. Variation of temperatures in bridge components significantly influences the overall deflection and deformation of the bridge. In addition, thermal stresses are usually induced due to redundancy of the structure, the restraint of movement from bearings and expansion joints, and the non-uniform distribution of temperature within the bridge. It has also been observed that changing temperature conditions may have a significant effect on structural vibration properties. Therefore temperature, including structural temperature and ambient air temperature, is one of the frequent measurands in bridge monitoring systems.

As number of sensors is always limited, a numerical thermal analysis is required to obtain the detailed temperature information within the bridge that is not available from the monitoring system. In addition, a quantitative assessment of the temperature effects on the bridge needs a thorough temperature distribution of the bridge components. At the same time, the numerical model and its analytical results need verification with the real data. Consequently it is imperative to integrate the numerical analysis and field monitoring for providing a better and thorough understanding of the temperature distribution of large-scale bridges. This approach is also applied to the monitoring of other effects.

In this regard, this chapter will first review the temperature monitoring exercises in bridge engineering and the basic temperature monitoring systems. The heat-transfer analysis and the associated thermodynamic models will then be introduced for obtaining the temperature distribution of a bridge structure with known thermal boundary conditions, including air temperature and solar radiation. The temperature distribution of the bridge components will be applied to the structural model to obtain the thermal responses including strain and displacement. The predicted responses can be verified with the field monitoring data. Finally, this integrated approach will be applied to the Tsing Ma Bridge to quantitatively monitor and assess the thermal effects on the displacement, strain, vibration properties of the bridge.

7.2 TEMPERATURE MONITORING

The solar radiation and ambient air temperature are the two main factors affecting the structural temperature behaviour. The temperature sensors can be either embedded in the bridge to measure the structural temperature or placed in the air to measure the ambient temperature. Temperature sensors cannot be exposed under sunshine directly to measure the ambient temperature.

The most often used temperature sensors include thermocouples, thermistors, and resistance temperature detectors. Thermocouples are inexpensive but less stable than the other two kinds. Thermistors and resistance temperature detectors are based on the principle that resistance of a metal increases when temperature goes up. They are generally more accurate and stable than thermocouples.

Besides temperature measurements, monitoring of environmental conditions also includes the measurement of air pressure, rainfall, humidity, and solar radiation. These measurement results are used to correlate with each other to provide complete environmental conditions. Barometric pressure sensors can be used to measure the atmospheric pressure of the environment whereas rain gauges can be used to measure intensity of amount of rainfall. The solar radiation can be measured continuously with a pyranometer. Figure 7.1 illustrates a set of pyranometers for measuring the diffuse solar radiation, reflected solar radiation, and total solar radiation. These quantities can also be estimated with thermal models as detailed in the next section.

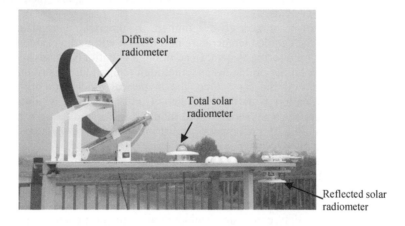

Figure 7.1 Configuration of a set of pyranometers

Thermal effects have been studied for a relatively long time. Zuk (1965) studied the thermal behaviour of several highway bridges. Capps (1968) measured temperature and the longitudinal movement of a steel box bridge in the UK. Based on the temperature distribution obtained from numerical or experimental studies, approximate analytical methods have been developed to estimate the thermal stresses of concrete bridges (Priestley, 1978; Elbadry and Ghali, 1983). Bridge design codes such as AASHTO (1989) include provisions on simple methods for predicting the longitudinal thermal movement of bridges. However, AASHTO's method is inadequate to predict thermal responses of complicated bridges (Moorty and Roeder, 1992).

Recently, with the development of SHM technologies, field monitoring of the effect of temperature on bridge performance has been widely employed for concrete bridges. For example, over 140 temperature sensors have been installed in sections of the Confederation Bridge to monitor its time-temperature history hourly (Cheung *et al.*, 1997; Li *et al.*, 2008). The measured movement at the expansion joints (with distance of 500 m) during a three-year period was 226 mm. The temperature-induced displacement of long-span bridges can be more significant. For example, the seasonal longitudinal displacement of the Runyang Suspension Bridge (a main span of 1490 m) was measured about 500 mm (Deng *et al.*, 2010).

Temperature effects on structural vibration properties (frequencies, damping ratios, and mode shapes) have also been studied for decades. In their pioneer study, Adams *et al.* (1978) studied the relation between temperature and the axial resonant frequency of a bar. Researchers from Los Alamos National Laboratory, USA (Cornwell *et al.*, 1999) found that the first three natural frequencies of the Alamosa Canyon Bridge varied about 4.7%, 6.6% and 5.0% during a 24 hours period as temperature of the bridge deck changed by approximately 22°C. The Z24-Bridge in Switzerland was continuously monitored for nearly a year (Peeters and De Roeck, 2001) and it was reported that the first four vibration frequencies varied 14~18% during the period. Desjardins *et al.* (2006) found that the frequencies of the Confederation Bridge during a six-month period varied by about 5%, with the temperature of the girders varying from −20°C to +25 °C. Xia *et al.* (2006) have studied the relation between environmental effects and structural dynamic properties of a continuous concrete slab for nearly two years. The effect of temperature on variation in frequency, mode shape, and damping ratio has been investigated. Other field studies include those of Fu and DeWolf (2001), Song and Dyke (2006), Ni *et al.* (2007), Liu and DeWolf (2007), Breccolotti *et al.* (2004), Macdonald and Daniell (2005), and Nayeri *et al.* (2008).

Quite a few laboratory experiments and field monitoring exercises have observed that changing temperature conditions may have a more significant effect on structural vibration properties than have operational loads or structural damage. For example, the researchers from Los Alamos National Laboratory also found that a significant artificial cut in the I-40 Bridge over the Rio Grande caused only insignificant frequency changes and a moderate cut even led to a slight increase in frequency (Farrar *et al.*, 1994). If the temperature effects are not well monitored and considered in the SHM, it may result in a false assessment (Xia *et al.*, 2006).

In SHM systems, the temperature is usually measured at a few limited points. However, the temperature distribution of a bridge is generally non-uniform and time dependent. As structural responses are associated with the temperature distribution of the entire bridge, using limited temperature data may not be able to capture the true quantitative relation between the structural responses and temperature, especially for large-scale bridges. In this regard, extensive temperature information is required. A realistic numerical heat-transfer analysis can be conducted.

7.3 HEAT-TRANSFER ANALYSIS

7.3.1 General Formulation

With the rapid development of computational methods and computer technology, a number of one-dimensional to three-dimensional finite element models have been developed since the 1970s (Elbadry and Ghali, 1983; Wang, 1994). Finite difference heat flow models have been employed to determine the temperature distribution on bridge members as well (Riding *et al.*, 2007).

The flow of heat in an isotropic solid is governed by the following Fourier partial differential equation (Whitaker, 1977):

$$\rho c \frac{\partial T}{\partial t} = k \left(\frac{\partial^2 T}{\partial x^2} + \frac{\partial^2 T}{\partial y^2} + \frac{\partial^2 T}{\partial z^2} \right) \tag{7.1}$$

where x, y, z are the Cartesian coordinates; t is the time; T is the temperature of the solid; ρ is the density; c is the specific heat of the solid (J/kg/°C); and k is the thermal conductivity (W/m/°C). For a bridge subjected to solar radiation, it can be assumed that thermal variation in the direction of the longitudinal axis is normally not significant (Elbadry and Ghali, 1983; Fu *et al.*, 1990). Consequently, the temperature field T of a bridge cross section at any time t can be expressed by a two-dimensional heat flow equation as

$$\rho c \frac{\partial T}{\partial t} = k \left(\frac{\partial^2 T}{\partial x^2} + \frac{\partial^2 T}{\partial y^2} \right) \tag{7.2}$$

7.3.2 Thermal Boundary Conditions

When the bridge temperature at the boundary is known, the condition is referred to as the First Kind thermal boundary condition (Lienhard and Lienhard, 2001). When heat flow at the boundary is known, the Third Kind thermal boundary condition can be employed as (Elbadry and Ghali, 1983):

$$k \left(\frac{\partial T}{\partial x} n_x + \frac{\partial T}{\partial y} n_y \right) + q = 0 \tag{7.3}$$

where n_x and n_y are the direction cosines of the unit outward vector normal to the boundary surfaces, and q is the boundary heat input or loss per unit area (W/m²). For a bridge surface subjected to solar radiation, q consists of convection q_c, thermal irradiation q_r, and solar radiation q_s, as illustrated in Figure 7.2 (Branco and Mendes, 1993).

The rate of heat transfer by convection q_c is associated with the movement of the air particles. It depends on the convection heat transfer coefficient h_c and difference between the air temperature T_a and the bridge surface temperature T_s as

$$q_c = h_c(T_a - T_s) \tag{7.4}$$

The convection heat transfer coefficient (W/m²/°C) is related to many variables such as wind speed, surface roughness, and geometric configuration of the exposed surface and is usually determined experimentally or calculated based on empirical formulae (Branco and Mendes, 1993; Froli *et al.*, 1996).

The heat transfer between the bridge surface and the surrounding environment due to thermal irradiation can be expressed in a quasi-linear form as

$$q_r = h_r(T_a - T_s) \tag{7.5}$$

The radiation heat transfer coefficient h_r can be approximately expressed as (Branco, 1986)

$$h_r = \varepsilon[4.8 + 0.075(T_a\text{-}5)] \tag{7.6}$$

where ε is the emissivity coefficient of the bridge surface. Thus, the effects of heat flow by convection and irradiation can be combined in terms of an overall heat transfer coefficient h (W/m²/°C) as

$$q = h(T_a - T_s), \text{ and } h = h_c + h_r \tag{7.7}$$

The solar radiation is detailed in the next section.

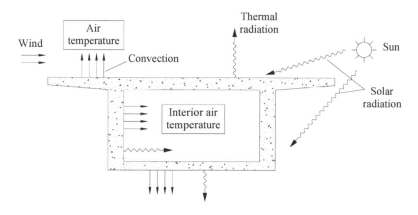

Figure 7.2 Energy exchange of a bridge section

7.3.3 Solar Radiation

The solar radiation is affected by several factors such as the hour of the day, the day of the year, the latitude and the altitude of the bridge, the orientation of the bridge, and the cloudiness of the sky. The spatial position of the sun and a bridge is illustrated in Figure 7.3, where ψ is the solar altitude, θ is the incidence angle of sun rays, γ_s is the azimuth angle, vector $\vec{n}$ is perpendicular to the surface, surface azimuth angle γ is measured from the south and is positive toward the west, β is the tilt angle from the horizontal and is positive for south-facing surface, solar incidence angle θ is defined as the angle between the surface normal $\vec{n}$ and a line collinear with the sun's rays. The rate of heat absorbed by the bridge surface due to solar radiation q_s is (Rohsenow, 1988; Branco *et al.*, 1992)

$$q_s = \alpha I \tag{7.8}$$

where α ($0 < \alpha < 1$) is the absorptivity coefficient of the surface material; I is the total solar radiation (W/m²) including direct solar radiation I_d, diffuse solar radiation I_i, and reflected solar radiation I_r on a surface, respectively, which can be obtained through the field measurement or empirical formulae.

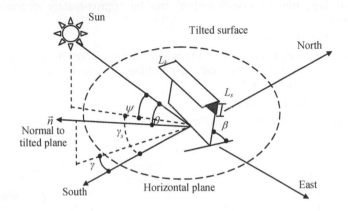

Figure 7.3 Spatial position of the sun and structure

7.3.3.1 Direct Solar Radiation

The direct radiation energy to heat up the bridge structure depends on the solar constant (I_{sc}) and the absorption of the solar energy in the atmosphere. The solar constant is the amount of solar electromagnetic radiation received by a unit area of a plane perpendicular to the sun, at the mean distance from the sun to the Earth. As the Earth's orbit around the sun is elliptical, solar constant I_{sc} is suggested to be (Rohsenow, 1988):

$$I_{sc} = 1367\left[1 + 0.033\cos\left(360°\frac{N}{365}\right)\right] \qquad (7.9)$$

where N is the number of the day of the year, counting from January 1. The solar constant varies by about 7.0% during a year (from 1.412 kW/m^2 in early January to 1.321 kW/m^2 in early July) due to the Earth's varying distance from the sun.

The Earth's atmosphere acts as a filter for the solar radiation so that only a fraction of the total solar radiation reaches the surface of the Earth. The energy reaching the Earth's surface by direct radiation can be expressed as

$$I_{d0} = I_{sc}K_T \qquad (7.10)$$

where K_T is the dimensionless transmittance factor, which depends on the atmospheric conditions such as the scatter of the light in a pure atmosphere and the absorption of certain wave lengths by the atmosphere. K_T can be calculated as

$$K_T = 0.9^{k_a t_u m} \qquad (7.11)$$

where k_a is the ratio of the atmospheric pressure to the pressure at sea level (see Table 7.1), t_u is the turbidity factor indicating the attenuation of radiation in different atmospheric conditions, and m is the air mass factor.

Table 7.1 Ratio of the atmospheric pressure to the pressure at sea level (k_a).

Elevation to sea level (m)	0	500	1000	1500	2000	2500	3000
k_a	1.0	0.94	0.89	0.84	0.79	0.74	0.69

Air pollution increases the absorption of the solar radiation and the turbidity factor t_u can be expressed as

$$t_u = A_{tu} - B_{tu}\cos\left(360°\frac{N}{365}\right) \tag{7.12}$$

where parameters A_{tu} and B_{tu} are suggested to take values in Table 7.2. The air mass factor, m, gives the relative path length of the radiation through the atmosphere, which is expressed in terms of the solar altitude ψ as

$$m = \frac{1}{\sin\psi + 0.15(\psi + 3.885)^{-1.253}} \tag{7.13}$$

The above equation can be simplified as

$$m = 1/\sin\psi \tag{7.14}$$

Table 7.2 Parameters of the turbidity factor.

	Mountain areas	Village	City	Industrial areas
A_{tu}	2.2	2.8	3.7	3.8
B_{tu}	0.5	0.6	0.5	0.6

When the sun has an incidence angle θ normal to the surface, the direct solar radiation is

$$I_d = I_{d0}\cos\theta = I_{sc}K_T\cos\theta \tag{7.15}$$

In the zone where an overhanging cantilever slab of length L_k provides shade, no direction radiation is present. The length of the shadow, L_s, can be calculated as

$$L_s = L_k \frac{\tan\psi}{\sin(90° + \gamma - \gamma_s)\sin\beta - \cos\beta\tan\psi} \tag{7.16}$$

7.3.3.2 Diffuse Solar Radiation

The direct solar radiation is accompanied with radiation that has been scattered or remitted, called the diffuse solar radiation. The energy reaching the Earth's surface by the diffuse solar radiation can be estimated as (Berlage, 1928)

$$I_{i0} = \frac{1}{2} \cdot \frac{1 - P^m}{1 - 1.4\ln P} I_{sc}\sin\psi \tag{7.17}$$

where

$$P = 0.9^{k_a t_u} \tag{7.18}$$

The diffuse solar radiation on a surface with a tilt angle β is

$$I_i = \frac{1+\cos\beta}{2}I_{i0}$$ (7.19)

7.3.3.3 Reflected Solar Radiation

The direct and diffuse solar radiation may also be reflected onto a nearby surface through the reflected solar radiation (Berlage, 1928)

$$I_{r0} = r_e(I_d + I_{i0})$$ (7.20)

where r_e is the reflected coefficient of the ground, which is usually set as 0.2 for ordinary earth surfaces and 0.7 for snow covered surfaces. The reflected solar radiation on a surface with a tilt angle β is

$$I_r = \frac{1-\cos\beta}{2}I_{r0}$$ (7.21)

7.3.4 Equivalent Temperature

The total solar radiation on a surface is then the summation of the direct solar radiation, diffuse solar radiation, and reflected solar radiation, that is

$$I = I_d + I_i + I_r$$ (7.22)

Based on Equations (7.7) and (7.8), the heat flow of a surface considering the solar radiation is

$$q = h(T_a - T_s) + \alpha I$$ (7.23)

or a more convenient form

$$q = h(T^* - T_s)$$ (7.24)

where

$$T^* = T_a + \frac{\alpha I}{h}$$ (7.25)

T^* includes both the air temperature and solar radiation and is thus usually termed 'equivalent air temperature'.

7.4 TEMPERATURE EFFECTS ON LONG-SPAN SUSPENSION BRIDGES

The temperature change affects a structure in a complicated way. It affects geometry, mechanical properties, and thus changes constraint conditions and boundary conditions. These make it difficult to quantify the temperature effect on structural behaviour. In the following sections temperature effects on both structural static behaviour and dynamic behaviour will be described.

7.4.1 Temperature Effects on Static Behaviour of Bridges

Variation in temperature results in movement of bridges, and stresses when the movement is restrained. Temperature-induced stresses depend not only on the temperature distribution but also the boundary conditions. The thermal stresses can be categorized into the primary thermal stress, induced by the continuity of the structure, and the secondary thermal stress, due to the nonlinear thermal gradient in cross sections (Fu *et al.*, 1990).

Given a thermal gradient, thermal stresses can be analysed based on the one-dimensional beam theory (Priestley, 1978), which is based on the assumptions of linear stress-strain relation, linear temperature-strain relation, and planar section. The longitudinal thermal effects and transverse thermal effects are considered independently and then combined. The thermal effects can be superimposed by other loading effects.

For cable-supported bridges, the one-dimensional beam theory may be insufficient as the cable system and the towers affect the bridge's behaviours as well. To obtain more accurate and detailed results, three-dimensional FE methods can be employed. The three-dimensional strain-stress relationship for linear and isotropic materials subjected to both mechanical stresses and temperature changes can be written in the following equation:

$$
\begin{Bmatrix} \varepsilon_1 \\ \varepsilon_2 \\ \varepsilon_3 \\ \gamma_{12} \\ \gamma_{23} \\ \gamma_{31} \end{Bmatrix} = \begin{bmatrix} 1/E & -v/E & -v/E & & & \\ -v/E & 1/E & -v/E & & & \\ -v/E & -v/E & 1/E & & & \\ & & & 1/G & & \\ & & & & 1/G & \\ & & & & & 1/G \end{bmatrix} \begin{Bmatrix} \sigma_1 \\ \sigma_2 \\ \sigma_3 \\ \tau_{12} \\ \tau_{23} \\ \tau_{31} \end{Bmatrix} + \theta_T \Delta T \begin{Bmatrix} 1 \\ 1 \\ 1 \\ 0 \\ 0 \\ 0 \end{Bmatrix} \quad (7.26)
$$

where E is the modulus of elasticity, G shear modulus, v Poisson's ratio, and θ_T the thermal coefficient of linear expansion.

7.4.2 Temperature Effects on Dynamic Behaviour of Bridges

It is believed that the variation in modal frequencies of structures with temperature is due to the change in the material properties, in particular the modulus of elasticity (Lee *et al.*, 1988).

To simplify the problem, we take a simply-supported uniform beam of length l, height h, density ρ, and modulus of elasticity E as the example. Mass and the boundary conditions are assumed unchanged, and temperature will affect only the geometry and mechanical properties. The undamped flexural vibration frequency of order n is (Clough and Penzien, 1993)

$$
f_n = \frac{n^2 \pi h}{2l^2} \sqrt{\frac{E}{12\rho}} \quad (7.27)
$$

It can be shown that

$$
\frac{\delta f_n}{f_n} = \frac{\delta h}{h} - 2\frac{\delta l}{l} + \frac{1}{2}\frac{\delta E}{E} - \frac{1}{2}\frac{\delta \rho}{\rho} \quad (7.28)
$$

where δ represents an increment in the corresponding parameter. Assuming the thermal coefficient of linear expansion of the material is θ_T, and the temperature coefficient of modulus is θ_E, one obtains

$$\frac{\delta h}{h} = \theta_T \delta T, \quad \frac{\delta l}{l} = \theta_T \delta T, \quad \frac{\delta \rho}{\rho} = -3\theta_T \delta T, \quad \frac{\delta E}{E} = \theta_E \delta T \tag{7.29}$$

Here we assume that the variation of Young's modulus with temperature is linear for small changes in temperature. Consequently Equation (7) yields

$$\frac{\delta f_n}{f_n} = \frac{1}{2}(\theta_T + \theta_E)\delta t \tag{7.30}$$

For prismatic multi-span beams, the natural frequencies can be expressed in a similar form to those of a simply-supported beam as (Blevins, 1979):

$$f_n = \frac{\lambda_n^2 h}{2\pi l^2}\sqrt{\frac{E}{12\rho}} \tag{7.31}$$

where λ_n is a dimensionless parameter which is a function of the boundary conditions. Consequently Equation (7.30) is still applicable for multi-span beam structures.

Equation (7.30) estimates the dimensionless rate of frequency change with temperature change. For some materials in civil engineering, θ_E is much higher than θ_T. For example, concrete's θ_E is about $-3.0\times10^{-3}/°C$ and θ_T about $1.0\times10^{-5}/°C$ at the temperature 20°C (CEB-FIP, 1993); and steel's θ_E is about $-3.6\times10^{-4}/°C$ and θ_T is about $1.1\times10^{-5}/°C$ at the temperature 38°C (Brockenbrough and Merritt, 1994). Therefore, variation of Young's modulus with temperature is much more significant than thermal expansion. From Equation (7.30), all frequencies will decrease 0.15% per degree for concrete structures regardless of lower modes or higher modes. If a concrete structure experiences temperature change of 20°C in one day, its natural frequencies will vary by 3.0%, which agrees with some field observations and laboratory experimental results as described in Section 7.2. It also indicates that steel structures are not sensitive to the temperature changes as the temperature coefficient of modulus is much smaller.

In the above quantitative analysis, the effect of internal force on the frequency is not considered. When the thermal stress is not negligible, its effect on the frequency should also be taken into account.

7.5 MONITORING OF TEMPERATURE EFFECTS ON THE TSING MA BRIDGE

7.5.1 Temperature Sensors in the Tsing Ma Bridge

The position of the temperature sensors in the Tsing Ma Bridge is shown in Figure 3.8, in which the numbers in parentheses are the sensor numbers at different positions. The collected temperature data from WASHMS can be grouped into three categories: (1) ambient temperature (P3, 6 in number); (2) section temperature (P1, P2, P4, and P6, 86 in number); and (3) cable temperature (TC, 23 in number). The TC type is a thermocouple sensor, and P1 to P6 are PT100 Platinum resistance temperature sensors.

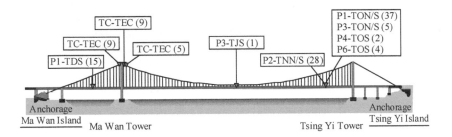

Figure 7.4 Distribution of temperature sensors in the Tsing Ma Bridge

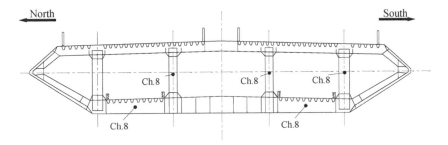

Figure 7.5 Ambient temperature sensors close to the cross frame (P3-TON and P3-TOS)

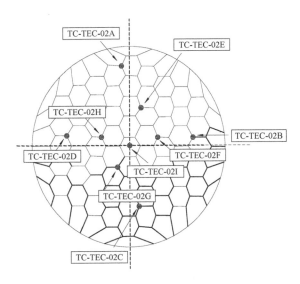

Figure 7.6 Temperature sensors at the main cable (TC-TEC)

One air temperature sensor (Ch. 81, P3-TJS) is approximately at the middle section of the main span and attached on a sign gantry which stands on the upper deck. The other five air temperature sensors (Ch. 82 to Ch. 86, P3-TON and P3-TOS), as shown in Figure 7.5, measure the ambient temperature around the bridge deck section of the main span near the Tsing Yi tower. 23 thermocouples (TC-TEC) were embedded inside the main cables at three different locations to measure the cable temperature and those of one section are shown in Figure 7.6. As the cross section of the main cable is relatively small, the dense sensors can capture the temperature of the cable sufficiently and thus no numerical analysis is required.

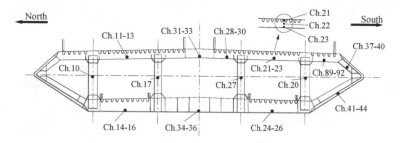

Figure 7.7 Temperature sensors on the cross frame (P1-TON and P1-TOS)

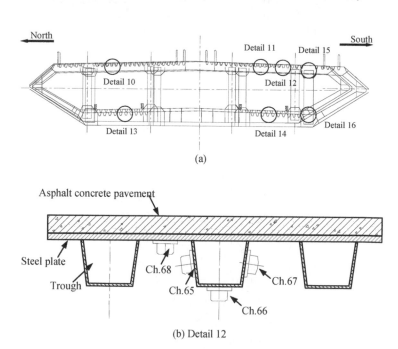

(a)

(b) Detail 12

Figure 7.8 Temperature sensors on the deck plate (P2-TNN and P2-TNS)

The deck section in the Ma Wan side span is equipped with 15 sensors (P1-TDS) and another deck section close to the Tsing Yi Tower is equipped with 71 sensors. The latter can be found in Figure 7.7 and Figure 7.8. In particular, Figure 7.7 shows the position of 37 sensors installed on the cross frame (P1-TON and P1-

TOS), four of which are located on the vertical members (Nos. 10, 17, 20 and 27) and others on the chord members. Figure 7.8 illustrates seven groups of temperature sensors (Details 10 to 16, P2-TNN and P2-TNS, 28 sensors in total) mounted on the orthotropic deck plates 2.25 m away from the section. At each Detail, one sensor measures the temperature of the steel plate and the other three measure the temperature distribution over the deck trough. In addition, four temperature sensors (P6-TOS) are embedded in the south carriageway and two temperature sensors (P4-TOS) are installed on the corrugated sheets on the south side of the section, which are not shown here for brevity. The sampling frequency of all of the temperature sensors is 0.07 Hz.

7.5.2 Heat-transfer Analysis of the Tsing Ma Bridge

Because of the complicated geometrical configuration of the bridge, it is very difficult to establish one single model for the entire bridge and conduct transient thermal analysis. Thermal variation of the bridge deck in the longitudinal direction of the bridge can be assumed insignificant and thus one typical segment is studied. This assumption is based on the fact that air temperature, solar radiation, and segment configuration along the longitudinal direction of the bridge are similar. Similarly, thermal variation of the bridge towers in the vertical direction is assumed small and consequently one section of the tower is analysed. By this means, different components are modelled separately to simplify the modelling and analysis. Consequently, FE models of a typical deck plate, a cross frame, and a segment of the bridge tower are established using three dimensional solid elements with the aid of commercial FE package ANSYS (ANSYS, 2004).

7.5.2.1 Deck Plate Models

The orthotropic bridge deck plate consists of steel deck plates, asphalt concrete pavement, and deck troughs. The upper and the lower deck plates have the same thickness of 13 mm. The upper deck troughes and lower deck troughes are 8 mm thick and 10 mm thick, respectively. All the steel deck plates are covered by a 40 mm thick asphalt concrete cover. It is noted that the top surface of the deck plates is subjected to direct solar radiation while the other parts of the deck plates are not. The deck plates, troughs, and deck covers are modelled using thermal type solid elements (SOLID90 in ANSYS). Figure 7.9(a) shows the FE model of Detail 12 for heat transfer analysis, consisting of 8,932 nodes and 6,720 three dimensional SOLID90 elements. The FE models of deck plates at other positions are constructed similarly.

Appropriate thermal boundary condition is critical for transient thermal analysis. Considering the complicated configuration of the large-scale bridge, different surfaces are subjected to different heat flows and air temperatures. In particular, the upper surface of the deck is subject to direct solar radiation and thus Equation (7.23) should be applied. For other boundary surfaces without solar radiation, only convection and irradiation as in Equation (7.7) are employed to determine the surface heat flow. The temperature measured from P3-TJS is used as the air temperature. The temperatures measured at the steel plate and deck troughs are utilized to verify the numerical results.

7.5.2.2 Cross Frame Model

A detailed three-dimensional FE model of one cross frame consisting of 24,242 nodes and 3,259 three dimensional SOLID90 elements is constructed as shown in Figure 7.9(b). The cross frame is modelled on the basis of the design with minor simplifications by disregarding accessory components. An enlarged detail of one connection of the cross frame is shown in Figure 7.9(c).

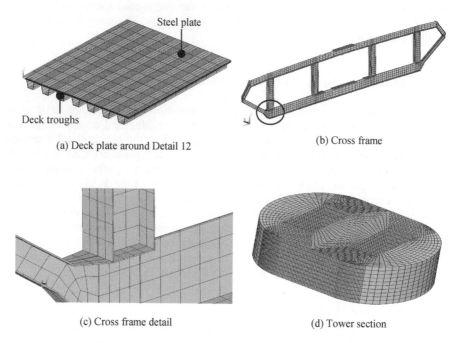

(a) Deck plate around Detail 12   (b) Cross frame

(c) Cross frame detail   (d) Tower section

Figure 7.9 Finite element models of the bridge components

Because the cross frame is enclosed by the deck plates and corrugated sheets, heat exchange occurs between the plates and the cross frame. Consequently, the temperature measured at the interface between the frame and enclosed plates (sheets) is used as the thermal boundary condition (the First Kind thermal boundary condition). Other surfaces of the frame that are exposed to air do not receive solar radiation directly. Consequently Equation (7.7) can be employed as the surface heat flow, while the air temperatures measured by the sensors close to the frame (P3-TON and P3-TOS in Figure 7.5) are used as the ambient temperature (the Third Kind boundary condition). Take the upper chord of the frame on which sensors Ch. 21 to Ch. 23 (P1-TOS) are mounted as an example. Sensor Ch. 21 was installed on the upper side of the member adjacent to the deck plate, and Ch. 23 was installed on the lower side of the member exposed to air. Two cases of boundary conditions are employed here. In the first case (Case 1), monitoring data from both Ch. 21 and Ch. 23 are used as the boundary temperature. Consequently, the temperature at Ch. 22 point can be calculated through the heat-conductivity analysis and compared with the measurement data.

In this situation, using the heat convection condition is not necessary. In the second case (Case 2), the field monitoring data from Ch. 21 are the boundary temperature of the upper surface, and the heat flow condition and air temperature measured from Ch. 86 are the boundary conditions of the bottom surface. In the case, temperature recorded at Ch. 22 and Ch. 23 will be employed to verify the numerical results.

7.5.2.3 Tower Section Model

A detailed FE model of one tower section is constructed using 23,969 nodes and 20,000 three dimensional solid elements as shown in Figure 7.9(d). The cross section of the bridge tower and its orientation are illustrated in Figure 7.10. The tower section is enclosed by two semicircles and one rectangle with two rectangular holes in the middle. The angle between the x axis and the south is 17°. The height of the tower segment model is 5.0 m. As the outer surface of the tower is a curve, different points receive different levels of solar radiation. In this regard, the outer surface of the tower is divided into 18 segments indicated by the 18 points (N1~N16, NN, and NS) . Each segment receives the same level of solar radiation calculated according to its orientation. The heat flow determined according to Equation (7.23) and the air temperature measured at P3-TJS are applied to the outer surfaces of the tower section as the thermal boundary conditions.

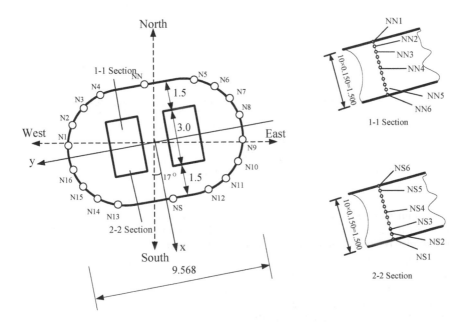

Figure 7.10 Configuration of the tower section (unit: m)

7.5.3 Temperature Distribution of the Tsing Ma Bridge

With the FE models and corresponding boundary conditions described above, each component is analyzed for transient heat transfer to obtain the temperature distribution among the components. Table 7.3 lists the material parameters for the thermal analysis. Note that convection heat transfer coefficient h_c depends on the wind speed and the geometric configuration. An empirical formulae (Froli *et al.*, 1996) is employed here to determine h_c, in which the wind speed measured just above the deck is used.

Table 7.3 Thermal properties of materials.

Parameters		Asphalt	Concrete	Steel
k (W/(m·C))	Thermal conductivity	2.50	1.54	55
ρ (kg/m^3)	Density	2100	2400	7800
c (J/(kg·C))	Heat capacity	960	950	460
α	Absorptivity coefficient	0.90	0.65	0.685
ε	Emissivity coefficient	0.92	0.88	0.8
E (Pa)	Elasticity modulus	1.3×10^9	3.25×10^{10}	2.01×10^{11}
α_T	Thermal expansion coefficient	1.0×10^{-5}	1.0×10^{-5}	1.3×10^{-5}

7.5.3.1 Deck Plate

Figure 7.11 compares the temperature calculated at the deck plate at Detail 12 on 17 July 2005 with the measurement counterparts (Ch. 66 to Ch. 68). It is observed that the structural temperature decreases slightly in the early morning before increasing to the maximum in the afternoon at about 16:00. The temperatures predicted for the three points agree with the measurements very well at all times, validating the effectiveness of the numerical model and the heat transfer analysis. The calculated temperatures of the steel plate and pavement at Detail 12 are plotted in Figure 7.12. It is observed that the temperature at the asphalt concrete pavement peaks at about 61°C in the summer. In addition, the temperature at the asphalt concrete pavement is much higher than that at the steel plate in the afternoon, with a maximum difference of 11°C. The temperature difference between the top and bottom of the steel plate is slight throughout the day. This is because the thermal conductivity of steel is much higher than that of asphalt.

Temperatures at the deck plate in the other three seasons are also calculated and illustrated in Figure 7.13. The parameters of the model are the same as those adopted for the summer calculation, except that the radiation intensity varies in different seasons. Similar to the results in the summer, the temperatures estimated for the other seasons agree well with the measurements.

Table 7.4 lists the daily peak temperatures at the top of the pavement (Detail 12) in different seasons. The temperature distribution of the bridge plate at other deck details has also been obtained but is not shown here for brevity.

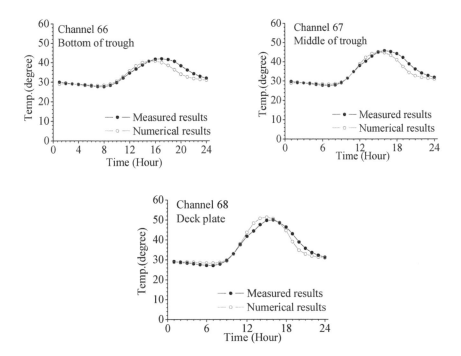

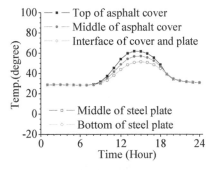

Figure 7.11 Comparison of measured and simulated temperature at Detail 12 on 17 July 2005

Figure 7.12 Comparison of measured and simulated temperature of the steel plate and pavement on 17 July 2005

Table 7.4 Daily peak temperature at top of the pavement (Detail 12) in different seasons (Unit: °C).

Season	Winter	Spring	Summer	Autumn
Date	17 January 2005	18 April 2005	17 July 2005	17 October 2005
Maximum	35.41	56.52	61.85	51.38
Minimum	12.67	21.31	28.22	24.23

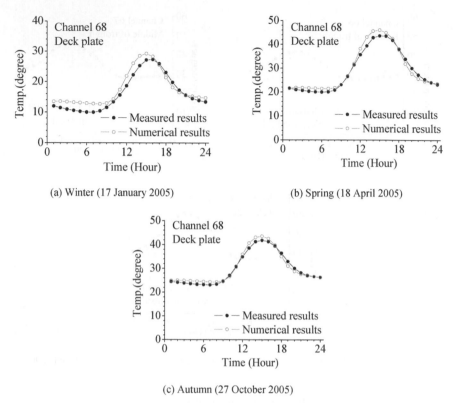

(a) Winter (17 January 2005)   (b) Spring (18 April 2005)

(c) Autumn (27 October 2005)

Figure 7.13 Comparison of measured and simulated temperature at Detail 12 in different seasons

7.5.3.2 Cross Frame

The temperature distribution across the cross frame is investigated in this section under two types of boundary conditions (Case 1 and Case 2) described in Section 7.5.2.2. The temperature variations at Ch. 22 (the upper right chord member) on 17 July 2005 calculated under the two cases are compared with the measured results in Figure 7.14. It can be seen that the numerical temperature results in both cases closely match the measurement counterparts at different times, and that the temperature using the first boundary condition (Case 1) can predict the temperature better than Case 2. This is because the temperature at the bottom of the member (Ch. 23) has been known already in Case 1 and it is more accurate than the numerical calculation in Case 2. Similar observations can be made for the upper left chord, lower chord, and vertical truss members, as illustrated in Figure 7.14, Figure 7.15, and Figure 7.16, respectively.

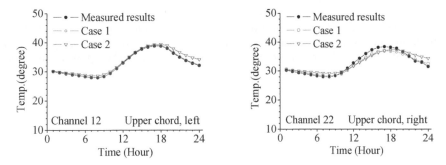

Figure 7.14 Comparison of measured and simulated temperature at the upper chord of the cross frame on 17 July 2005

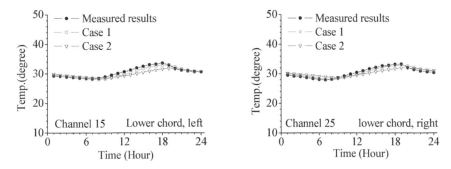

Figure 7.15 Comparison of measured and simulated temperature at the lower chord of the cross frame on 17 July 2005

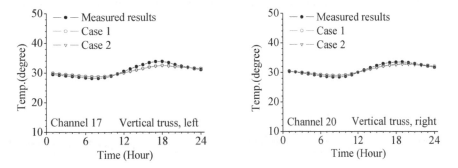

Figure 7.16 Comparison of measured and simulated temperature at the truss members of the cross frame on 17 July 2005

7.5.3.3 Bridge Tower

Solar intensity on the tower surfaces differs from that on the deck surface, and different facets receive different levels of solar radiation. Figure 7.17 shows the solar intensity on the west and south surfaces on 17 July 2005, indicating that peak solar intensity occurs at about 9:00 on the east surface and about 16:00 on the west. As far as the east surface is concerned, there is no direct solar radiation in the afternoon due to the sun being blocked by the bridge tower and only a small amount of diffuse solar radiation therefore being received. The opposite situation applies to the west surface.

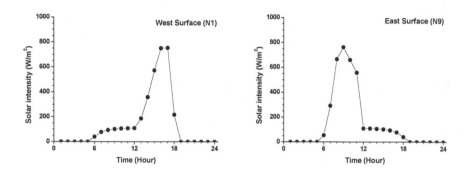

Figure 7.17 Solar intensity of the surfaces of the bridge tower on 17 July 2005

Variation in temperature at the tower section on 17 July 2005 is calculated and shown in Figure 7.18. Only numerical results are reported here due to no temperature sensor being installed on the tower. The figure shows that the temperatures at outer surface points are quite different from each other because of the different levels of solar radiation intensity they receive. The temperatures at inner points fall dramatically as the distance to the surface of the tower increases. No obvious temperature variation can be observed at the middle points (NN4 and NS4) throughout the day, while temperature variations among points 30 cm from the surface are insignificant because concrete has a low level of heat conductivity.

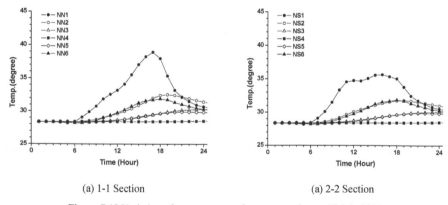

(a) 1-1 Section   (a) 2-2 Section

Figure 7.18 Variation of temperature at the tower section on 17 July 2005

7.5.3.4 Summary

As mentioned above, the cross section of the cables is relatively small. The densely distributed sensors across the cable section can capture the temperature distribution adequately and no numerical analysis has been conducted in the present study. The measured and simulated temperatures at different components on a summer's day are compared in Table 7.5. It is observed that the simulation results agree well with the measurement data, with a maximum difference of only 1.3°C. Therefore, the numerical models and heat transfer analysis are able to simulate the temperature distribution throughout the bridge.

Table 7.5 Peak temperature at different components on 17 July 2005 (Unit: °C).

Component	Minimum		Maximum	
	Measurement	Simulation	Measurement	Simulation
Deck plate (Ch. 68)	27.12	28.25	49.95	51.28
Cross frame (Ch. 12)	27.98	28.04	38.84	39.05
Tower section (NE)	N.A.	28.37	N.A.	43.86
Cable (Ch. 115)	29.36	N.A.	36.75	N.A.

7.5.3.5 Long-term Temperature Variation

The temperature variation in the long term (for example, over years) can be obtained from the monitoring system (HKPU, 2007; Xu *et al.* 2010). Here numerical calculation is not conducted. For brevity, the monthly effective temperature over the bridge deck and the ambient temperature during 1999 to 2005 are illustrated in Figure 7.19, where the minimum, average, and maximum in each month are shown. The effective temperature is calculated as the averaged temperatures of various locations within the cross section weighted by the area around the temperature point (Emerson, 1979). Here it refers to the deck section at sensors P2 (see Figure 3.8). The ambient temperature is measured at Ch. 82 (see Figure 7.5).

It is clear that the effective temperature of the bridge deck has a similar variation pattern to the ambient temperature. The minimum effective temperature is almost the same as the minimum ambient temperature in each month. The differences between the mean values of the two temperatures are also slight, that is, the mean effective temperature of the bridge deck is normally 2°C higher than the mean ambient temperature. Regarding the maximum monthly temperature, the effective temperature is significantly higher than the ambient temperature, by about 6°C in the winter to 10°C in the summer. It should be pointed out that the range of ambient temperature is well within the design values of −2°C to 46°C.

The maximum daily or monthly temperature variation for each month is also obtained and shown in Figure 7.20. It shows that the maximum variation in effective temperature within one day is about 14°C for all months while that in the ambient temperature is roughly 7°C. Except in a few extreme cases, the maximum daily variation is very uniform and seems independent of the month and year. The maximum variation in effective temperature within one month is about 15~28°C while that in the ambient temperature is about 6~20°C.

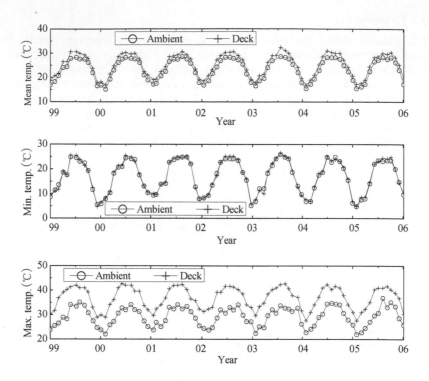

Figure 7.19 Monthly ambient temperature and deck effective temperature during 1999 to 2005

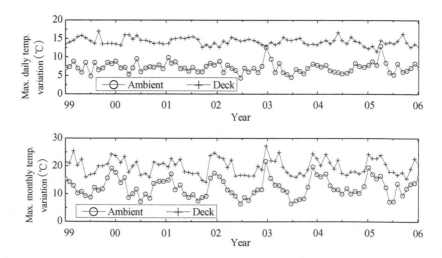

Figure 7.20 Maximum daily and monthly temperature variation during 1999 to 2005

7.5.4 Temperature-induced Displacement of the Tsing Ma Bridge

Figure 7.21 shows a global FE model of the entire bridge constructed to facilitate structural analysis and calculation of temperature-induced displacement and strain. Components are simplified appropriately without ignoring details of temperature variations and are modelled with relatively coarse meshes in the structural analysis in comparison with those employed in the thermal analysis. In particular, the towers are modelled using solid elements to take different temperatures at different facets into consideration. The horizontal plane is divided into six different temperature zones — south, north, southeast, southwest, northeast, and northwest — each of which has a uniform temperature along the vertical direction of the facet at any point in time. As the steel deck plate temperatures are uniformly distributed, the deck plates are modelled using shell elements in the global model (the asphalt cover is excluded from the structural analysis). For components with relatively small temperature variations over the cross section, beam-type elements are used and the effective temperature is applied to the joints. In particular, the cross frame members, the longitudinal trusses, and the bridge piers are modelled using three-dimensional beam elements, and the suspenders and main cables are modelled using bar elements. The global FE model comprises 23,960 nodes and 28,856 elements. A total of 4,808 MPCs are used for connections between the orthotropic deck plates and the chords of cross frames.

The temperature distribution calculated at one point in time is input into the global model to facilitate static analysis. Displacement and strain in various components can then be calculated hourly. The temperature, strain, and displacement at midnight (00:00) are set as the initial reference values, with the results calculated at later times indicating relative changes to the reference values. The numerical results are compared with the recorded measurements, which also show the relative changes in the corresponding reference values. To reduce random measurement error, the temperatures and responses measured each hour are averaged. Without further clarification, the displacements measured by the pairs of GPS receivers located at the north and south ends of the bridge are averaged and regarded as the displacements of the corresponding locations.

Figure 7.21 Finite element model of the Tsing Ma Bridge

7.5.4.1 Displacement of the Bridge Deck

Figure 7.22 compares the numerical displacement of the bridge deck in the middle of the main span (MDMS) on 17 July 2005 with measurements taken using GPS at the section. It can be observed that displacements in the longitudinal (x) and vertical (z) directions show a good agreement with those measured in the field. However, the calculated lateral (y) displacements do not agree well with the measurements, possibly because the displacements are affected not only by the temperature loading but also by other loadings such as traffic. Measurement noise is another factor as the lateral displacements are small throughout the day.

It can also be seen that the deck moves slightly towards Ma Wan Island (negative values) in the early morning as the temperature decreases slightly during this period. It then moves towards Tsing Yi Island (positive values) as the temperature increases later in the day. The daily movement in the longitudinal direction is about 170 mm. In the vertical direction, the deck first moves upwards during 0:00 to 8:00 before moving downwards as the temperature increases. In the evening, the deck goes up again as the temperature decreases. The daily movement in the vertical direction is about 220 mm. The displacements at other deck sections also exhibit similar trends.

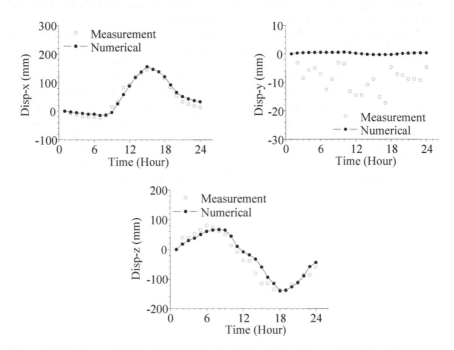

Figure 7.22 Displacement of the bridge deck in the middle of the main span (MDMS) on 17 July 2005

7.5.4.2 Displacement of the Bridge Towers

Displacements at the top of the Ma Wan Tower on 17 July 2005 are shown in Figure 7.23. It can be seen that only the displacements calculated in the longitudinal direction (x) are correlated with the measurements in the longitudinal direction whereas the displacements calculated in the lateral and vertical directions are not. In addition, the measured displacements in the lateral and vertical directions are relatively scattered and irregular in comparison with the longitudinal one. Similar observations are made for the Tsing Yi Tower. One reason for these results might be that the displacements of the towers are not only induced by temperature variation of the tower themselves, but also affected by movement of the main cables and bridge deck.

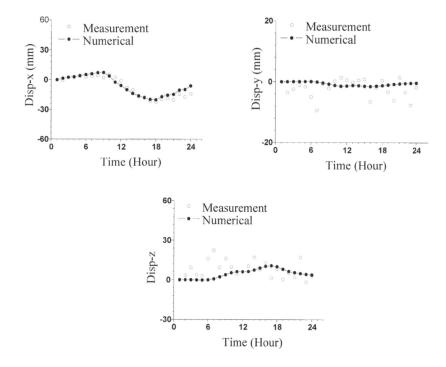

Figure 7.23 Displacement of the Ma Wan Tower on 17 July 2005

7.5.4.3 Displacement of the Bridge Abutment

The hourly calculations of longitudinal displacement of the bridge abutment on the Tsing Yi side on 17 July 2005 are drawn in Figure 7.24 as compared with the displacement measurements recorded by the displacement transducer, showing a good agreement except for the peak values. The daily temperature-induced variation in the longitudinal displacement of the abutment is about 300 mm on a summer's day.

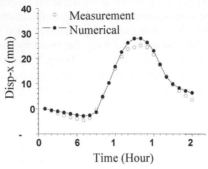

Figure 7.24 Longitudinal displacement of the Tsing Yi abutment on 17 July 2005

7.5.4.4 Displacement of the Main Cable

Displacements of the main cables are monitored by a pair of GPS stations that are placed in the middle of the main cables. The hourly mean displacements of the main cables in three directions are shown in Figure 7.25. Similar to the bridge deck results, the calculated displacements in the longitudinal and vertical directions agree with the field monitoring results whereas the calculated lateral displacements do not.

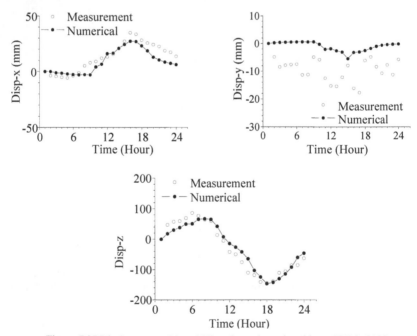

Figure 7.25 Displacement of the middle point at the main cable on 17 July 2005

It can also be seen that displacement of the main cable is in phase with that of the deck, that is, the main cable first moves upwards as the temperature decreases slightly from 0:00 to 8:00 before moving downwards as the temperature increases. In the evening, the main cables go up again as the temperature decreases. The daily vertical displacement of the main cables is about 220 mm peak-to-peak, close to the degree of movement of the deck; whereas its longitudinal movement is about 40 mm, much smaller than that of the deck.

7.5.4.5 Displacement Responses in Long-term

The temperature-induced bridge displacement over the long term can be obtained from the monitoring system (HKPU, 2007; Xu *et al.*, 2010). No numerical analysis is conducted due to the large calculations involved. The daily mean displacements of the bridge deck in the middle of the main span in 2003 are illustrated in Figure 7.26. It is noted that the displacements in the figure are the displacements relative to those when the GPS stations were installed and thus differ from those in the last section.

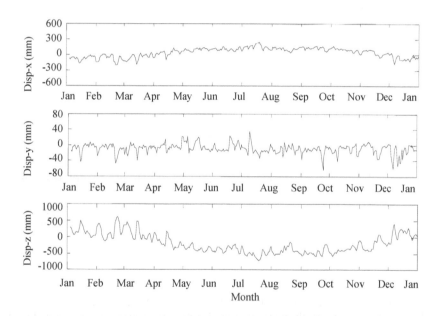

Figure 7.26 Daily mean displacement of bridge deck in the middle of the main span (MDMS) in 2005

The monthly mean displacements of the bridge deck in the middle of the main span during 2003 to 2005 are illustrated in Figure 7.27. Seasonal variation in the longitudinal and vertical displacements can be clearly observed from the figure. In particular, the bridge deck moves downwards and to Tsing Yi Island in the summer, and upwards and to Ma Wan Island in the winter. In addition, a fairly constant seasonal variation in the longitudinal and vertical displacements is observed. The lateral displacements of the deck section, however, do not present a seasonal variation trend. Similar results can be observed from other deck points in the main span.

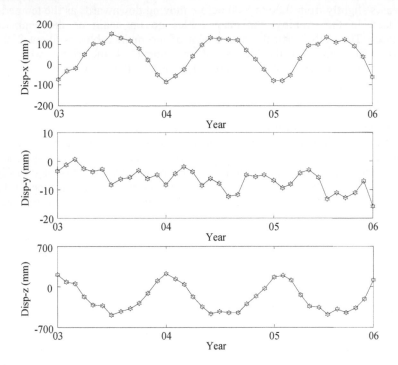

Figure 7.27 Monthly mean displacement of the bridge deck in the middle of the main span (MDMS) during 2003 to 2005

The daily mean displacement during 2003 to 2005 is also plotted with respect to the deck effective temperature for exploring the temperature-displacement relationship, as shown in Figure 7.28. As described before, both the daily vertical and longitudinal displacements present a very good linear relationship with the deck effective temperature. The lateral displacements are not shown in the figure as no clear relation with the temperature can be found. The corresponding linear regression model is taken as

$$x = b + at + \varepsilon_x \tag{7.32}$$

where x is the displacement, b (intercept) and a (slope) are regression coefficients, and ε_x is the error. With least-squares fitting, the regression coefficients are obtained as $a = 15.50$ mm/°C and $b = -349.82$ mm for the longitudinal displacements, and $a = -39.32$ mm/°C and $b = 837.88$ mm for the vertical displacements.

The regression coefficients for the displacement at other components, not shown here for brevity, can be used to quantify the temperature-induced movement of the bridge. In particular, slope a is 3.58, 11.56, 15.50, 18.55, and 26.11 mm/°C for the longitudinal displacements at the five deck sections (middle of Ma Wan side span, ¼ of the main span, middle of the main span, ¾ of the main span, and Tsing Yi abutment) respectively. The ratio of the slope to the free distance of the section is respectively 1.35×10^{-5}, 1.46×10^{-5}, 1.36×10^{-5}, 1.25×10^{-5}, and $1.31 \times 10^{-5}/^{\circ}C$, which are close to the thermal coefficient of linear expansion of steel $\theta_T = 1.1 \times 10^{-5}/^{\circ}C$ (Brockenbrough and Merritt, 1994). This error might be attributed to using the effective temperature of the deck section, while the global responses of the bridge are associated with the temperatures over the entire section.

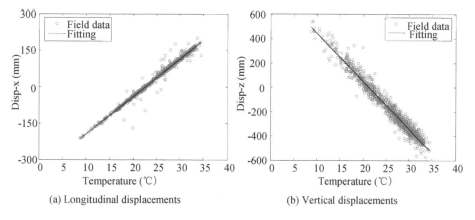

(a) Longitudinal displacements   (b) Vertical displacements

Figure 7.28 Relationship between daily mean displacement and deck effective temperature of the bridge deck (MDMS) during 2003-2005

7.5.4.6 Summary of the Displacement Responses

The maximum peak-to-peak ranges of the longitudinal and vertical displacements at different components during 2005 are summarized in Figure 7.29, in which the monitored maximum displacements are used rather than mean values. It can be seen that the maximum range of the longitudinal displacement of the deck sections gradually increases along the longitudinal direction toward Tsing Yi Island. In the middle of the main span, the vertical displacement range of the bridge section is very close to that of the main cables. Therefore, the temperature-induced vertical displacement of the deck section in the middle of the main span is dominated by deformation of the main cables.

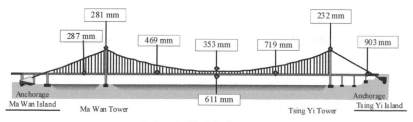

(a) Longitudinal displacement

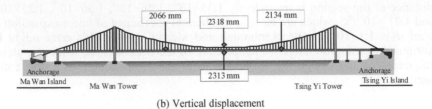

(b) Vertical displacement

Figure 7.29 Maximum displacement ranges of the Tsing Ma Bridge in 2005

7.5.5 Temperature-induced Stress of the Tsing Ma Bridge

The variation in stresses of bridge members is investigated in this section. Considering that temperature tends to cause static strain of a bridge, whereas traffic loadings cause dynamic strain, the hourly mean strain is calculated and variation therein is regarded as being due to temperature changes. The stress is calculated from the measured strain and temperature according to Equation (7.26) where only axial stress and strain are involved.

110 strain gauges are distributed at three sections of the Tsing Ma Bridge as shown in Figure 7.30. Taking Section E (Chainage 23623) shown in Figure 7.31 as an example, 32 strain gauges were installed at the cross frame, near the Ma Wan Tower. The north outer truss, north inner truss, south outer truss, and south inner truss were equipped with four pairs of gauges (SP type) and two single gauges (SS type). Figure 7.31b illustrates the position of six strain gauges (four SP type and two SS type) at the north inner longitudinal truss.

The strains at the upper chord, diagonal braces, and bottom chord are measured by gauges. Their hourly mean stresses determined from the measurement data on 17 July 2005 are compared their numerical counterparts in Figure 7.32. It is noted that the figure illustrates changes in stress from the initial values at 0:00. The figure also demonstrates that the calculated hourly stress variation agrees well with the field measurement, thus indicating that it is mainly caused by temperature changes.

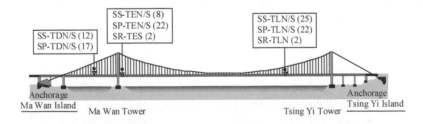

Figure 7.30 Distribution of strain gauges in the Tsing Ma Bridge

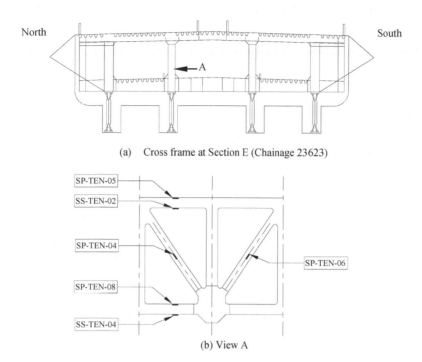

(a)   Cross frame at Section E (Chainage 23623)

(b) View A

Figure 7.31 Distribution of strain gauges at North inner longitudinal truss (Chainage 23623)

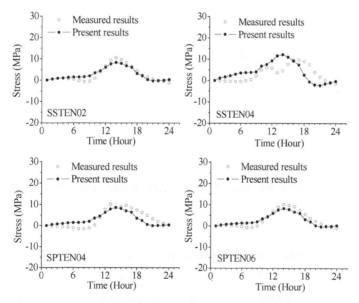

Figure 7.32 Axial stresses at upper chord, diagonal braces, and bottom chord at Section E on 17 July 2005

7.5.6 Temperature Effect on Vibration Properties of the Tsing Ma Bridge

Variation in vibration properties of the Tsing Ma Bridge is investigated in this section. Only the first few measured frequencies of the global structure in different seasons are studied here.

The frequencies are identified from the ambient excited acceleration data measured by the sensors installed on the main span deck, as illustrated in Figure 3.15. The vertical acceleration data measured by sensor AS-TIN-01 and AS-TJN-01 are employed to extract the vertical and torsional modal frequencies. The horizontal acceleration data by AB-TIS-01 and AB-TJS-01 are employed to extract the lateral modal frequencies. The Stochastic Subspace Identification method (Van Overschee and De Moor, 1996) is used as the modal identification algorithm and the vertical acceleration data by AS-TFN-01 are employed as the reference.

Figure 7.33 shows the variation in the first few frequencies (four vertical, one torsional, and two lateral modes) and temperatures on 17 January 2005. In the figure, the frequency data at each hour is obtained from the acceleration data recorded during that hour (3600 seconds × 51.2 Hz = 184,320 points in total), and the effective temperature of the deck section at each hour is adopted. From the figure, one can find that

1)   The variations in the frequencies are very small. This is because the bridge was made of steel, whose modulus is insensitive to the temperature change.
2)   All of the frequencies of the bridge generally decrease when the temperature goes up before noon while they increase as the temperature drops down in the afternoon, as observed by many other researchers.
3)   The minimum frequencies and maximum temperatures do not occur at the same time, and the time difference is about three hours. This confirms that the frequencies are global properties and associated with the temperature distribution of the entire structure.

As modal frequencies are different, the variation percentage of the frequencies with respect to the effective temperature is shown in Figure 7.34, where frequencies of each mode are divided by their corresponding maximum values. It can be seen that the variation trends of the vertical modal frequencies are very similar with a maximum variation of about 1%. The variation percentages of the first two lateral modes are larger with a maximum variation of about 2.5%.

The relation of the frequencies to the temperature in the other three seasons (18 April, 17 July, and 26 October in 2005) is also studied. Similar results can be observed but not shown here for brevity. The first four vertical modal frequencies on the four days are plotted in Figure 7.35(a) against the effective temperature, where the frequencies are normalized for each mode. The figure shows a clear linear relation between the frequency ratio and the temperature. A linear regression model has the form of (Kottegoda and Rosso, 1997)

$$f = \beta_0 + \beta_T T + \varepsilon_f \tag{7.33}$$

where f is the frequency, β_0 (intercept) and β_T (slope) are the regression coefficients, and ε_f is the regression error. The slope (β_T) is calculated as -1.7×10^{-4}, very close to half of $\theta_E = -3.6 \times 10^{-4}/°C$, as described in Equation (7.30). This confirms that the variation in the vertical frequencies of the bridge is mainly caused by the modulus change under different temperature conditions.

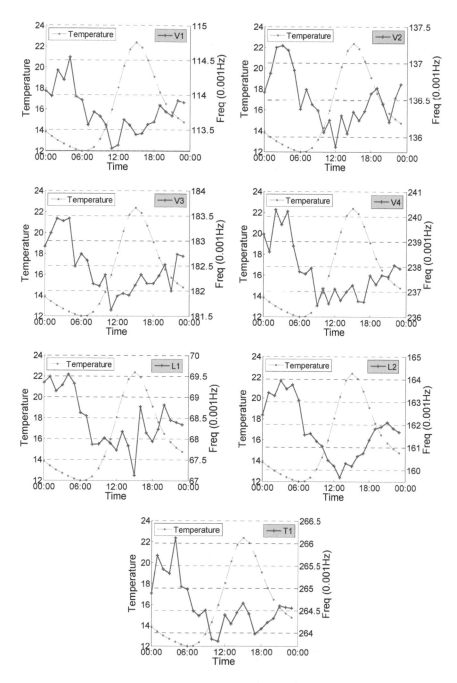

Figure 7.33 Variation in frequencies on 17 January 2005

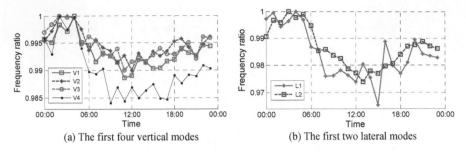

(a) The first four vertical modes   (b) The first two lateral modes

Figure 7.34 Variation percentage of frequencies on 17 January 2005

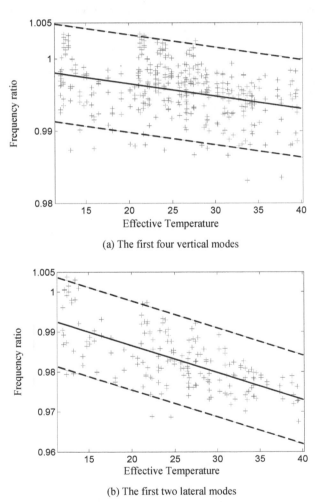

(a) The first four vertical modes

(b) The first two lateral modes

Figure 7.35 Normalized frequencies versus temperature

Similarly the first two lateral modal frequencies are plotted in Figure 7.35(b) against the effective temperature. With the same linear regression model, the slope (β_T) is calculated as -6.7×10^{-4}, much larger than half of θ_E of steel ($-3.6 \times 10^{-4}/^{\circ}C$). This might be because the lateral modal frequencies are also affected by the lateral bending rigidity (y-direction) of the towers. The bridge towers are made of concrete, whose temperature coefficient of modulus ($\theta_E = -3.0 \times 10^{-3}/^{\circ}C$) is much higher than that of the steel.

7.6 SUMMARY

Calculation of temperature distribution is the first step to study the thermal effects on bridges. For a long-span bridge, estimation of the temperature distribution is a major challenge for practitioners due to the complex configuration of the bridge. In this chapter, the heat-transfer analysis and the associated thermodynamic models have been described and applied to the Tsing Ma Bridge. The temperature distribution of each component was calculated and applied to a global FE model of the bridge to calculate the temperature-induced displacement and strain/stress in each component. The temperature distribution and the temperature-induced displacement and strain results are then compared with their field monitoring counterparts. Conclusions can be drawn as follows:

1. The numerical models can successfully predict the structural temperature of various components at different times. The assumption of a constant temperature distribution along the longitudinal direction of the deck and the towers is numerically accurate and computationally effective.
2. In general, the temperature-induced displacements calculated for the deck, main cables, and tower in the longitudinal (x) and vertical (z) directions show good agreement with those of the field measurement data. However, the calculated lateral (y) displacements are usually small and do not agree well with the measurement data, possibly because of other loading effects and measurement noise.
3. At the measured positions, the vertical displacements are much larger than the lateral ones. The longitudinal displacements depend on the section location and show a good linear relation with the free distance to the Ma Wan abutment.
4. The calculated temperature-induced strains and stresses generally agree well with the measurement.
5. The variation in the vertical frequencies of the bridge is mainly caused by the change in modulus of steel under different temperature conditions, although the magnitude is small.

The temperature monitoring provides an approach to quantify the temperature effects on the deformation, internal forces, and vibration properties of the bridge. The results have a few benefits for the SHM:

1. The results can be used to evaluate the safety and serviceability of the bridge by comparing them with the design criteria.
2. The results can be regarded as a set of reference data for elimination of temperature effects in bridge condition evaluation. For example, in assessing

the effects of traffic loading, the temperature effect should be removed from the original measured displacements.

3. The quantitative effects of temperature on structural responses can be employed as damage indicators. For example, if the Tsing Yi abutment is completely or partially restrained from longitudinal movement, then the measured displacements and vibration properties will show different patterns.

Many temperature sensors have been installed on the Tsing Ma Bridge. The analytical study shows that the layout of these sensors provides sufficient boundary conditions for the numerical analysis to be carried out. Comparison between the numerical results and the field monitoring data demonstrates that the quality of the measurement data is very high. After more than ten years of operation, the Tsing Ma Bridge SHM system is still quite reliable and effective.

7.7 REFERENCES

AASHTO, 1989, *AASHTO Guide Specifications: Thermal Effects in Concrete Bridge Superstructures*, (Washington, D.C.: American Association of State Highway and Transportation Officials).

Adams, R.D., Cawley, P., Pye, C.J. and Stone, B.J., 1978, A vibration technique for non-destructively assessing the integrity of structures. *Journal of Mechanical Engineering Science*, **20(2)**, pp. 93-100.

ANSYS, 2004, *Thermal Analysis Guide*, Version 8.1, (Southpointe: ANSYS Inc.).

Berlage, H.P., 1928, Zur theorie der beleuchtung einer horizontalen flache durch tageslicht. *Meteorologische Zeitschrift*, **45(5)**, pp. 174-180.

Blevins, R.D., 1979, *Formulas for Natural Frequency and Mode Shape*, (New York: Van Nostrand Reinhold).

Branco, F.A., 1986, Thermal effects on composite box girder bridges during construction. In *Proceedings of 2nd International Conference on Short and Medium Span Bridges*, (Montreal: Canadian Society for Civil Engineering), pp. 215-226.

Branco, F.A. and Mendes, P.A., 1993, Thermal actions for concrete bridge design. *Journal of Structural Engineering, ASCE*, **119(8)**, pp. 2313-2331.

Branco, F.A., Mendes, P.A. and Mirambell, E., 1992, Heat of hydration effects in concrete structures. *ACI Materials Journal*, **89(2)**, pp. 139-145.

Breccolotti, M., Franceschini, G. and Materazzi, A.L., 2004, Sensitivity of dynamic methods for damage detection in structural concrete bridges. *Shock and Vibration*, **11**, pp. 383-394.

Brockenbrough, R.L. and Merritt, F.S., 1994, *Structural Steel Designer's Handbook*, 2nd ed., (New York: McGraw-Hill).

Capps, M.W.R., 1968, *The Thermal Behavior of the Beachley Viaduct/Wye Bridge*. Report LR 234, (London: Road Research Laboratory, Ministry of Transport).

CEB-FIP, 1993, *Model Code 1990*, (London: Thomas Telford).

Cheung, M.S., Tadros, G.S., Brown, T., Dilger, W.H., Ghali, A. and Lau, D.T., 1997, Field monitoring and research on performance of the Confederation Bridge. *Canadian Journal of Civil Engineering*, **24(6)**, pp. 951-962.

Clough, R.W. and Penzien, J., 1993, *Dynamics of Structure*, 2nd ed., (New York: McGraw-Hill).

Cornwell, P., Farrar, C.R., Doebling, S.W. and Sohn, H., 1999, Environmental variability of modal properties. *Experimental Techniques*, **23(6)**, pp. 45-48.

Deng, Y., Ding, Y.L. and Li, A.Q., 2010, Structural condition assessment of long-span suspension bridges using long-term monitoring data. *Earthquake Engineering and Engineering Vibration*, **9(1)**, pp. 123-131.

Desjardins, S.L., Londoño, N.A., Lau, D.T. and Khoo, H., 2006, Real-time data processing, analysis and visualization for structural monitoring of the Confederation Bridge. *Advances in Structural Engineering*, **9(1)**, pp. 141-157.

Doebling, S.W., Farrar, C.R., Prime, M.B. and Shevitz, D.W., 1996, *Damage Identification and Health Monitoring of Structural and Mechanical Systems from Changes in their Vibration Characteristics: A Literature Review*. Report LA-13070-MS, (Los Alamos: Los Alamos National Laboratory).

Elbadry, M.M. and Ghali, A., 1983, Temperature variation in concrete bridges. *Journal of Structural Engineering, ASCE*, **109(10)**, pp. 2355-2374.

Emerson, M., 1979, *Bridge Temperatures for Settling Bearings and Expansion Joints*, TRRL Supplementary Report 479, (Berkshire, England: Department of Transport).

Farrar, C.R., Baker, W.E., Bell, T.M., Cone, K.M., Darling, T.W., Duffey, T.A., Eklund, A. and Migliori, A., 1994, *Dynamic Characterization and Damage Detection in the I-40 Bridge Over the Rio Grande*. Report LA-12767-MS, (Los Alamos: Los Alamos National Laboratory).

Froli, M., Hariga, N. and Nati, G., 1996, Longitudinal thermal behavior of a concrete box girder bridge. *Structural Engineering International*, **6(4)**, pp. 237-242.

Fu, H.C., Ng, S.F. and Cheung, M.S., 1990, Thermal behavior of composite bridges. *Journal of Structural Engineering, ASCE*, **116(12)**, pp. 3302-3323.

Fu, Y. and DeWolf, J.T., 2001, Monitoring and analysis of a bridge with partially restrained bearings. *Journal of Bridge Engineering, ASCE*, **6(1)**, pp. 23-29.

HKPU, 2007, *Establishment of Bridge Rating System for Tsing Ma Bridge: Statistical Modelling and Simulation of Predominating Temperature Loading Effects*. Report No. 4, (Hong Kong: Department of Civil and Structural Engineering, The Hong Kong Polytechnic University).

Kottegoda, N.T. and Rosso, R., 1997, *Statistics, Probability, and Reliability for Civil and Environmental Engineers*, (New York: McGraw-Hill).

Lee, G.C., Shih, T.S. and Chang, K.C., 1988, Mechanical properties of concrete at low temperature. *Journal of Cold Regions Engineering, ASCE*, **2(1)**, pp. 13-24.

Li, D.N., Maes, M.A. and Dilger, W.H., 2008, Evaluation of temperature data of Confederation bridge- thermal loading and movement at expansion joint. In *Proceedings of the 2008 Structures Congress*, Vancouver, BC, Canada, edited by Don, A., Carlos, V., David, H. and Marc, H., (Reston, Virginia: ASCE), pp. 1-10.

Lienhard IV, J.H. and Lienhard V, J.H., 2001, *A Heat Transfer Textbook*, (Massachusetts: Phlogiston Press).

Liu, C.Y. and DeWolf, J.T., 2007, Effect of temperature on modal variability of a curved concrete bridge under ambient loads. *Journal Structural Engineering, ASCE*, **133(12)**, pp. 1742-1751.

Macdonald, J.H.G. and Daniell, W.E., 2005, Variation of modal parameters of a cable-stayed bridge identified from ambient vibration measurements and FE modelling. *Engineering Structures*, **27(13)**, pp. 1916-1930.

Moorty, S. and Roeder, C.W., 1992, Temperature-dependent bridge movements. *Journal of Structural Engineering, ASCE*, **118(4)**, pp. 1090-1105.

Nayeri, R.D., Masri, S.F., Ghanem, R.G. and Nigbor, R.L., 2008, A novel approach for the structural identification and monitoring of a full-scale 17-story building based on ambient vibration measurements. *Smart Materials and Structures*, **17(2)**, pp. 1-19.

Ni, Y.Q., Ko, J.M., Hua, X.G. and Zhou, H.F., 2007, Variability of measured modal frequencies of a cable-stayed bridge under different wind conditions. *Smart Structures and Systems*, **3(3)**, pp. 341-356.

Peeters, B. and De Roeck, G., 2001, One-year monitoring of the Z24-Bridge: environmental effects versus damage events. *Earthquake Engineering and Structural Dynamics*, **30**, pp. 149-171.

Priestley, M.J.N., 1978, Design of concrete bridges for thermal gradients. *ACI Journal*, **75(5)**, pp. 209-217.

Riding, K.A., Poole, J.L., Schindler, A.K., Juenger, M.CG and Folliard, K.J., 2007, Temperature boundary condition models for concrete bridge members. *ACI Materials Journal*, **104(4)**, pp. 379-387.

Rohsenow, W.M., 1988, *Handbook of Heat Transfer Applications*, (New York: McGraw-Hill).

Song, W. and Dyke, S.J., 2006, Ambient vibration based modal identification of the Emerson bridge considering temperature effects. In *Proceedings of the 4th World Conference on Structural Control and Monitoring*, San Diego, USA, edited by Benzoni, G., (Los Angeles, California: International Association for Structural Control and Monitoring).

Van Overschee, P. and De Moor, B., 1996, *Subspace Identification for Linear Systems: Theory-Implementation-Applications*, (Boston, USA: Kluwer Academic Publishers).

Wang, C., 1994, *Temperature Effects in Hardening Concrete*, Ph.D. thesis, (Calgary, Canada: University of Calgary).

Whitaker, S., 1977, *Fundamental Principles of Heat Transfer*, (New York: Pergamon Press).

Xia, Y., Hao, H., Zanardo, G. and Deeks, A.J., 2006, Long term vibration monitoring of a RC Slab: temperature and humidity effect. *Engineering Structures*, **28(3)**, pp. 441-452.

Xu, Y.L., Chen, B., Ng, C.L., Wong, K.Y. and Chan, W.Y., 2010, Monitoring temperature effect on a long suspension bridge. *Journal of Structural Control and Health Monitoring*, **17**, pp. 632-653.

Zuk, W., 1965, Thermal behaviour of composite bridges-insulated and uninsulated. *Highway Research Record*, **76**, pp. 231-253.

7.8 NOTATIONS

a, b, ε_x	Regression coefficients: slope, intercept, error.
c	Specific heat of the solid (J/kg/°C).

E, G	Modulus of elasticity, shear modulus of elasticity.
f_n	Vibration frequency.
h_c, h_r	Heat transfer coefficient of convection and radiation.
I, I_d, I_i, I_r	Total solar radiation, direct solar radiation, diffuse solar radiation, and reflected solar radiation on a surface.
I_{d0}	Energy reaching the earth's surface by direct radiation.
I_{sc}	Solar constant.
K_T	Dimensionless transmittance factor.
k	Thermal conductivity (W/m/°C).
k_a	Ratio of the atmospheric pressure to the pressure at sea level.
L_k, L_s	Length of an overhanging cantilever slab, length of the shadow of the slab.
l, h	Length and height of a simply-supported uniform beam.
m	Air mass factor.
N	Number of the day of the year, counting from January 1.
n_x, n_y	Direction cosines of the unit outward vector normal to the boundary surfaces.
$\vec{n}$	Vector perpendicular to the surface.
q	Boundary heat input or loss per unit area (W/m²), heat flow of a surface considering solar radiation.
q_c, q_r, q_s	Convection, thermal irradiation and solar radiation.
r_e	Reflected coefficient of the ground.
T	Temperature of the solid.
T^*	Equivalent air temperature.
T_a, T_s	Air temperature and the bridge surface temperature.
t	Time.
t_u	Turbidity factor indicating the attenuation of radiation.
x, y, z	Cartesian coordinates.
α	Absorptivity coefficient of the surface material.
β	Tilt angle from the horizontal.
$\beta_T, \beta_0, \varepsilon_f$	Regression coefficients: slope, intercept, regression error.
γ_s	Azimuth angle.
δ	An increment in the corresponding parameter.
ε	Emissivity coefficient of the bridge surface.
θ	Incidence angle of sun rays.
θ_T, θ_E	Thermal coefficient of linear expansion of the material, the temperature coefficient of modulus.
ρ	Density.
ψ	Solar altitude.
ν	Poisson's ratio.

CHAPTER 8

Monitoring of Wind Effects

8.1 PREVIEW

When a long-span suspension bridge is built in a wind-prone region, wind environment around and wind effects on the bridge must be monitored to ensure the functionality and safety of the bridge in high winds. Wind environment for a long-span suspension bridge is often described in terms of boundary layer wind characteristics which are directly related to wind loads acting on the bridge. The wind effects which must be considered for a long-span suspension bridge are mainly due to buffeting excitation caused by fluctuating forces induced by turbulence, flutter instabilities of several types that occur at very high wind speeds as a result of the dominance of self-excited aerodynamic forces, and vortex shedding excitation which usually occurs in low wind speed and low turbulence conditions.

This chapter first describes sensors and sensing systems used for wind monitoring of long-span suspension bridges. Boundary layer wind characteristics are then introduced in terms of mean wind speed, mean wind direction, turbulence components, turbulence intensities, integral scales, wind spectra and other parameters. A non-stationary wind model is also presented to deal with wind data recorded during typhoons over a complex terrain. Aerodynamic analyses of buffeting-induced bridge vibration, bridge flutter instability, and vortex shedding-induced bridge vibration are concisely introduced. Finally, the Tsing Ma suspension bridge is taken as an example to illustrate the monitoring of wind environment around and wind effects on the bridge.

8.2 WIND MONITORING SYSTEM

Similar to other loading effect monitoring, wind monitoring of a long-span suspension bridge also starts with the selection and arrangement of a sensing system and the associated data acquisition system. A variety of sensors is applicable for wind monitoring of long-span suspension bridges. The traditional sensors widely used for wind monitoring of a suspension bridge include anemometers for measuring wind velocity, accelerometers for acceleration responses, and strain gauges for strain responses. Occasionally, pressure transducers are installed to measure wind pressures and pressure distribution over a particular part of the bridge envelope, and displacement transducers and tiltmeters are used to measure displacement responses. In recent years, Doppler radar, GPS drop-sonde, and Doppler sodar have become powerful devices for measuring boundary layer wind profiles whilst Real-Time Kinematic GPS has been used to measure the total displacement of bridges and buildings during high winds, e.g., Tamura *et al.* (1999), Nakamura (2000), and Kijewski-Correa and Kochly (2007).

Simultaneous measurements of wind speed and direction with bridge responses are imperative for the assessment of the variations of structural dynamic characteristics with wind speed and for the determination of the relationship between wind and bridge responses. The wind records from nearby observatory stations or airports in many cases cannot be directly adopted for this purpose because the bridge is often located in a unique wind environment and these wind records may not embed the effects of bridge surroundings on boundary layer wind characteristics at the bridge site. Propeller and ultrasonic anemometers are the most commonly used instruments for measuring wind velocity on the site. A propeller anemometer, directly recording wind speed and direction, is convenient and relatively reliable and sustainable in harsh environments but it is not sensitive enough to capture the nature of turbulent winds at higher frequencies. This is particularly true in situations when the wind speed or direction changes rapidly. An ultrasonic anemometer, measuring wind speed through its two or three orthogonal components, is quite sensitive but not sustainable in harsh environments. The accuracy requirement for wind velocity measurement of the anemometers must be maintained under heavy rains, that is, no occurrence of spikes during heavy rainstorms. For a long-span suspension bridge, the anemometers are often installed at a few bridge deck sections on both sides and along the height of the towers so that not only wind characteristics at key points can be measured but also the correlation of wind velocity in both horizontal and vertical directions can be determined. The positions of the anemometers must be selected so as to minimize the effect of the adjacent edges of the bridge deck and towers on the airflow towards them. To meet this requirement, anemometer booms or masts are often needed so that the anemometer can be installed at a few metres from the bridge edges. The boom or mast must be equipped with a retrievable device to enable retracting of the anemometer in an unrestricted and safe manner for inspection and maintenance.

For measuring wind-induced acceleration responses of a long-span suspension bridge, piezoelectric accelerometers are sometimes used. One drawback of this type of accelerometers is that it cannot accurately detect low-frequency vibration. In cases where the low-frequency performance of a long-span suspension bridge in high winds needs to be considered, servo-type accelerometers can be adopted since they are capable of recording acceleration at frequency as low as DC. The selection of the number and location of the anemometers must consider not only the measurement of wind-induced acceleration responses but also the measurement of other loading effects such as traffic-induced acceleration responses and the identification of bridge dynamic characteristics. Other requirements of accelerometers for wind monitoring are similar to those for railway monitoring.

The dynamic displacement response of the bridge can be measured using displacement transducers or obtained by integrating the acceleration records twice with time. However, the absolute static or quasi-static displacement responses of the bridge deck and towers caused by the long-period component in the wind cannot be captured using these devices. To circumvent the difficulties, the GPS technique is preferred. The GPS technique provides a powerful ability to track dynamic as well as static displacements of long-span suspension bridges in high winds (Ashkenazi and Roberts, 1997; Fujino *et al.*, 2000; Miyata *et al.*, 2002). The measured total displacement is imperative for monitoring the integrity and safety of the bridge in conjunction with the disaster prevention system.

The requirements of strain gauges and measurements for wind-induced strain responses are similar to those for traffic-induced strain responses of the bridge. Wind-induced strain responses are smaller than railway-induced strain responses in most cases.

8.3 MONITORING OF BOUNDARY LAYER WIND ENVIRONMENT

The Earth's surface exerts on the moving air a horizontal drag force, which reduces wind speed and introduces turbulence near the ground. The wind region affected by the roughness of the terrain is referred to as the atmospheric boundary layer. The depth of the boundary layer may range from a few hundred metres to several kilometres, depending upon wind intensity, roughness of terrain, and angle of latitude (Holmes, 2007). Since long-span suspension bridges are placed on the ground, it is important to deal only with boundary layer winds, that is, winds near the ground surface.

Within the boundary layer, the wind velocity can be decomposed as a mean component and three perpendicular turbulence components, which cause the mean and dynamic responses of the bridge, respectively. Long-span suspension bridges are often located in a unique wind environment. One major objective of SHM of a long-span suspension bridge under wind action is to determine mean wind speed levels and turbulent wind characteristics by analysing wind data recorded by the anemometers. In this section, the traditional wind model for analysis of stationary wind data to obtain boundary layer wind characteristics is touched upon and followed by the presentation of a relatively new wind model for handling both stationary and non-stationary wind data.

8.3.1 Stationary Wind Model

In the analysis of wind characteristics using the traditional stationary wind model, the mean wind direction during a given time is first determined. The measured wind components are then resolved with respect to the mean wind direction to find the longitudinal wind speed $U(t)$ in the mean wind direction, the lateral turbulence component $v(t)$ which is horizontal and normal to the mean wind direction, and the vertical turbulence component $w(t)$ which is upward and perpendicular to both $U(t)$ and $v(t)$. Procedures for resolving the wind components from the measured wind data can be found in Xu and Zhan (2001) and Xu and Chen (2004) for propeller anemometers and ultrasonic anemometers, respectively.

In current practice, boundary layer longitudinal wind speed $U(t)$ at a given height is assumed as an ergodic random process consisting of a constant mean wind speed $\bar{U}$ and a longitudinal turbulence component $u(t)$ (ESDU, 2001).

$$U(t) = \bar{U} + u(t) \tag{8.1}$$

The mean wind speed $\bar{U}$ can be estimated over a designated time duration T by

$$\bar{U}_T = \frac{1}{T}\int_0^T U(t)\,dt \tag{8.2}$$

The typical value used in practice for time duration T is ten minutes or one hour, leading to the so-called 10-minute or hourly mean wind speed. The mean values of $v(t)$ and $w(t)$ are assumed to be zero, i.e., $\bar{v} = \bar{w} = 0$. The boundary layer wind characteristics, including the turbulence variance, turbulence intensity, wind spectrum, probability density function, integral length scale, and root coherence function, are then calculated by assuming $u(t)$, $v(t)$ and $w(t)$ as zero-mean stationary random processes. The formulas for calculating these wind characteristics by using wind data recorded by the anemometers installed in the bridge are summarized in Table 8.1. In addition to the wind characteristics listed in Table 8.1, the mean wind speed profile, turbulence intensity profile, cross spectra and coherence functions are also important for determining wind effects on a suspension bridge. For a properly designed wind monitoring system, these wind characteristics could be estimated by analysing wind data recorded by the monitoring system.

Table 8.1 Definitions of boundary layer wind characteristics in two wind models.

Parameter	Stationary	Non-stationary
Turbulent wind components	$u(t) = U(t) - \bar{U}$ $v(t)$ $w(t)$	$u^*(t) = U(t) - \bar{U}^*(t)$ $v^*(t) = v(t) - \bar{v}^*(t)$ $w^*(t) = w(t) - \bar{w}^*(t)$
Mean wind speed	$\bar{U} = \frac{1}{T}\int_0^T U(t)\,dt$	$\bar{U}_T^* = \frac{1}{T}\int_0^T \bar{U}^*(t)$
Variance	$\sigma_\alpha^2 = \frac{1}{T}\int_0^T \alpha^2(t)\,dt$	$\sigma_{\alpha^*}^2 = \frac{1}{T}\int_0^T \alpha^{*2}(t)\,dt$
Turbulence intensity	$I_a = \dfrac{\sigma_a}{\bar{U}}$	$I_a^* = \dfrac{\sigma_{a^*}}{\bar{U}_T}$
Probability density	$p(u) = \dfrac{1}{\sqrt{2\pi}\sigma_u}e^{-\frac{u^2}{2\sigma_u^2}}$	$p(u^*) = \dfrac{1}{\sqrt{2\pi}\sigma_{u^*}}e^{-\frac{u^{*2}}{2\sigma_{u^*}^2}}$
Wind spectrum (longitudinal, von-Karman for instance)	$\dfrac{fS_u(f)}{\sigma_u^2} = \dfrac{4\dfrac{L_u}{\bar{U}}f}{\left[1+70.8\left(\dfrac{L_u}{\bar{U}}f\right)^2\right]^{\frac{5}{6}}}$	$\dfrac{fS_{u^*}(f)}{\sigma_{u^*}^2} = \dfrac{4\dfrac{L_{u^*}}{\bar{U}_T}f}{\left[1+70.8\left(\dfrac{L_{u^*}}{\bar{U}_T}f\right)^2\right]^{\frac{5}{6}}}$
Wind spectrum (lateral and vertical)	$\dfrac{fS_\beta(f)}{\sigma_\beta^2} = \dfrac{4\dfrac{L_\beta}{\bar{U}}\left[1+755\left(\dfrac{L_\beta f}{\bar{U}}\right)\right]}{\left[1+283\left(\dfrac{L_\beta}{\bar{U}}\right)^2\right]^{\frac{11}{6}}}$	$\dfrac{fS_{\beta^*}(f)}{\sigma_{\beta^*}^2} = \dfrac{4\dfrac{L_{\beta^*}}{\bar{U}_T}\left[1+755\left(\dfrac{L_{\beta^*}f}{\bar{U}_T}\right)\right]}{\left[1+283\left(\dfrac{L_{\beta^*}}{\bar{U}_T}\right)^2\right]^{\frac{11}{6}}}$
Root coherence	$coh_{\alpha\alpha}(\Delta, f)$	$coh_{\alpha^*\alpha^*}(\Delta, f)$

Note: $\alpha = u, v, w$; $\beta = v, w$; $\alpha^* = u^*, v^*, w^*$; and $\beta^* = v^*, w^*$

8.3.2 Non-stationary Wind Model

The boundary layer wind measured during typhoons over complex terrain may not comply with the assumption of an ergodic or stationary random process (Kareem, 2008). The probability distribution of the longitudinal turbulence component obtained by subtracting the constant mean wind speed from the longitudinal wind speed may not follow the Gaussian distribution. Thus, when the stationary wind model is applied to wind data collected during typhoons, the accuracy of resulting wind characteristics is doubtful. The turbulence intensity may be overestimated (Schroeder *et al.*, 1998). To overcome this problem, a non-stationary wind model was proposed, which can be applied to the wind data to obtain proper wind characteristics (Xu and Chen, 2004).

In the non-stationary wind model, the longitudinal wind speed is modelled as a deterministic time-varying mean wind speed plus a zero-mean stationary random process for a longitudinal turbulent wind component.

$$U(t) = \bar{U}^*(t) + u^*(t) \tag{8.3}$$

where $\bar{U}^*(t)$ is the deterministic time-varying mean wind speed reflecting the temporal trend of wind speed; $u^*(t)$ is the longitudinal turbulent wind component that can be modelled as a zero-mean stationary process. The asterisk denotes that the quantity is obtained from the non-stationary wind model. Clearly, if the total longitudinal wind speed $U(t)$ is a strictly stationary random process, the time-varying mean wind speed $\bar{U}^*(t)$ will be reduced to a constant mean wind speed $\bar{U}$ as defined in Equation (8.1) for the traditional wind model.

Furthermore, when wind velocity is non-stationary the lateral and vertical turbulence components may also be subject to a time-varying trend around the zero level. The non-stationary wind model is thus extended to the lateral and vertical turbulent wind components as

$$v(t) = \bar{v}^*(t) + v^*(t) \tag{8.4}$$

$$w(t) = \bar{w}^*(t) + w^*(t) \tag{8.5}$$

where $v^*(t)$ and $w^*(t)$ are respectively the lateral and vertical turbulent wind components after time-varying trends $\bar{v}^*(t)$ and $\bar{w}^*(t)$ having been removed from the original turbulent wind components of $v(t)$ and $w(t)$. Since the time-varying mean wind speed is introduced in the non-stationary wind speed model, the definitions of turbulent wind characteristics should be revised accordingly. The formulas for calculating turbulent wind characteristics by using the non-stationary wind model and wind data recorded by the anemometers installed in the bridge are summarized in Table 8.1 and compared with those of the traditional stationary wind model.

8.3.3 Time Varying Mean Wind Speed

One of the key issues for using the above non-stationary wind model is how to extract an appropriate time-varying wind speed from a given wind time history recorded in the field. To circumvent this problem, a relatively new data processing method, termed the empirical mode decomposition (EMD), is employed (Huang *et al.*, 1998). The EMD can decompose any complicated data set into a finite but often small number of intrinsic mode functions (IMF). This decomposition procedure is adaptive because it is based on the local characteristic time scale of the data.

Let $U(t)$ represent a longitudinal wind speed time history of one-hour duration. It can be decomposed by EMD as the sum of a series of IMFs plus a final residue.

$$U(t) = \sum_{j=1}^{N} c_j(t) + r_u(t)_N \tag{8.6}$$

where N is the number of IMF components $c_j(t)$; and $r_u(t)_N$ is the final residue. If there is a trend in the wind speed $U(t)$, the finial residue $r_u(t)_N$ then represents that trend and it can be defined as the time-varying mean wind speed $\bar{U}^*(t)$. Similarly, the time-varying trend of the lateral and vertical turbulent wind components is defined as the residue obtained after applying EMD to $v(t)$ and $w(t)$.

$$\bar{v}^*(t) = r_v(t)_N \quad \text{and} \quad \bar{w}^*(t) = r_w(t)_N \tag{8.7}$$

Clearly, EMD naturally extracts the time-varying mean wind speed (trend) from the recorded wind speed time history without any prior information.

8.4 WIND EFFECTS ON LONG-SPAN SUSPENSION BRIDGES

Wind effects on long-span suspension bridges are mainly due to static forces induced by mean winds, buffeting excitation caused by fluctuating forces induced by turbulence, flutter instabilities of several types that occur at very high wind speeds as a result of the dominance of self-excited aerodynamic forces, and vortex shedding excitation which usually occurs in low wind speed and low turbulence conditions. The analytical backgrounds of bridge aerodynamics are introduced in this section to serve two purposes: (1) use the measurement data from an SHM system to verify the existing analytical methods or to develop new analytical methods if necessary; and (2) use the verified analytical methods to assess fatigue damage, strength and integrity of the bridge under action of extreme wind. In some cases, the number and location of sensors in the SHM system may have to be refined based on the assessment results.

8.4.1 Buffeting Analysis of Long-span Suspension Bridges

When a long-span suspension bridge is immersed in a wind field, the bridge will be subjected to static and dynamic wind forces caused by mean and fluctuating wind speeds, respectively. Buffeting action on a long-span suspension bridge is a random vibration caused by fluctuating winds that appear within a wide range of wind speeds. In wind resistance design of a long-span suspension bridge, the buffeting responses are normally dominant to determine the size of structural members. In addition to buffeting action, the self-excited forces induced by wind-structure interaction are also important for predicting the buffeting response of long-span suspension bridges, because the additional energy injected into the oscillating structure by self-excited forces increases the magnitude of vibrations. To model the action of buffeting wind load, the buffeting forces resulting from turbulent wind and the self-excited forces due to the wind-bridge interaction should be taken into account. The buffeting response prediction can be performed in both the frequency domain (Davenport, 1962; Scanlan, 1978) and the time domain (Bucher and Lin, 1988; Xiang *et al.*, 1995; Chen *et al.*, 2000). The theoretical backgrounds of the frequency domain buffeting response prediction (Ding and Xu, 2004a) are introduced in this section.

In the 3-D FE based buffeting analysis, the governing equation for the buffeting response of a long-span suspension bridge in terms of a displacement vector X can be expressed as

$$[M]\{\ddot{X}\}+[C]\{\dot{X}\}+K\{X\}=\{f_{se}\}+\{f_b\}+\{f_s\} \tag{8.8}$$

where $[M]$, $[C]$, and $[K]$ are the mass, damping, and stiffness matrices of the bridge, respectively; $\{X\}$, $\{\dot{X}\}$, $\{\ddot{X}\}$ are the nodal displacement, velocity, and acceleration vector, respectively; $\{f\}$ indicates the nodal equivalent force vector, and the subscripts *se*, *b*, and *s* represent the self-excited force, buffeting force, and mean wind force components, respectively. The mean deformation of the bridge caused by the static wind forces can be readily determined using the mean wind speeds measured from the anemometers installed in the bridge and the other parameters determined by wind tunnel tests through an iterative solution scheme (Zhang *et al.*, 2002).

The self-excited vertical and lateral forces and self-excited moment acting on the bridge deck per unit length are often expressed in Scanlan's format as follows:

$$L_{se}(t)=\frac{1}{2}\rho\bar{U}^2(2B)(kH_1^*\frac{\dot{h}}{\bar{U}}+kH_2^*\frac{B\dot{\alpha}}{\bar{U}}+k^2H_3^*\alpha+k^2H_4^*\frac{h}{B}+kH_5^*\frac{\dot{p}}{\bar{U}}+k^2H_6^*\frac{p}{B}) \tag{8.9}$$

$$D_{se}(t)=\frac{1}{2}\rho\bar{U}^2(2B)(kP_1^*\frac{\dot{p}}{\bar{U}}+kP_2^*\frac{B\dot{\alpha}}{\bar{U}}+k^2P_3^*\alpha+k^2P_4^*\frac{p}{B}+kP_5^*\frac{\dot{h}}{\bar{U}}+k^2P_6^*\frac{h}{B}) \tag{8.10}$$

$$M_{se}(t)=\frac{1}{2}\rho\bar{U}^2(2B^2)(kA_1^*\frac{\dot{h}}{\bar{U}}+kA_2^*\frac{B\dot{\alpha}}{\bar{U}}+k^2A_3^*\alpha+k^2A_4^*\frac{h}{B}+kA_5^*\frac{\dot{p}}{\bar{U}}+k^2A_6^*\frac{p}{B}) \tag{8.11}$$

where ρ is the air density; $B = 2b$ is the bridge deck width; $k = \omega B/\bar{U}$ is the reduced frequency; ω is the circular frequency of vibration; h, p, and α are the vertical, lateral, and torsional displacements of the bridge deck, respectively, with respect to the centre of the deck cross section; the over-dot denotes the partial differentiation with respect to time t; and H_i^*, P_i^*, A_i^* ($i = 1\sim6$) are the non-dimensional flutter derivatives. In the 3-D FE based buffeting analysis, the distributed self-excited forces acting on one element of the bridge deck are converted into the equivalent forces at the two nodes of the element in the local coordinate. The equivalent nodal forces are then transformed into the global coordinate and assembled for all the elements in a complex domain, leading to

$$f_{se}=\omega^2\left[\mathbf{A}_{se}\right]\{X\} \tag{8.12}$$

where $[\mathbf{A}_{se}]$ is a complex self-excited force matrix of the bridge. Since the self-excited forces are non-conservative, $[\mathbf{A}_{se}]$ is generally unsymmetrical and a function of the reduced frequency. The details of $[\mathbf{A}_{se}]$ will be given in section 8.4.2. The vertical and lateral buffeting forces and buffeting moment acting on the bridge deck per unit length due to wind fluctuations are given as

$$\{L_b\}=\frac{1}{2}\rho\bar{U}^2B\left[2C_L\chi_{Lu}\frac{u}{\bar{U}}+(C_L'+C_D)\chi_{Lw}\frac{w}{\bar{U}}\right] \tag{8.13}$$

$$\{D_b\} = \frac{1}{2}\rho\bar{U}^2B\left[2C_D\chi_{Du}\frac{u}{\bar{U}} + C_D'\chi_{Dw}\frac{w}{\bar{U}}\right] \tag{8.14}$$

$$\{M_b\} = \frac{1}{2}\rho\bar{U}^2B^2\left[2C_M\chi_{Mu}\frac{u}{\bar{U}} + C_M'\chi_{Mw}\frac{w}{\bar{U}}\right] \tag{8.15}$$

where C_L, C_D, and C_M are the lift, drag, and moment coefficients referring to deck width B, respectively; $C_L' = dC_L/d\alpha$, $C_D' = dC_D/d\alpha$, and $C_M' = dC_M/d\alpha$; χ_{Lu}, χ_{Lw}, χ_{Du}, χ_{Dw}, χ_{Mu}, χ_{Mw} are the aerodynamic admittance functions, which are functions of the reduced frequency and dependent on the geometrical configuration of the cross section of the bridge deck.

If the aerodynamic admittance functions are taken as units, the buffeting forces aforementioned can be expressed as follows:

$$\{P_b\} = 0.5\rho\bar{U}(C_{bu}\{u\} + C_{bw}\{w\}) \tag{8.16}$$

where

$$\mathbf{P}_b = \begin{Bmatrix} L_b \\ D_b \\ M_b \end{Bmatrix}, \quad C_{bu} = B\begin{Bmatrix} 2C_L \\ 2C_D \\ 2BC_M \end{Bmatrix}, \quad C_{bw} = B\begin{Bmatrix} C_L' + C_D \\ C_D' \\ BC_M' \end{Bmatrix} \tag{8.17}$$

When the element is small enough, it can be assumed that the longitudinal and vertical wind fluctuations are distributed linearly on the element:

$$\{u\} = \begin{bmatrix} 1-\dfrac{x}{L} & \dfrac{x}{L} \end{bmatrix}\begin{Bmatrix} u_1 \\ u_2 \end{Bmatrix} = [\mathbf{A}]\{u^e\} \tag{8.18}$$

$$\{w\} = \begin{bmatrix} 1-\dfrac{x}{L} & \dfrac{x}{L} \end{bmatrix}\begin{Bmatrix} w_1 \\ w_2 \end{Bmatrix} = [\mathbf{A}]\{w^e\} \tag{8.19}$$

where x and L are the axial location and the length of the element, respectively; and subscripts 1 and 2 indicate the two ends of the element.

The consistent buffeting forces at the element ends in the local coordinate system can be obtained by the following definite integral:

$$f_b^e = \int_L \mathbf{B}^T p_b \, dx = 0.5\rho\bar{U}(\int_L\left[\mathbf{B}^T C_{bu}\mathbf{A}\right]dx\{u^e\} + \int_L\left[\mathbf{B}^T C_{bw}\mathbf{A}\right]dx\{w^e\})$$

$$= 0.5\rho\bar{U}(\left[\mathbf{A}_{bu}^e\right]\{u^e\} + \left[\mathbf{A}_{bw}^e\right]\{w^e\}) \tag{8.20}$$

where $\left[\mathbf{A}_{bu}^e\right]$ and $\left[\mathbf{A}_{bw}^e\right]$ are the buffeting force matrices of the element corresponding to the longitudinal and vertical wind fluctuations, respectively; and $[\mathbf{B}]$ is the matrix of interpolated functions

$$[\mathbf{B}] = \begin{bmatrix} 0 & -N_1 & 0 & 0 & 0 & -N_3 & 0 & -N_2 & 0 & 0 & 0 & N_4 \\ 0 & 0 & -N_1 & 0 & N_3 & 0 & 0 & 0 & -N_2 & 0 & -N_4 & 0 \\ 0 & 0 & 0 & -N_5 & 0 & 0 & 0 & 0 & 0 & -N_6 & 0 & 0 \end{bmatrix} \tag{8.21}$$

where $N_1 = 1 - 3\left(\dfrac{x}{L}\right)^2 + 2\left(\dfrac{x}{L}\right)^3$; $N_2 = 3\left(\dfrac{x}{L}\right)^2 - 2\left(\dfrac{x}{L}\right)^3$; $N_3 = x\left(1-\dfrac{x}{L}\right)^2$;

$N_4 = \dfrac{x^2}{L}\left(1-\dfrac{x}{L}\right)$; $N_5 = 1-\dfrac{x}{L}$; and $N_6 = \dfrac{x}{L}$

Matrices $\left[\mathbf{A}_{bu}^e\right]$ and $\left[\mathbf{A}_{bw}^e\right]$ can be derived as

$$\left[\mathbf{A}_{bu}^e\right]=\frac{-BL}{30}\begin{bmatrix}0 & 21C_L & 21C_D & 20BC_M & -3LC_D & 3LC_L & 0 & 9C_L & 9C_D & 10BC_M & 2LC_D & -2LC_L \\ 0 & 9C_L & 9C_D & 10BC_M & -2LC_D & 2LC_L & 0 & 21C_L & 21C_D & 20BC_M & 3LC_D & -3LC_L\end{bmatrix}^T$$

(8.22)

$$\left[\mathbf{A}_{bw}^e\right]=\frac{-BL}{60}\begin{bmatrix}0 & 21(C_L'+C_D) & 21C_D' & 20BC_M' & -3LC_D' & 3L(C_L'+C_D) \\ 0 & 9(C_L'+C_D) & 9C_D' & 10BC_M' & -2LC_D' & 2L(C_L'+C_D)\end{bmatrix}$$

$$\begin{matrix}0 & 9(C_L'+C_D) & 9C_D' & 10BC_M' & 2LC_D' & -2L(C_L'+C_D) \\ 0 & 21(C_L'+C_D) & 21C_D' & 20BC_M' & 3LC_D' & -3L(C_L'+C_D)\end{matrix}\Bigg]^T$$

(8.23)

The local nodal buffeting forces can be converted into the global coordinate system using the coordinate transformation matrix. As a result, the global nodal buffeting force vector can be obtained as

$$f_b = 0.5\rho\bar{U}(\left[\mathbf{A}_{bu}\right]\{u\}+\left[\mathbf{A}_{bw}\right]\{w\})$$

(8.24)

where $[\mathbf{A}_{bu}]$ and $[\mathbf{A}_{bw}]$ are the global buffeting force matrices; and $\{u\}$ and $\{w\}$ are the *r*-row nodal fluctuating wind vectors for the longitudinal and vertical components, respectively, where *r* is the number of nodes subjected to wind fluctuations.

Apart from the bridge deck, the buffeting forces also act on the bridge towers, cables, and other components. These buffeting forces can be determined using a similar way to the determination of the forces acting on the bridge deck. It is thus possible to have a buffeting analysis of the bridge as a whole rather than the bridge deck only.

Based on the preceding discussion, the governing equation of motion of the bridge as a whole can be written as

$$[M]\{\ddot{X}\}+[C]\{\dot{X}\}+[K]\{X\}-\omega^2\left[\mathbf{A}_{se}\right]\{X\}=\{f_b\}$$

(8.25)

The buffeting response of the bridge is dominant by the first *m* vibration modes and thus a linear transformation is introduced as

$$\{X\}=[\mathbf{\Phi}]\{q\}$$

(8.26)

where $[\mathbf{\Phi}]$ is an $n\times m$ matrix of mode shapes; $\{q\}$ is an $m\times 1$ vector of generalized coordinates; and *n* is the total number of DOFs of the bridge. By using the linear transformation, the equation of motion of the bridge can be expressed as

$$\{\ddot{q}\}+\left[\bar{C}\right]\{\dot{q}\}+[\mathbf{\Lambda}]\{q\}-\omega^2\left[\bar{\mathbf{A}}_{se}\right]\{q\}=\{Q_b\}$$

(8.27)

where $[\mathbf{\Lambda}]$ is the diagonal eigenvalue matrix, $\left[\bar{C}\right]=\left[\mathbf{\Phi}^T\right][C][\mathbf{\Phi}]$, $\left[\bar{\mathbf{A}}_{se}\right]=\left[\mathbf{\Phi}^T\right][\mathbf{A}_{se}][\mathbf{\Phi}]$, $\{Q_b\}=\left[\bar{\mathbf{A}}_{bu}\right]\{u\}+\left[\bar{\mathbf{A}}_{bw}\right]\{w\}$, and $\left[\bar{\mathbf{A}}_{bu}\right]=\left[\mathbf{\Phi}^T\right][\mathbf{A}_{bu}]$ and $\left[\bar{\mathbf{A}}_{bw}\right]=\left[\mathbf{\Phi}^T\right][\mathbf{A}_{bw}]$ are the generalized buffeting force vector.

According to the random vibration theory, the power spectral density (PSD) matrices of the generalized modal response $\{q\}$ and nodal displacement $\{X\}$ can be obtained as

$$S_q(\omega) = H^*(\omega)S_{Q_b}(\omega)H^T(\omega)$$

(8.28)

$$S_X(\omega) = \mathbf{\Phi}H^*(\omega)S_{Q_b}(\omega)H^T(\omega)\mathbf{\Phi}^T$$

(8.29)

where $H(\omega)$ is the transfer function matrix

$$H(\omega) = [-\omega^2(\mathbf{I}+\bar{\mathbf{A}}_{se})+i\omega\bar{C}+\mathbf{\Lambda}]^{-1}$$

(8.30)

I is a unit matrix; and superscript * and *T* denote the complex conjugate and transpose, respectively.

The PSD matrix of the generalized buffeting forces is given by

$$S_{Q_b}(\omega) = 0.25\rho^2\overline{U}^2(\overline{\mathbf{A}}_{bu}S_{uu}\overline{\mathbf{A}}_{bu}^T + \overline{\mathbf{A}}_{bw}S_{ww}\overline{\mathbf{A}}_{bw}^T) \tag{8.31}$$

where S_{uu} and S_{ww} are the PSD matrices of $\{u\}$ and $\{w\}$ components, respectively. If the aerodynamic admittance functions are taken into consideration, the above PSD matrix includes the aerodynamic admittance functions.

The power spectra of wind components $\{u\}$ and $\{w\}$ are functions of circular frequency ω. The cross-spectral density functions of the wind component between two points can be expressed in a conventional form.

$$S_{uu(ww)}(z_1, z_2, \omega) = \sqrt{S_{uu(ww)}(z_1, \omega)S_{uu(ww)}(z_2, \omega)}e^{-\hat{f}_{u(w)}} \tag{8.32}$$

where z_1 and z_2 denote the two points; $S_{uu(ww)}(z, \omega)$ is the auto spectra identified from the wind characteristics study; and $e^{-\hat{f}_{u(w)}}$ is the coherence function of fluctuating winds given by the wind characteristics study.

The components of the matrices S_q and S_X can be expressed as

$$S_{q_{ij}}(\omega) = \sum_{k=1}^{m}\sum_{l=1}^{m} H_{ik}^*(\omega)S_{Qb_{kl}}(\omega)H_{jl}(\omega) \tag{8.33}$$

$$S_{X_i}(\omega) = \sum_{k=1}^{m}\sum_{l=1}^{m} \phi_{ik}S_{q_{kl}}(\omega)\phi_{il} \tag{8.34}$$

The variances of the *i*th generalized modal response and nodal displacement are thus given by

$$\sigma_{q_{ii}}^2 = \int_0^\infty S_{q_{ii}}(\omega)d\omega \tag{8.35}$$

$$\sigma_{X_i}^2 = \int_0^\infty S_{X_i}(\omega)d\omega = \sum_{k=1}^{m}\sum_{l=1}^{m}\phi_{ik}\left(\int_0^\infty S_{q_{kl}}(\omega)d\omega\right)\phi_{il} \tag{8.36}$$

8.4.2 Flutter Analysis of Long-span Suspension Bridges

The theoretical backgrounds of coupled flutter analysis of a long-span suspension bridge (Ding *et al.*, 2001; Ding and Xu, 2004b) are introduced in this section. It is assumed that the buffeting forces have no influence on flutter instability and are excluded in the flutter analysis. Thus, the governing equation of motion of a long suspension bridge in the flutter analysis is given by

$$[M]\{\ddot{X}\}+[C]\{\dot{X}\}+[K]\{X\} = \{f_{se}\} \tag{8.37}$$

At present, a two DOFs section model of the bridge deck is widely used to identify the flutter derivatives H_i^* and A_i^* ($i = 1\sim4$). The drag components associated with the lateral motion and some coupling terms are generally negligible. If the flutter derivatives are not available from the buffeting wind tunnel testing assignment, the following empirical expressions based on the quasi-steady theory may be used in the analysis.

$$P_1^* = -\frac{1}{K}C_D, \quad P_2^* = \frac{1}{2K}C_D', \quad P_3^* = \frac{1}{2K^2}C_D' \tag{8.38}$$

$$P_5^* = \frac{1}{2K}C_D', \quad H_5^* = \frac{1}{K}C_L, \quad A_5^* = -\frac{1}{K}C_M \tag{8.39}$$

$$P_4^* = P_6^* = H_6^* = A_6^* = 0 \tag{8.40}$$

Using a complex notation, the self-excited forces expressed by Equations (8.9-8.11) may be expressed as

$$L_{se}(t) = \omega^2 \rho B^2 (C_{Lh}h + C_{Lp}p + BC_{L\alpha}\alpha) \tag{8.41}$$

$$D_{se}(t) = \omega^2 \rho B^2 (C_{Dh}h + C_{Dp}p + BC_{D\alpha}\alpha) \tag{8.42}$$

$$M_{se}(t) = \omega^2 \rho B^2 (BC_{Mh}h + BC_{Mp}p + B^2 C_{M\alpha}\alpha) \tag{8.43}$$

where C_{rs} ($r = L, D, M$; $s = h, p, \alpha$) are the complex flutter derivatives of self-excited forces. The relationships between the real and complex flutter derivatives can be found as

$$C_{Lh} = H_4^* + iH_1^*, \quad C_{Lp} = H_6^* + iH_5^*, \quad C_{L\alpha} = H_3^* + iH_2^* \tag{8.44}$$

$$C_{Dh} = P_6^* + iP_5^*, \quad C_{Dp} = P_4^* + iP_1^*, \quad C_{D\alpha} = P_3^* + iP_2^* \tag{8.45}$$

$$C_{Mh} = A_4^* + iA_1^*, \quad C_{Mp} = A_6^* + iA_5^*, \quad C_{M\alpha} = A_3^* + iA_2^* \tag{8.46}$$

In the 3-D FE based flutter analysis, the distributed self-excited forces acting on one element of a bridge deck are converted into the equivalent nodal loads at two ends of the element.

$$\{f_{se}^e\} = \omega^2 \left[\mathbf{A}_{se}^e\right]\{X^e\} \tag{8.47}$$

where subscript e represents the local coordinates of the element, and $\left[\mathbf{A}_{se}^e\right]$ is a 12 × 12 aeroelastic matrix. For an element of length L, the aeroelastic matrix is

$$\left[\mathbf{A}_{se}^e\right] = \begin{bmatrix} \mathbf{A}_1 & \mathbf{0} \\ \mathbf{0} & \mathbf{A}_1 \end{bmatrix} \tag{8.48}$$

where

$$\mathbf{A}_1 = \frac{1}{2}\rho B^2 L \begin{bmatrix} 0 & 0 & 0 & 0 & 0 & 0 \\ 0 & C_{Lh} & C_{Lp} & BC_{L\alpha} & 0 & 0 \\ 0 & C_{Dh} & C_{Dp} & BC_{D\alpha} & 0 & 0 \\ 0 & BC_{Mh} & BC_{Mp} & B^2 C_{M\alpha} & 0 & 0 \\ 0 & 0 & 0 & 0 & 0 & 0 \\ 0 & 0 & 0 & 0 & 0 & 0 \end{bmatrix} \tag{8.49}$$

Since aeroelastic forces are non-conservative, the aeroelastic matrix of the element is generally unsymmetrical and is a function of the reduced frequency. The element aeroelastic matrices can be transformed into the global coordinate system and then assembled to the global matrix as

$$f_{se} = \omega^2 \left[\mathbf{A}_{se}\right]\{X\} \tag{8.50}$$

The governing equation of motion of the bridge for the flutter analysis can then be expressed as

$$[M]\{\ddot{X}\} + [C]\{\dot{X}\} + [K]\{X\} = \omega^2 \left[\mathbf{A}_{se}\right]\{X\} \tag{8.51}$$

Let $\{x\}$ equal $\{Re^{st}\}$, where $\{R\}$ is the complex modal response amplitude vector of the system. Denote the complex frequency $s = (-\xi + i)\omega$ (where ξ and ω are the damping ratio and circular frequency of the complex mode of vibration, respectively, and $i^2 = -1$). The governing equation of motion can be written as

$$(s^2[M] + s[C] + [K] - \omega^2 \mathbf{A}_{se})Re^{st} = 0 \tag{8.52}$$

The complex mode response of the system can be given approximately by the first m modes of vibration.

$$\{R\} = \mathbf{\Phi}\{q\} \tag{8.53}$$

The above linear transformation yields

$$[s^2\mathbf{I} - \omega^2\overline{\mathbf{A}}_{se} + s\overline{C} + \mathbf{\Lambda}]\{qe^{st}\} = 0 \tag{8.54}$$

where $\left[\overline{\mathbf{A}}_{se}\right] = \mathbf{\Phi}^T[\mathbf{A}_{se}]\mathbf{\Phi}$ and $\left[\overline{C}\right] = \mathbf{\Phi}^T[C]\mathbf{\Phi}$.

Considering the fact that the damping ratios of the system (positive or negative) are small, the approximate relation $\omega^2 = -s^2$ exists. As a result, one may have

$$[s^2(\mathbf{I} + \overline{\mathbf{A}}_{se}) + s\overline{C} + \mathbf{\Lambda}]qe^{st} = 0 \tag{8.55}$$

The above equation can be further expressed in the following state-space format (Ding *et al.*, 2001):

$$(\mathbf{A} - s\mathbf{I})\mathbf{Y}e^{st} = 0 \tag{8.56}$$

where

$$\{\mathbf{Y}\} = \begin{Bmatrix} q \\ sq \end{Bmatrix}, \quad [\mathbf{A}] = \begin{bmatrix} \mathbf{0} & \mathbf{I} \\ -\overline{\mathbf{M}}\mathbf{\Lambda} & -\overline{\mathbf{M}}\overline{C} \end{bmatrix} \tag{8.57}$$

$$\left[\overline{\mathbf{M}}\right] = (\mathbf{I} + \overline{\mathbf{A}}_{se})^{-1} \tag{8.58}$$

Thus, to have a nontrivial solution, the following equation must exist, leading to a standard eigenvalue problem.

$$[\mathbf{A}]\{\mathbf{Y}\} = s\{\mathbf{Y}\} \tag{8.59}$$

where the characteristic matrix [**A**] is a $2m \times 2m$ complex matrix and a function of reduced frequency k (or reduced velocity) only. Thus, the above equation can be solved for only two variables, s and k.

For a given k, standard linear eigensolvers are available to find the $2m$ sets of eigenvalues s and corresponding eigenvectors $\{\mathbf{Y}\}$ from the above equation.

$$s = (-\xi + i)\omega, \quad \{q\} = \mathbf{a} + \mathbf{b}i \tag{8.60}$$

The m eigenvalues with positive imaginary part are the complex frequencies of the system, and the upper half vector $\{q\}$ in the corresponding eigenvector $\{\mathbf{Y}\}$ is the complex mode shape of the system. In a prescribed complex mode shape, the magnitude and phase of the kth natural mode are given as

$$|q_k| = \sqrt{a_k^2 + b_k^2}, \quad \varphi_k = \tan^{-1}(b_k/a_k) \tag{8.61}$$

If the damping ratios of all complex modes are positive, the system is stable; if at least one damping ratio is equal to zero, the system is neutrally stable; and if at least one damping ratio is negative, the system is unstable. Therefore, the flutter analysis described above is able to find the critical state through searching the reduced frequency k. The corresponding circular frequency is the flutter circular frequency ω_f and the critical wind speed U_{cr} is then equal to $B\omega_f/k$. At the critical wind speed, the generalized modal coordinate vector $q(t)$ and the nodal displacement vector of the bridge can be expressed as

$$q(t) = \{|q_i| \sin(\omega_f t + \varphi_i)\} \tag{8.62}$$

$$x(t) = \sum_{i=1}^{m} \phi_i |q_i| \sin(\omega_f t + \varphi_i) = x_0 \sin(\omega_f t + \overline{\varphi}) \tag{8.63}$$

where ϕ_i is the ith natural mode shape; ω_f is the flutter circular frequency; x_0 and $\overline{\varphi}$ are the amplitude and phase of $x(t)$; m is the number of participating modes. It is clear that the coupled flutter motion is three-dimensional and that the phase shift exists among mode components.

The total energy in the characteristic motion (flutter motion) of the bridge at the lowest critical wind speed is

$$E = \frac{1}{2}\{\dot{X}_{max}\}^T M\{\dot{X}_{max}\} = \frac{1}{2}\omega_f^2 \sum_{i=1}^{m} |q_i|^2 \tag{8.64}$$

and the energy in the ith mode is expressed as

$$E_i = \frac{1}{2}\omega_f^2 |q_i|^2 \tag{8.65}$$

The ratio of the ith modal energy over the total energy E_i/E is defined as the modal energy ratio e_i. Clearly, the modal energy ratio provides a uniform measurement to the contribution of a particular vibration mode to the flutter instability of the whole bridge.

8.4.3 Vortex Shedding Analysis of Long-span Suspension Bridges

Vortex shedding excitation can induce significant, but limited, amplitude of vibration of a long-span suspension bridge in low wind speed and low turbulence conditions. For simplicity, Scanlan's model can be used for calculating the vortex-shedding force at "lock-in" when the vortex shedding frequency matches one of the natural frequencies of the bridge (Simiu and Scanlan, 1996).

$$L_{VS}(t) = \frac{1}{2}\rho U^2 D\left[Y_1(K)\left(1 - \varepsilon\frac{h^2}{D^2}\right)\frac{\dot{h}}{U} + Y_2(K)\frac{h}{D} + \tilde{C}_L(K)\sin(\omega_n t + \varphi)\right] \tag{8.66}$$

where ω_n is the circular lock-in frequency; D is the bridge deck depth; Y_1 and ε are two aeroelastic damping parameters; Y_2 is the aeroelastic stiffness parameter; and $\tilde{C}_L$ is the root mean squares of the lift coefficient. They are all functions of reduced frequency, k, at lock-in.

The aeroelastic damping parameters, Y_2 and $\tilde{C}_L$, are usually ignored since they have negligible effects on the response. Y_1 and ε are functions of the Scruton number and can be extracted from wind tunnel observations of steady-state amplitudes of models at "lock-in" based on the following equation:

$$\frac{h_0}{D} = 2\left[\frac{Y_1 - 8\pi \cdot Sc \cdot St}{\varepsilon \cdot Y_1}\right]^{1/2} \tag{8.67}$$

where h_0/D is the reduced amplitude; h_0 is the vibration amplitude of the vertical displacement; $S_C = m_1\xi/\rho D^2$ is the Scruton number; m_1 is the mass per unit length; ξ is the damping ratio; $S_t = f_n D/\overline{U}$ is the Strouhal number; and f_n is the vortex-shedding frequency or the natural frequency. Once these parameters are obtained, the vortex shedding force acting on the bridge deck per unit length can be obtained by Equation (8.66). In a similar way to the buffeting analysis, the vortex shedding analysis of the bridge can be performed and vortex shedding-induced response of the bridge can then be determined.

8.5 MONITORING OF WIND EFFECTS ON THE TSING MA BRIDGE

Hong Kong is situated at latitude N22.2° and longitude E114.1° and it is just on the south eastern coast of China facing the South China Sea. The weather system of Hong Kong is influenced by the land mass to its north as well as by the ocean to its south and east. Two types of wind conditions dominate Hong Kong: monsoon wind prevailing in the months from November to April and typhoon wind predominating in the summer. The local topography surrounding the bridge is quite unique and complex, which includes sea, islands, and mountains of 69 to 500 m high (see Figure 8.1).

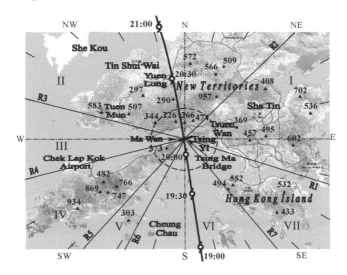

Figure 8.1 Local topography of Hong Kong

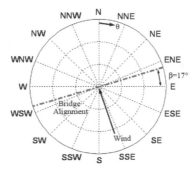

Figure 8.2 Alignment of Tsing Ma Bridge

The alignment of the bridge deck deviates by 17° in counter-clockwise from the east-west axis (see Figure 8.2). The complex topography makes wind characteristics at the bridge site very complicated. The WASHMS of the Tsing Ma Bridge makes it possible to investigate wind characteristics on the bridge site and to assess the wind resistance performance of the bridge (HKPU, 2008).

8.5.1 Monitoring of Wind Environment

Wind data recorded by the WASHMS were analysed, and wind characteristics such as mean wind speed and direction, turbulence components and intensities, integral scales, and wind spectra were obtained for both monsoon wind and typhoon wind. Due to limitation of length, only wind characteristics of typhoons and the joint PDF of monsoons at the bridge site are briefly introduced in this chapter.

8.5.1.1 Anemometers in WASHMS

The WASHMS of the bridge includes a total of six anemometers with two at the middle of the main span, two at the middle of the Ma Wan side span, and one of each on the Tsing Yi Tower and Ma Wan Tower (see Figure 8.3). To prevent disturbance from the bridge deck, the anemometers at the deck level were respectively installed on the north side and south side of the bridge deck via a boom of 8.965 m long from the leading edge of the deck (see Figure 8.4).

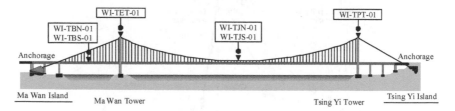

Figure 8.3 Distribution of anemometers in Tsing Ma Bridge

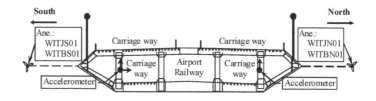

Figure 8.4 Deck cross-section and sensor positions

The anemometers installed on the north side and south side of the bridge deck at the middle of the main span, respectively specified as WI-TJN-01 and WI-TJS-01, are the digital type Gill Wind Master ultrasonic anemometers. The anemometers located at the two sides of the bridge deck near the middle of the Ma

Wan approach span, specified as WI-TBN-01 on the north side and WI-TBS-01 on the south side, are the analogue mechanical (propeller) anemometers. Each analogue anemometer consists of a horizontal component (RM Young 05106) with two channels, giving the horizontal resultant wind speed and its azimuth, and a vertical component (RM Young 27106) with one channel, providing the vertical wind speed. The other two anemometers arranged at 11 m above the top of the bridge towers, respectively specified as WI-TPT-01 for the Tsing Yi tower and WI-TET-01 for the Ma Wan tower, are analogue mechanical anemometers of a horizontal component only. The sampling frequency of measurement of wind speeds was set as 2.56 Hz.

8.5.1.2 Typhoon Victor

On 31 July 1997, less than three months after the opening of the Tsing Ma Bridge, the tropical depression Victor formed in the middle of the South China Sea, and became a real typhoon when it entered the region of 250 km south of Hong Kong at 8:00 on 2 August 1997 Hong Kong Time. The WASHMS installed on the bridge recorded the wind speed and bridge response time histories in time. The measured wind data of seven-hour duration from 17:00 to 24:00 on 2 August 1997 were used in the analysis to investigate typhoon wind characteristics. The 10-minute averaged mean wind speed and mean wind direction at anemometers WITJN01 and WITJS01 were computed using the traditional approach and plotted against time in Figures 8.5.

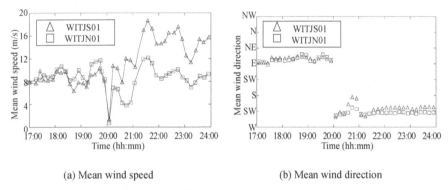

(a) Mean wind speed   (b) Mean wind direction

Figure 8.5 Variations of 10-min mean wind speed and direction

It can be seen that there is a sudden change of wind direction from north-east to south-west and the mean wind speed becomes very small around 20:00. This indicates that the eye of Typhoon Victor just crossed over the bridge at that time. The measurement results also showed that turbulence intensities of winds during Typhoon Victor were higher than those due to monsoon winds and that some of wind samples were not stationary as assumed traditionally. Further information can be found in the literature (Xu *et al.*, 2000).

8.5.1.3 Non-stationary Wind Model

A new model for characterizing non-stationary wind speed was proposed in Xu and Chen (2004). In this model, a measured wind sample of one-hour duration is decomposed into the sum of a deterministic time-varying mean wind speed plus a stationary fluctuating wind component by using the empirical mode decomposition (EMD) (Huang *et al.*, 1998). The time-varying mean wind speed identified by the EMD at a designated intermittency frequency level is more natural than the traditional time-averaged mean wind speed over the certain time interval. The field measurement wind data recorded from the anemometers installed in the bridge during Typhoon Victor were used to test the new model. It was found that the new model could be used to check whether a wind sample is stationary or not. Most of the non-stationary wind samples recorded during Typhoon Victor could be decomposed into a time-varying mean wind speed plus a well-behaved fluctuating wind speed admitted as a stationary random process. The fluctuating wind components obtained by the EMD from the non-stationary wind samples complied with the standard Gaussian distribution well while those gained from the traditional approach deviated significantly from the Gaussian distribution. The spectral density functions of fluctuating wind components obtained by the EMD method had a slightly lower amplitude in the low frequency range than those gained by the traditional approach. The turbulence intensities calculated by the EMD were more appropriate than the traditional approach. The gust factors obtained by the EMD were similar to those obtained by the traditional approach. It can be concluded that the proposed approach is more appropriate than the traditional approach for characterizing wind speed. The detailed information can be found in the literature (Xu and Chen, 2004).

8.5.1.4 Typhoon Wind Characteristics

To understand typhoon wind characteristics at the bridge site, typhoons with signal No. 3 and above issued by the Hong Kong Observatory during the period from July 1997 to September 2005 were studied. A total of 247 hourly typhoon data records was correspondingly retrieved. Four major steps were then taken for pre-processing the original data records: (1) to eliminate unreasonable data with abnormal magnitude for the propeller anemometer at the Ma Wan tower; (2) to decide which anemometer, on the south or on the north, at the deck level must be considered for each typhoon and monsoon events; (3) to eliminate unreasonable data with abnormal magnitude for ultrasonic anemometers at the middle of the main span; and (4) to obtain 10-minute and hourly mean wind speeds and wind turbulence components from the original wind data recorded by the ultrasonic anemometers. After the data pre-processing, a total of 147 hourly typhoon wind records is of acceptable quality for subsequent statistical analysis.

To understand the mean wind speed and mean wind direction of typhoon events experienced at the Tsing Ma Bridge, the wind records were further split into four groups in terms of mean wind speed. These wind speed groups included: (1) less than 10 m/s; (2) between 10 and 18 m/s; (3) between 18 and 45.8 m/s; and (4) greater than 45.8 m/s. The latter three groups represent stages 1, 2 and 3 specified in the high wind management system for the bridge (Xu et al., 2007). Figure 8.6 shows the polar plot of the 10-minute mean wind direction for typhoon events. It can be seen that almost 23% records are taken from 'N' direction. This observation is consistent with that made by the Hong Kong Observatory.

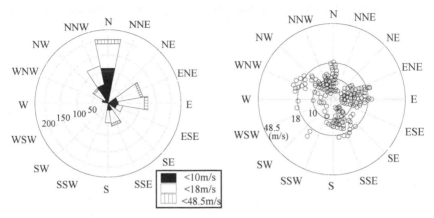

Figure 8.6 Polar plot of mean wind direction   **Figure 8.7** Polar plot of mean wind speed

Figure 8.7 displays the polar plot of the 10-minute mean wind speed for typhoon events. It can be observed that wind speeds within the range from 10 to 18 m/s are dominant for typhoon events. This indicates that stage 1 of the high wind management system was issued for most typhoons for the bridge during the period concerned. Amongst the 16 direction sectors, the SW direction shows the maximum 10-minute mean wind speed of 22.67 m/s.

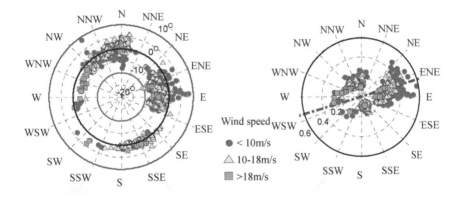

Figure 8.8 Polar plot of mean wind incidence   **Figure 8.9** Longitudinal turbulence intensity

The mean wind incidence is defined as the angle between the mean wind velocity and the horizontal plane. The positive mean wind incidence means the wind blowing upward. Figure 8.8 displays the polar plot of the 10-minute mean wind incidence recorded at the deck level of the bridge for typhoon events. All the wind incidences are within ±10° at the 95% upper and lower limits of wind incidences under a mean wind speed of 20 m/s. However, the wind incidences measured in the easterly directions are much more scattered than those in other directions. This may be because easterly wind directions are almost parallel to the longitudinal direction of the bridge deck. There is a possibility that the flow of air could be disrupted by the bridge deck. It can also be seen that wind incidences tend to be approximately zero for the open-sea area. The mean values of 10-minute mean wind incidences were recorded as 0.33°, 0.09° and 0.52° in the SE, SSE and S directions, respectively.

Figure 8.9 shows the polar plot of the 10-minute turbulence intensity in the longitudinal direction at the deck level for typhoon events. It can be seen that the turbulence intensities measured in the NE and E directions vary within a larger range than those measured in other directions. The NE direction shows the most turbulent winds for typhoon events. The mean values of the longitudinal, lateral and vertical turbulence intensities in this direction are 38.6, 36.2 and 26.1%, respectively. Contrarily, the least turbulent winds are in the S direction which has the mean longitudinal, lateral and vertical turbulence intensities of 8.6, 8.4 and 5.3%, respectively. The average ratio of the lateral to longitudinal turbulence intensities is 0.903 and the ratio of the vertical to longitudinal turbulence intensities is 0.703.

The alongwind, crosswind, and upwind power spectra and the integral length scales derived by the curve fitting method from four selected typhoon samples were investigated in detail. Figure 8.10 shows the alongwind and crosswind power spectra for one selected typhoon sample. The measured integral length scales for the over-land exposure vary between 94.81 m and 188.52 m (the average value being 136.75 m) for L_u^x, between 31.34 m and 54.55 m (the average value being 44.96 m) for L_v^x, and between 33.18 m and 43.04 m (the average value being 37.93 m) for L_w^x. Contrarily, for the open-sea fetch the measured integral length scales range between 110.66 m and 539.66 m (the average value being 280.76 m) for L_u^x, between 43.74 m and 149.18 m (the average value being 83.85 m) for L_v^x, and between 29.15 m and 60.04 m (the average value being 40.91 m) for L_w^x. It seems that the integral length scales from the open-sea fetch are larger than those from the over-land fetch. This observation is consistent with the comment in the literature that the length scale is a decreasing function of the terrain roughness (Simiu and Scanlan, 1996).

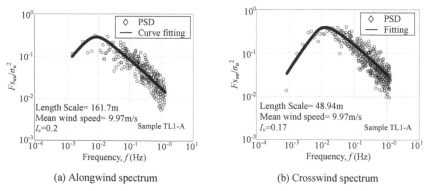

(a) Alongwind spectrum   (b) Crosswind spectrum

Figure 8.10 Wind spectra of one typhoon sample from over-land fetch

8.5.1.5 Monsoon Wind and Joint Probability Density Function

The joint PDF of wind speed and wind direction is essential when assessing wind-induced fatigue damage to the bridge. A practical joint PDF was adopted for a complete population of wind speed and wind direction based on two assumptions: (1) the distribution of the component of wind speed for any given wind direction follows the Weibull distribution; and (2) the interdependence of wind distribution in different wind directions can be reflected by the relative frequency of occurrence of wind.

$$P_{u,\theta}(U,\theta) = P_\theta(\theta)\left(1 - \exp\left[-\left(\frac{U}{c(\theta)}\right)^{k(\theta)}\right]\right) = \iint f_\theta(\theta) f_{u,\theta}(U, k(\theta), c(\theta)) \, du \, d\theta \qquad (8.68)$$

$$f_{u,\theta}(U, k(\theta), c(\theta)) = \frac{k(\theta)}{c(\theta)}\left(\frac{U}{c(\theta)}\right)^{k(\theta)-1} \exp\left[-\left(\frac{U}{c(\theta)}\right)^{k(\theta)}\right] \qquad (8.69)$$

$$P_\theta(\theta) = \int_0^\theta f_\theta(\theta) \, d\theta \qquad (8.70)$$

where $0 \le \theta < 2\pi$, and $P_\theta(\theta)$ is the relative frequency of occurrence of wind in wind direction θ. The occurrence frequency $P_\theta(\theta)$ as well as the distribution parameters $k(\theta)$ and $c(\theta)$ can be estimated using wind data recorded at the bridge site. Wind records of hourly mean wind speed and direction within the period between 1 January 2000 and 31 December 2005 from the anemometer installed on the top of the Ma Wan tower were used to ascertain the joint PDF of hourly mean wind speed and direction. The height of the anemometer is 214 m above sea level. Wind records having an hourly mean wind speed lower than 1 m/sec were removed in order to avoid any adverse effects on the statistics. As a result, 19,775 hourly monsoon records were available for calculation of the joint PDF of the wind speed and direction. The number of hourly typhoon wind records during the period concerned was so small that the corresponding joint PDF could not be obtained at present.

All the monsoon records were classified into 16 sectors of the compass with an interval of $\Delta\theta = 22.5^0$ according to the hourly mean wind direction (see Figure 8.2). In each sector, mean wind speed was further divided into 16 ranges from zero to 32 m/sec with an interval of $\Delta U = 2$ m/sec. This leads to a total of 256 cells, and the relative frequency of hourly mean wind speed and wind direction in each cell was calculated. Based on the relative frequencies of wind speed and wind direction calculated, the theoretical expression of the joint PDF was deduced based on Equation (8.68). The Weibull function was used to fit the histogram of hourly mean wind speed for each wind direction, and the typical results in the east, south and west directions are depicted in Figures 8.11(a) to 8.11(c), respectively. The Weibull function was also applied to the complete wind records without considering wind direction, as shown in Figure 8.11(d). The results show that the Weibull function also fits the complete wind data adequately. The relative frequency of wind direction and the scale and shape parameters of the Weibull function obtained are given in polar plot in Figure 8.12. It can be seen that the dominant monsoon direction is the east, and the scale and shape parameters do not vary significantly with wind direction. The details can be found in the literature (Xu *et al.*, 2008).

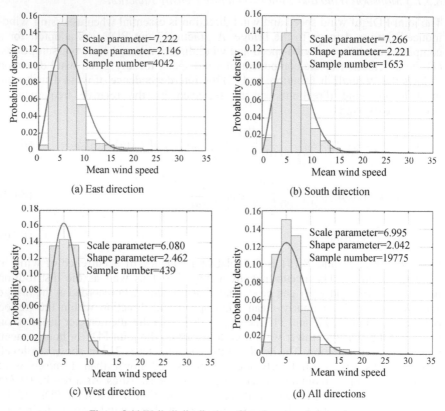

Figure 8.11 Weibull distribution of hourly mean wind speed

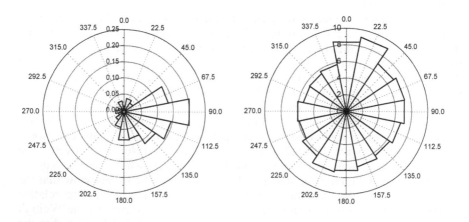

(a) Relative frequency of wind direction   (b) Weibull scale parameter

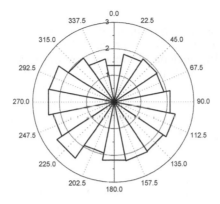

(c) Weibull shape parameter

Figure 8.12 Relative frequency of wind direction and Weibull scale and shape parameters

8.5.2 Monitoring of Bridge Displacements

Bridge displacement data recorded by the WASHMS were analyzed. The statistical relationships between the strong wind and wind-induced bridge displacement responses at the locations of GPS stations are then established. The statistical relationships obtained shed light on displacement behaviour of the bridge under strong winds and provide the information for comfort and deflection assessment of the bridge under winds as well as the comfort of road vehicles and passengers. The statistical relationships can also be taken as reference to verify wind tunnel test results, and will be extended to assess the wind-resistant performance of the bridge under extreme wind speeds through computer simulations.

8.5.2.1 GPS and Data Processing

The displacements of the Tsing Ma Bridge in longitudinal, lateral and vertical directions are mainly monitored by GPS. There are in total 14 GPS monitoring stations installed on the bridge towers, main cables and bridge deck as shown in Figure 8.13.

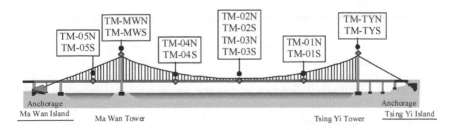

Figure 8.13 Distribution of GPS receivers in Tsing Ma Bridge

The original bridge displacement data recorded by GPS stations at a sampling frequency of 10 Hz are pre-processed first. The main steps of the pre-processing of GPS data include (1) to convert the GPS data from the HK80 geographic coordinate to the Universal Transverse Mercator grid coordinate; (2) to compute the bridge displacement coordinate with respect to the reference coordinate measured during the installation of the GPS systems; (3) to eliminate unreasonable data with abnormal magnitude caused by an abrupt change in the number of satellites or unsatisfactory geometric configurations; (4) to obtain bridge displacement responses in three orthogonal directions; (5) to filter displacement response time histories using a low-pass filter of upper frequency 1 Hz to maintain high quality data for subsequent analysis; and (6) to decompose mean and dynamic displacements from the total response measurements recorded by GPS.

The mean displacement response of the bridge recorded by GPS may be affected by GPS background noise, temperature, and mean wind. Furthermore, the dynamic displacement response of the bridge recorded by GPS may also be affected by moving road vehicles, moving trains, GPS background noise, and wind.

First calibration tests using a 2D motion simulation table were carried out on the site of Tsing Ma Bridge, to find the possible effects of GPS background noise on the bridge displacement responses. The calibration results show that the ten-minute mean value of GPS background noise averaged in one hour is 0.07 mm in the horizontal direction and 0.11 mm in the vertical direction only. The averaged ten-minute standard deviation of GPS background noise is 1.62 mm in the horizontal direction and 3.67 mm in the vertical direction. Therefore, the effect of GPS background noise on the mean displacement response is small and can be disregarded when strong wind events are considered. The dynamic displacement responses with the standard deviation less than 4.5 mm and 9.7mm in the respective horizontal and vertical directions are not included in the statistical analysis to avoid the effect of GPS background noise on the dynamic displacement response.

To assess the effect of temperature on the mean displacement response of the bridge, the measurement data from both the temperature sensors and GPS stations were selected during forty days with only light wind. It has been observed that temperature mainly affects the deck and cable mean displacements in the vertical and longitudinal directions, the tower mean displacement in the longitudinal direction, whereas the mean wind mainly affects the lateral mean displacements of the bridge deck, cables and towers. These observations provide a way of distinguishing temperature effects from wind effects on bridge mean displacements.

The statistical analysis on the dynamic displacement responses has revealed that the dynamic displacement responses of the bridge deck in the vertical direction, of the cables in the vertical and longitudinal directions, and of the bridge towers in the longitudinal direction within the frequency range less than 0.08 Hz are dominated by railway load effects (see Chapter 5). Therefore, the wavelet decomposition is applied to the time histories of the relevant dynamic displacement responses to eliminate the railway-load effects. The remaining concern is how to eliminate highway load effects on the dynamic displacement responses. In this regard, the measurement data from both the weight-in-motion (WIM) sensors and GPS stations were examined. Part of the measurement data which meet the requirements of weak wind and no passing trains were selected. The variation of dynamic displacement response of the bridge with the traffic flow rate was then assessed. It was found, for instance, that the average standard deviation of the dynamic displacement response of the bridge deck in the mid-main span over the period of high traffic flow rates is about 5 mm in the lateral direction and about 11 mm in the vertical direction. By assuming wind- and traffic-induced displacements are uncorrelated with each other, the standard deviation of the wind-induced dynamic displacement response (σ_W) can be determined by:

$$\sigma_W = \sqrt{\sigma_{WT}^2 - c^2} \qquad (8.71)$$

where σ_{WT} is the standard deviation of the total dynamic displacement response induced by both highway load and wind load; and c is the highway-induced displacement standard deviation. The highway-induced standard deviation is often small or even can be disregarded when the total displacement standard deviation is large enough. By Equation (8.71), such disregard causes less than 5% error in wind-induced displacement standard deviation if the total displacement standard deviation is greater than or equal to $3.28c$.

8.5.2.2 Statistical Relationship between Wind and Wind-induced Bridge Displacement

The relationships between ten-minute mean wind speed and ten-minute wind-induced mean, dynamic, and total displacement responses of the bridge were explored, but only the relationship between ten-minute mean wind speed and ten minute wind-induced total displacement of the bridge deck and main cables is presented in this book.

All valid 10-minute mean wind speeds higher than 5 m/s and the corresponding 10-minute total displacements of the bridge deck and main cable are considered for the time period from June 2003 to September 2005. The power law function expressed by Equation (8.72) is adopted to fit the measurement data.

$$\hat{D} = \bar{D} \pm m_p \sigma_D = a\bar{U}^b \qquad (8.72)$$

where $\hat{D}$ is the wind-induced total displacement of the bridge deck or main cable at a given section in a given direction; $\bar{D}$ and σ_D represent the corresponding mean and standard deviation, respectively; m_p is the statistical peak factor; $\bar{U}$ denotes the corresponding mean wind speed; a and b are the two parameters to be determined. Because of the nature of the problem, only the positive and non-zero value of b in Equation (8.72) reflects the fact that a larger mean wind speed leads to a larger total displacement. Hence, in the curve fitting, the one-tail testing hypothesis concerning parameter b and the two-tail testing hypothesis concerning parameter a are performed both with a t-test significance 0.05 (p-value). The relationships between the 10-minute mean wind speed and the lateral total displacement recorded at the four positions of the bridge deck and the main cable are explored by using the regression model given in Equation (8.72). The results of the regressions for the mid-main span and the main cable in the lateral direction in the N direction sector are shown in Figure 8.14.

It is noted from the figures that the estimated exponents in the power law function for the lateral direction of the mid-main span and the main cable are 1.822 and 1.741, respectively. This means that the wind-induced total displacement in the lateral direction is approximately proportional to the squares of wind speed. When the mean wind speed is taken as 10 m/s, the lateral total displacements of the bridge deck at the mid-main span and the main cable are estimated as −191.29 and −179.73 mm, respectively, using the power law functions. These results are similar to the summarization of the lateral mean displacement and the triple of the lateral standard deviation.

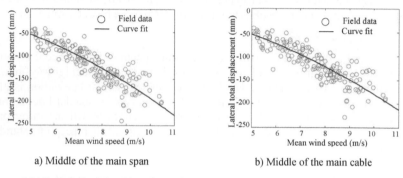

a) Middle of the main span   b) Middle of the main cable

Figure 8.14 Statistical relationships of ten-minute mean wind speed and lateral total displacement of the bridge in N direction sector

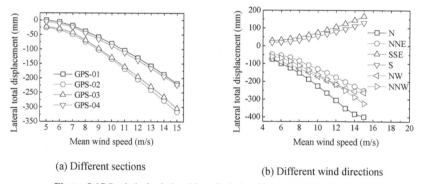

(a) Different sections   (b) Different wind directions

Figure 8.15 Statistical relationships of wind and lateral total displacement

The statistical relationships between wind and wind-induced total displacement responses at four locations on the bridge deck and main cable are also explored for wind direction sector NNW. The regression coefficients determined using the least-squares method with p-value less than 0.05 and R^2 greater than 0.6 are accepted. The curve fitting fails for the measurement data from the deck section in the Ma Wan side span because the R^2 is less than 0.6. Figure 8.15(a) illustrates the statistical relationships obtained by Equation (8.72) with the regression parameters obtained for the three locations on the bridge deck and the main cable in the lateral direction. It is noted from the figure that the total displacements are almost symmetric with respect to the middle of the main span. It is also observed that the total displacement of the main cable is similar to that of the mid-main span.

The statistical relationships between wind and wind-induced total displacement response of the bridge deck at the middle of the main span for different wind directions are also investigated. Similarly, only the responses with the p-value less than 0.05 and R^2 greater than 0.6 are accepted for six wind directions. The total displacement response at the middle of the main span in the lateral direction for the six wind direction sectors is plotted in Figure 8.15(b). From the figure, it can be noted that the total displacement responses reach their most negative values in the N direction sector, whereas the most positive total displacements come from the SSE direction. It can also be observed that the relationship between wind and wind-induced total displacement response varies with the wind direction.

8.5.3 Wind-resistant Performance Assessment of Bridge

8.5.3.1 Background

There are three levels of wind-resistant performance assessments for the Tsing Ma Bridge. The first level is related to comfort and safety of road vehicles and trains running on the bridge in crosswind. In this regard, the comfort and deflection of the bridge must be examined, respectively, in terms of the maximum acceleration and displacement responses of the bridge against the maximum allowable acceleration and displacement of the roadway and the railway. A high wind management system is also introduced for traffic management of the bridge for the safety of road vehicles. The operation guidelines adopted for the bridge during a high wind period specify that both the upper and lower decks must be closed to all road vehicles except trains when the hourly-mean wind speed recorded on site is in excess of 165 km/hour. The second level of the wind-resistant performance assessment is associated with the serviceability limit state (SLS) of the bridge, in which the bridge shall behave elastically and is expected to be serviceable immediately without the need for any repair. To examine the serviceability of the bridge, buffeting analysis of the bridge is necessary and the maximum total displacement response of the bridge under the serviceability design wind speed must be examined against the maximum allowable displacement. The third level refers to the ultimate limit state (ULS) of the bridge, in which the bridge may undergo large deformation in the post elastic range without substantial reduction in strength and the damage level of the bridge will be considered as economically and technically feasible to repair. In this level, the aerodynamic instability should be avoided and the critical wind speed should be confirmed.

Wind-resistant performance assessment of the bridge based on the data from the WASHMS has not been explored in detail. The number of sensors in the WASHMS is always limited for such a large structure and the locations of structural defects or degradation may not be at the same positions of the sensors. It is possibile that the worst condition may not be directly monitored by the sensors, and the structural performance assessment could not rely on the data from the sensors only. Therefore, for the accomplishment of a complete wind-resistant performance assessment of the bridge, the WASHMS must be combined with advanced computer simulation in an evolutionary way. For this purpose, a SHM-based computer simulation of wind-induced responses of the bridge has been presented and verified by the measurement data to some extent. Nevertheless, this section only discusses the application of GPS for wind-resistant performance assessment of the bridge in terms of bridge displacement response. Since the design hourly mean wind speed of the bridge is 50 m/s at the deck level for SLS and 65 m/s for ULS, the statistical relationships presented in the last section cannot be used to assess wind-resistant performance of the bridge under extreme wind speed because the maximum wind speed recorded by the WASHMS so far is much lower than the design wind speeds. The statistical relationship presented in the last section is therefore extended through the computer simulation to consider extreme wind cases. It is also extended for other deck sections of the bridge through the computer simulation in order to have a complete assessment of the comfort and deflection of the bridge associated with the comfort of road vehicles and passengers.

8.5.3.2 SHM-oriented Finite Element Model

To facilitate a complete wind-resistant performance assessment of the bridge, an SHM-oriented FE model is needed so that stresses/strains in all important bridge components can be directly computed and some of them can be compared with the measured ones for verification. However, the currently conducted buffeting analyses of long span bridges are often based on a simplified spine beam FE model of equivalent sectional properties (Xu *et al.*, 1997). Such a simplified model is effective to capture the dynamic characteristics and global structural behaviour of the bridge under strong winds without heavy computational effort. However, local structural behaviour linked to stress and strain, which is prone to cause local damage, could not be estimated directly. On the other hand, with the rapid development of information technology, the improvement of speed and memory capacity of personal computers has made it possible to establish an SHM-oriented FE model for a long suspension bridge. In this regard, an SHM-oriented FE model for the Tsing Ma Bridge is established and shown in Figure 8.16, with significant modelling features of the bridge deck included for the good replication of geometric details of the as-built complicated deck (Liu *et al.*, 2009). The bridge was modelled using a series of beam elements, plate elements, shell elements, and others. The FE model contains 12,898 nodes, 21,946 elements (2,906 plate elements and 19,040 beam elements) and 4,788 MPCs. The SHM-oriented model has also been updated using the measured first 18 natural frequencies and mode shapes of the bridge with the updated parameters being material properties only because the geometric features and supports of bridge deck have been modelled in great detail in the proposed model. It turns out that the updated complex FE model could provide comparable and credible structural dynamic characteristics. Detailed information about the model can be found in Chapter 4 of this book.

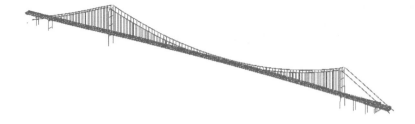

Figure 8.16 SHM-oriented finite element model of Tsing Ma Bridge

8.5.3.3 Computer Simulation of Buffeting-induced Displacement Responses

Based on the established SHM-oriented FE model, a numerical simulation procedure for buffeting induced responses of long suspension bridges has been proposed (Liu *et al.*, 2009). Significant improvements of the proposed procedure are that the effects of the spatial distribution of both buffeting forces and self-excited forces on a bridge deck structure are taken into account, as opposed to lumping all buffeting forces and self-excited forces at the centre of elasticity in an equivalent beam FE model. Local strains and stresses in structural members of the bridge deck, which are prone to cause local damage, are predicted directly using the mode superposition technique in the time domain. The field measurement data including wind, displacement, acceleration and stress recorded by the anemometers, GPS stations, accelerometers and strain gauges in the WASHMS installed on the Tsing Ma Bridge have been analysed. The buffeting-induced displacement and acceleration responses at the locations of GPS stations and accelerometers have been computed using the mode superposition method. The buffeting-induced stress responses at the locations of strain gauges have been computed through the modal stress analysis. The comparison shows that the computed displacement, acceleration and stress time histories are similar to the measured ones in terms of both pattern and magnitude. The SHM-oriented FE model of the bridge and the proposed computer simulation of buffeting-induced response offer a good tool for the extension of the statistical relationship presented in the last section to consider extreme wind cases.

8.5.3.4 Extension of Statistical Relationships and Performance Assessment

The statistical relationships between wind speed and wind-induced total displacement as developed in Section 8.4.2 are extended for the extreme wind cases through the computer simulation. Figures 8.17(a) and 8.17(b) illustrate the extended statistical relationships between wind speed and wind-induced total displacement in the lateral and vertical direction of the bridge at the mid-main span for the SSE wind direction (almost perpendicular to the bridge deck). As shown in the figure, the computed results match well with the measured data within the low wind speed range. When the mean wind speed is above about 15 m/s, only computed results are available. When the hourly-mean wind speed reaches 50 m/s, the total displacement response of the bridge at the mid-main span in the SSE wind direction is 2,305 mm in the lateral direction and 860 mm in the vertical direction. Figure 8.18 further depicts the variations of the lateral and vertical total displacement responses along the bridge deck for wind velocity normal to the bridge deck but under different mean wind speeds. The maximum total displacement response of the bridge occurs at the mid main span.

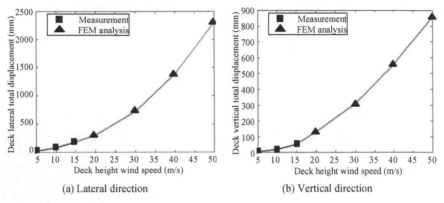

(a) Lateral direction   (b) Vertical direction

Figure 8.17 Statistical relationships of mean wind speed and total displacement response of bridge

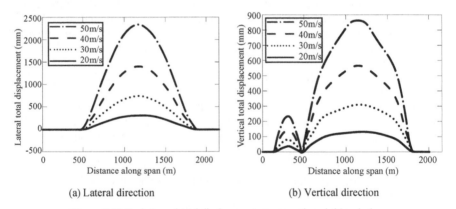

(a) Lateral direction   (b) Vertical direction

Figure 8.18 Variation of total displacement response along bridge deck

For the Tsing Ma Bridge, the design hourly-mean wind speed for SLS is 50 m/s for a 120 year return period. The maximum allowable total displacement of the bridge in the lateral direction under the serviceability design wind speed is 2.9 m. Since the time period for collecting wind data on the site is not longer enough to examine the design wind speed for a 120 year return period, the design wind speed of 50 m/s for a 120 year return period is used for the preliminary wind-resistant performance assessment. As shown in Figure 8.18(a), the total displacement response of the bridge at the mid-main span is 2,305 mm in the lateral direction under the design wind speed of 50 m/s at the deck level. To compare the computed maximum total displacement with the maximum allowable total displacement, the lateral total displacement movement of the bridge is within the tolerance movement according to the design criterion with respect to SLS. Similar exercises can be performed for other wind-resistant performance assessments of the bridge against other design criteria. One important point is that such exercises must be updated as more data are available so as to yield an evolutionary wind-resistant performance assessment of the bridge.

8.6 SUMMARY

The major sensors and sensing systems used for wind monitoring of long-span suspension bridges have been described in this chapter. Boundary layer wind characteristics related to aerodynamic forces acting on long-span suspension bridges have been also introduced based on SHM-recorded data. The FE method-based analytical backgrounds of bridge aerodynamics, including buffeting-induced bridge vibration, bridge flutter instability, and vortex shedding-induced bridge vibration, have been discussed. Finally, the Tsing Ma suspension bridge has been taken as an example to illustrate the monitoring of wind environment around and wind effects on the bridge. The wind environment monitoring includes the number and location of the anemometers in the WASHMS, the monitoring of Typhoon Victor with its eye passing through the bridge, the development of a non-stationary wind model, the major characteristics of typhoon wind, and the joint PDF of monsoon wind. The wind effect monitoring contains the number and location of GPS receivers in the WASHMS, the procedure of GPS data processing to find wind-induced displacement response of the bridge, the statistical relationship between wind and wind-induced displacement response of the bridge, the SHM-oriented FE model of the bridge, the computer simulation of wind-induced displacement response of the bridge, and the wind resistant performance assessment in terms of displacement response of the bridge.

8.7 REFERENCES

Ashkenazi, V. and Roberts, G.W., 1997, Experimental monitoring of the Humber Bridge using GPS. *ICE Proceedings,* **120(4),** pp. 177-182.

Bucher, G.C. and Lin Y.K., 1988, Stochastic stability of bridges considering coupled modes. *Journal of Engineering Mechanics, ASCE,* **114(12),** pp. 2055-2071.

Chen, X.Z., Matsumoto, M., and Kareem, A., 2000, Time domain flutter and buffeting response analysis of bridges. *Journal of Engineering Mechanics, ASCE,* **126(1),** pp. 7-16.

Davenport, A.G. ,1962, Buffeting of a suspension bridge by storm winds. *Journal of Structural Division, ASCE,* **88(3),** pp. 233-268.

Ding Q.S., Chen, A.R. and Xiang, H.F., 2001, A state space method for coupled flutter analysis of long span bridges. *Structural Engineering and Mechanics,* **14(4),** pp. 491-504.

Ding Q.S. and Xu, Y.L., 2004a, *3-D Finite Element Based Buffeting Analysis of Stonecutters Bridge.* Report No. 05, (Hong Kong: Department of Civil and Structural Engineering, The Hong Kong Polytechnic University).

Ding Q.S. and Xu, Y.L., 2004b, *3-D Finite Element Based Flutter Analysis of Stonecutters Bridge*. Report No. 03, (Hong Kong: Department of Civil and Structural Engineering, The Hong Kong Polytechnic University).

ESDU, 2001, *Characteristics of Atmospheric Turbulence Near the Ground, Part II: Single Point Data for Strong Winds*, Neutral atmosphere, (Houndsditch, London: IHS Engineering Sciences Data Unit).

Fujino, Y., Murata, M., Okano, S., and Takeguchi, M., 2000, Monitoring system of the Akashi Kaikyo Bridge and displacement measurement using GPS. In *Proceedings of Nondestructive Evaluation of Highways, Utilities, and Pipelines IV*, **3995**, (Bellingham, Washington: SPIE), pp. 229-236.

HKPU, 2008, *Establishment of Bridge Rating System for Tsing Ma Bridge: Statistical Modelling and Simulation of Predominating Wind Loading Effects*. Report No. 3, (Hong Kong: Department of Civil and Structural Engineering, The Hong Kong Polytechnic University).

Holmes, J.D., 2007, *Wind Loading of Structures*, 2nd ed., (London: Taylor & Francis).

Huang, N. E., Shen, Z., Long, S. R., Wu, M. C., Shih, H. H., Zheng, Q., Yen, N. C., Tung, C. C. and Liu, H. H., 1998, The Empirical mode decomposition and the Hilbert spectrum for nonlinear and non-stationary time series analysis. *Proceeding of Royal Society London A*, **454**, pp. 903-995.

Kareem, A., 2008, Numerical simulation of wind effects: A probabilistic perspective. *Journal of Wind Engineering and Industrial Aerodynamics*, **96(10-11)**, pp. 1472-1497.

Kijewski-Correa, T. and Kochly, M., 2007, Monitoring the wind-induced response of tall buildings: GPS performance and the issue of multipath effects. *Journal of Wind Engineering and Industrial Aerodynamics*, **95(9-11)**, pp. 1176-1198.

Liu, T.T., Xu, Y.L., Zhang, W.S., Wong, K.Y., Zhou, H.J. and Chan, K.W.Y., 2009, Buffeting-Induced stresses in a long suspension bridge: structural health monitoring orientated stress analysis. *Wind and Structures*, **12(6)**, pp. 479-504.

Miyata, T., Yamada, H., Katsuchi, H., and Kitagawa, M., 2002, Full-scale measurement of Akashi-Kaikyo Bridge during typhoon. *Journal of Wind Engineering and Industrial Aerodynamics*, **90(12)**, pp. 1517-1527.

Nakamura, S., (2000), GPS measurement of wind-induced suspension bridge girder displacements. *Journal of Structural Engineering, ASCE*, **126(12)**, pp. 1413-1419.

Scanlan, R.H., 1978, The action of flexible bridge under wind, II: buffeting theory. *Journal of Sound and Vibration*, **60(2)**, pp. 201-211.

Schroeder, J.L., Smith, D.A. and Peterson, R.E., 1998, Variation of turbulence intensities and integral scales during the passage of a hurricane. *Journal of Wind Engineering and Industrial Aerodynamics*, **77&78**, pp. 65-72.

Simiu, E. and Scanlan, R., 1996, *Wind Effects on Structures*, 3rd ed., (New York: John Wiley).

Tamura, Y., Suda, K., Sasaki, A., Iwatani, Y., Fujii, K., Hibi, K. and Ishibashi, R., 1999, Wind speed profiles measured over ground using Doppler sodars. *Journal of Wind Engineering and Industrial Aerodynamics*, **83**, pp. 83-93.

Xiang, H.F., Liu, C.H. and Gu, M., 1995, Time-domain analysis for coupled buffeting response of long span bridge. In *Proceeding of the 9th ICWE*, New Delhi, (Mumbai, Maharashtra: Wiley Eastern Limited), pp. 881-892.

Xu. Y. L., and Chen, J., 2004, Characterizing non-stationary wind speed using empirical mode decomposition. *Journal of Structural Engineering, ASCE*, **130**, pp. 912-920.

Xu, Y.I., Chen, J., Ng, C.L and Zhou, H.J., 2008, Occurrence probability of wind-rain-induced stay cable vibration. *Advances in Structural Engineering-An International Journal*, **11**, pp. 53-69.

Xu, Y.L., Guo, W.W., Chen, J., Shum, K.M. and Xia, H., 2007, Dynamic response of suspension bridge to typhoon and trains. I: field measurement results. *Journal of Structural Engineering, ASCE*, **133**, pp. 3-11.

Xu, Y.L., Ko, J. M. and Zhang, W.S., 1997, Vibration studies of Tsing Ma long suspension bridge. *Journal of Bridge Engineering*, **2**, pp. 149-156.

Xu, Y. L. and Zhan, S., 2001, Field measurements of Di Wang Tower during Typhoon York. *Journal of Wind Engineering and Industrial Aerodynamics*, **89**, pp. 73-93.

Xu, Y.L., Zhu, L.D., Wong, K.Y. and Chan, K.W.Y., 2000, Field measurement results of Tsing Ma suspension bridge during Typhoon Victor. *Structural Engineering and Mechanics - An International Journal*, **10(6)**, pp. 545-559.

Zhang, X., Xiang, H., Sun, B., 2002, Nonlinear aerostatic and aerodynamic analysis of long-span suspension bridges considering wind-structure interactions. *Journal of Wind Engineering and Industrial Aerodynamics*, **90**, pp. 1065-1080.

8.8 NOTATIONS

$[\mathbf{A}_{se}]$	Complex self-excited force matrix of the bridge.
$[\bar{\mathbf{A}}_{se}]$	Generalized complex self-excited force.
$[\mathbf{A}_{se}^e]$	Aeroelastic matrix.
$[\mathbf{A}_{bu}^e]$, $[\mathbf{A}_{bw}^e]$	Buffeting force matrices of an element corresponding to the longitudinal and vertical wind fluctuations.
$[\mathbf{A}_{bu}]$, $[\mathbf{A}_{bw}]$	Global buffeting force matrices corresponding to the longitudinal and vertical wind fluctuations.
$[\bar{\mathbf{A}}_{bu}]$, $[\bar{\mathbf{A}}_{bw}]$	Generalized buffeting force vectors.
a, b	Real and imagine part of complex modal response.
B	Bridge deck width.
$[\mathbf{B}]$	Matrix of interpolated functions.
C_L, C_D, C_M	Lift, drag, and moment coefficients.
C_{rs}	Complex flutter derivatives of self-excited forces.
$\tilde{C}_L$	Root mean squares of the lift coefficient.
$c_j(t)$	IMF components.
c	Highway-induced displacement standard deviation.
D	Bridge deck depth.

$\hat{D}$	Wind-induced total displacement of the bridge deck or main cable.
$\bar{D}$	Mean value of the wind-induced total displacement of the bridge deck or main cable.
D_{se}	Self-excited lateral forces acting on the bridge deck per unit length.
E	Energy of motion.
e_i	Modal energy ratio.
$e^{-\hat{f}_{u(w)}}$	Coherence function of fluctuating winds.
$\{f_{se}\}, \{f_b\}, \{f_s\}$	Self-excited force, buffeting force, and mean wind force.
$\{f_{se}^e\}$	Equivalent nodal loads at two ends of an element.
f_n	Vortex-shedding frequency.
$H(\omega)$	Transfer function matrix.
H_i^*, P_i^*, A_i^*	Non-dimensional flutter derivatives of the vertical, lateral, and torsional displacements.
h, p, α	Vertical, lateral, and torsional displacements of the bridge deck.
h_0	Vibration amplitude of the vertical displacement.
I_a	Turbulence intensity.
k	Reduced frequency.
$k(\theta), c(\theta)$	Distribution parameters of occurrence of wind.
L	Length of an element.
L_{VS}	Vortex-shedding force.
L_{se}	Self-excited vertical forces acting on the bridge deck per unit length.
$[M], [C], [K]$	Mass, damping, and stiffness matrices of the bridge.
M_{se}	Self-excited moment acting on the bridge deck per unit length.
m	Vibration modes of the bridge.
m_1	Mass per unit length.
m_p	Statistical peak factor.
N	Number of IMF components.
n	Total number of DOFs of the bridge.
$P_\theta(\theta)$	Relative frequency of occurrence of wind.
$\{\mathbf{P}_b\}$	Buffeting forces.
$p(u)$	Probability density.
$\{Q_b\}$	Generalized buffeting force.
$\{q\}$	Modal response.
$\{R\}$	Complex modal response amplitude vector.
r	Number of nodes subjected to wind fluctuations.
$r_u(t)_N, r_w(t)_N$	Final residue of IMF components.
$S_q(\omega), S_X(\omega)$	Power spectral density matrices of the generalized modal response $\{q\}$, nodal displacement $\{X\}$.
$S_{Q_b}(\omega)$	Power spectral density matrices of generalized buffeting forces.

S_{uu}, S_{ww}	Power spectral density matrices of $\{u\}$ and $\{w\}$.
$S_{uu(ww)}(z,\omega)$	Auto spectra.
$S_{uu(ww)}(z_1,z_2,\omega)$	Cross-spectral density functions of the wind component between two points.
S_c, S_t	Scruton number, Strouhal number.
s	Complex frequency.
T	Time duration.
t	Time point.
$\bar{U}$	Mean wind speed.
$\bar{U}^*(t)$	Deterministic time-varying mean wind speed.
$U(t)$	Longitudinal wind speed in the mean wind direction.
U_{cr}	Critical wind speed.
$u(t)$	Longitudinal turbulence wind speed.
$\{u\}$	The r-row nodal fluctuating wind vectors for the longitudinal components.
$v(t)$	Lateral turbulence component horizontal and normal to the mean wind direction.
$\bar{v}$	Mean wind speed of lateral turbulence.
$w(t)$	Vertical turbulence component upward and perpendicular to both $U(t)$ and $v(t)$.
$\{w\}$	The r-row nodal fluctuating wind vectors for the vertical components.
$\bar{w}$	Mean wind speed of vertical turbulence.
$\{X\}$, $\{\dot{X}\}$, $\{\ddot{X}\}$	Nodal displacement, velocity, and acceleration vectors.
x	Axial location of an element.
x_0	Amplitude of displacement vector $x(t)$.
$\{Y\}$	Complex mode shape.
Y_1	Aeroelastic damping parameter.
Y_2	Aeroelastic stiffness parameter.
z_1, z_2	Two points.
ρ	Air density.
θ	Wind direction.
ε	Aeroelastic damping parameter.
σ_D	Standard deviation of the wind-induced total displacement of the bridge deck or main cable.
σ_W	Standard deviation of wind-induced dynamic displacement response.
σ_{WT}	Standard deviation of the total dynamic displacement response induced by both highway load and wind load.
σ_q^2, σ_x^2	Variances of the modal response and nodal displacement.
ω, $[\Lambda]$,	Circular frequency of vibration, matrix of eigenvalues.
ξ	Damping ratio.
ω_f, ω_n	Flutter circular frequency, circular lock-in frequency.

χ_{Lu}, χ_{Du}, χ_{Mu} | Aerodynamic admittance functions corresponding to the longitudinal wind fluctuations.

χ_{Lw}, χ_{Dw}, χ_{Mw} | Aerodynamic admittance functions corresponding to the vertical wind fluctuations.

φ_i, $\overline{\varphi}$ | Phrase of displacement.

$\{\phi\}$, $[\Phi]$, | Mode shape, matrix of mode shapes.

CHAPTER 9

Monitoring of Seismic Effects

9.1 PREVIEW

When a long-span suspension bridge is built in a seismic region, seismic ground motions and seismic effects on the bridge must be monitored closely to ensure the functionality and safety of the bridge before or after an earthquake. A long-span suspension bridge often has multi-supports separated by certain distances. If the bridge is located in a complex terrain with different local soil conditions, the spatial variation and wave passage effect of seismic ground motions whose essential features are considerably different in amplitude and phase may result in quantitative and qualitative differences in seismic responses as compared with those produced by synchronous motion at all supports (Abdelghaffar, 1982; Abdelghaffar and Rubin, 1983). Therefore, the simulation of non-uniform seismic ground motions at multi-supports is essential for a proper seismic monitoring of a long-span suspension bridge in consideration that the number of seismometers in the SHM system is often limited. Furthermore, to make full use of measured ground motion time histories at limited points and retain more information on actual seismic ground motions, which can be provided by SHM systems, the conditional simulation of non-uniform seismic ground motions at multi-supports is more appropriate than the unconditional simulation. After having properly simulated ground motions, the method for predicting seismic responses of a long-span suspension bridge under non-uniform seismic ground motions becomes crucial.

This chapter first describes sensors and sensing systems used for seismic monitoring of long-span suspension bridges. The method proposed by Vanmarcke *et al.* (1993) is then introduced for simulating properly-correlated earthquake ground motions at an arbitrary set of closely-spaced points, which are compatible with known motions at other locations monitored by SHM systems. Two relatively simple methods, the relative motion method (RMM) and the large mass method (LMM), are described concisely for predicting seismic responses of a long-span suspension bridge under non-uniform seismic ground motions (Léger *et al.*, 1990). Finally, the Tsing Ma suspension bridge is taken as an example to illustrate the monitoring of seismic ground motions and seismic effects on the bridge.

9.2 SEISMIC MONITORING SYSTEM

Similar to other loading effect monitoring, seismic monitoring of a long-span suspension bridge also starts with the selection and arrangement of a sensing system and the associated data acquisition system. The commonly used sensors for seismic monitoring of a long-span suspension bridge include seismometers for measuring seismic ground motions, accelerometers for measuring earthquake-induced bridge acceleration responses, strain gauges for strain responses, and displacement transducers for displacement responses.

Seismometers are sensors that measure motions of the ground, including those of seismic waves generated by earthquakes, nuclear explosions, and other seismic sources. Modern seismometers are sensitive electromechanical devices. For short-period seismometers, the inertial force produced by a seismic ground motion deflects the mass from its equilibrium position, and the displacement or velocity of the mass is then converted into an electric signal as output proportional to the seismic ground motion. Long-period or broadband seismometers are built according to the force-balanced principle, in which the inertial force is compensated with an electrically generated force so that the mass moves as little as possible. The feedback force is generated with an electromagnetic force transducer through a servo loop circuit. The feedback force is strictly proportional to seismic ground acceleration and it is converted into an electric signal as output without depending on the precision of a mechanical suspension. Strong-ground seismometers, which are widely used for measuring earthquake-induced high ground accelerations that are particularly important for designing bridges and structures, are also built on the force-balanced principle and sometimes called accelerometers. The measured ground accelerations can be mathematically integrated later to give ground velocities and displacements. In order to observe ground motions in three dimensions, a triple set of seismometers oriented towards east-west, north-south and up-down has been the standard for a century. Location for a seismometer must be selected with low seismic noise. An aggressive atmosphere may cause corrosion, wind and short-term variations of temperature may induce noise, and seasonal variations of temperature may exceed the drift specifications. Seismometers must be protected against these influences, sometimes by hermetic containers. The foundation of a seismic station is also critical. A professional station is sometimes mounted on bedrock. The best mounting may be in deep boreholes, which avoid thermal effects, ground noise and tilting from weather and tides.

Similar to other loading effect monitoring, the dynamic displacement responses of a bridge to seismic loading can be measured using displacement transducers, tilt-meters and GPS or obtained by integrating the acceleration records twice with time. The dynamic acceleration responses of the bridge to seismic loading can be measured using piezoelectric or servo-type tri-axial accelerometers. The dynamic stress responses can be measured using strain gauges or optical fibre sensors. Nevertheless, attention must be paid to the effective frequency range of the sensors because the frequency range of dynamic responses of the bridge to seismic loading is quite different from that to railway, highway, and wind loadings. In some cases, even though the displacement transducers, accelerometers and strain gauges used for other load effect monitoring can be utilized for seismic effect monitoring, additional sensors may be needed.   The locations of the additional sensors must be selected according to the requirement of seismic monitoring.

9.3 CONDITIONAL SIMULATION OF SEISMIC GROUND MOTIONS

Observations during earthquakes have clearly demonstrated that seismic ground motions can vary significantly over distances of the same order of magnitude as the dimensions of a long span bridge. This is particularly true for a site with complex topography and different soil conditions. Three major factors are responsible for the spatial variation of seismic ground motion: (1) wave passage effect due to the finite velocity of propagation; (2) incoherence effects resulting from wave scattering or extended source effects; and (3) local site effect because of the spatial variation of soil characteristics (Der Kiureghian and Neuenhofer, 1992; Zerva, 1999). The differences in the multiple support ground motions can then significantly influence the dynamic responses of the bridge. Therefore, non-uniform seismic ground motions must be considered for the seismic response analysis of a long-span suspension bridge.

The most common approach for the generation of spatially variable ground motions caused by earthquake is the unconditional simulation of stochastic fields described by a power spectral density function and a spatial variability model. Deodatis (1996) proposed a spectral-representation-based simulation algorithm to generate sample functions of seismic ground motions at multiple supports of a bridge according to its prescribed cross-spectral density matrix. The wave passage effect, incoherence effect, local side effect, and non-stationary effect are all taken into consideration in the prescribed cross-spectral density matrix, by which non-uniform seismic ground acceleration time histories at multiple supports of a bridge can be generated. Although the unconditional simulation provides a valuable tool for the generation of seismic ground motions, the ground motions generated by the unconditional simulation bear limited association with the actually measured ground motion time histories recorded by the seismometers of an SHM system for the bridge. By contrast, the conditional simulation based on the limited measured ground motion time histories is more flexible and compatible with the site conditions of the area where the long-span suspension bridge is located. The conditional simulation permits the use of one or more measured time histories and generates spatially variable ground motions compatible with these measured time histories and the prescribed spatial variability model (Zerva, 2008). The seismic ground motion time histories generated by the conditional simulation can not only exhibit spatial variability features but also inherit the physical characteristics of non-stationary ground motions in both time and frequency domains.

The method developed by Vanmarcke *et al.* (1993) for conditional simulation of a non-uniform seismic ground motion field is concisely introduced here because of its simplicity. The other methods can be found in the literature (e.g. Heredia-Zavoni and Santa-Cruz, 2000). The simulation is based on multivariate linear prediction and ensures an unbiased covariance structure for the conditional field (Journel and Huijbregts, 1978). It is accomplished by generating Fourier coefficients for independent frequency-specific spatial fields whose component correlation structure is obtained in terms of the spectral density function and the real coherency spectrum of the space-time field. The FFT technique is then used to recompose the time histories from the frequency-specific Fourier coefficients.

9.3.1 Ground Motion Simulation

Let us consider a segment of the ground motion at a point x_i and assume it can be represented by a non-ergodic, zero-mean, homogeneous, mean-square continuous space time process $Y_i(t)$. The process $Y_i(t)$ can be expressed as a sum of independent frequency-specific spatial processes in consecutive constant-size frequency intervals as follows (Vanmarcke *et al.*, 1993),

$$Y_i(t) = \sum_{k=1}^{K} \left[A_{ik} \cos(\omega_k t) + B_{ik} \sin(\omega_k t) \right] \tag{9.1}$$

where coefficients A_{ik} and B_{ik} are zero-mean random variables; $\omega_k = (k-1)\Delta\omega$, ($k = 1, 2, ..., K$); $\Delta\omega = 2\pi / (t_f + \Delta t)$; and t_f is the length of the process. For a discrete-time process $Y_i(t_j)$, defined at times $t_j = (j-1)\Delta t$, ($j = 1, 2, ..., K$), coefficients A_{ik} and B_{ik} are related to $Y_i(t_j)$ through the discrete Direct Fourier Transform,

$$A_{ik} = \frac{1}{K} \sum_{j=1}^{K} Y_i(t_j) \cos(\frac{2\pi(k-1)(j-1)}{K}), \ B_{ik} = \frac{1}{K} \sum_{j=1}^{K} Y_i(t_j) \sin(\frac{2\pi(k-1)(j-1)}{K}) \tag{9.2}$$

where $\Delta t = t_f/(K-1)$. The following symmetry conditions about the Nyquist frequency, $\omega_{1+K/2} = \pi / \Delta t$, apply to the Fourier coefficients A_{ik} and B_{ik} when the process $Y_i(t)$ is real,

$$A_{ik} = A_{i(K-k+2)}, \quad B_{ik} = -B_{i(K-k+2)} \quad \text{for } k = 2,3,...,1+K/2 \tag{9.3}$$

It can be shown (Heredia-Zavoni, 1993) that by using Equation (9.2), the covariance between coefficients at point x_i and x_j, $C_{ij}(\omega_k) = E(A_{ik}A_{jk})$, can be written as follows,

$$C_{ij}(\omega_k) = \begin{cases} \frac{1}{2}\rho_{\omega_k}(\gamma_{ij})G(\omega_k)\Delta\omega, & \text{for} \quad k=1 \\ \frac{1}{4}\left\{\rho_{\omega_k}(\gamma_{ij})G(\omega_k) + \rho_{\omega_{K-k+2}}(\gamma_{ij})G(\omega_{K-k+2})\right\}\Delta\omega, & \text{for} \quad k=2,\cdots,K/2 \\ \rho_{\omega_k}(\gamma_{ij})G(\omega_k)\Delta\omega, & \text{for} \quad k=K/2+1 \end{cases} \tag{9.4}$$

where $\gamma_{ij} = x_i - x_j$ is the relative position vector; $G(\omega)$ is the one-sided "point" spectral density function; and $\rho_{\omega_k}(\gamma_{ij})$ is the frequency-dependent spatial correlation function. Note that $E(A_{ik}A_{jk}) = E(B_{ik}B_{jk}) = C_{ij}(\omega_k)$, for $k = 2,...,K/2$.

The spatially correlated ground motions can then be obtained by the following steps: (1) generate sets of Fourier coefficients A_{ik} and B_{ik}, for each frequency ω_k ($k = 1, 2, ..., 1+K/2$); (2) obtain the remainder using the symmetry conditions in Equation (9.3); and (3) use an FFT algorithm to perform the inverse discrete transform.

9.3.2 Conditional Simulation of Seismic Ground Motion

Consider the simulation of seismic ground motions at a set of m target points x_β , given that some motions have been recorded at a set of $n = N - m$ measurement points, where N is the total number of points. By using Equation (9.4), the covariance matrix $C_k = [C_{ij}(\omega_k)]$, $i, j = \{1, 2, ..., n+m\}$, for each Fourier frequency ω_k $(k = 1, 2, ..., 1+K/2)$ can be assembled and expressed as

$$C_k = \begin{bmatrix} C_{\alpha\alpha} & C_{\alpha\beta} \\ C_{\alpha\beta}^T & C_{\beta\beta} \end{bmatrix} \qquad (9.5)$$

Denote $A_s = \{A_{s\alpha}, A_{s\beta}\}$ as a set of simulated Fourier coefficients, where subsets $A_{s\alpha} = \{A_{1k}, A_{2k}, ... A_{nk}\}$ and $A_{s\beta} = \{A_{(n+1)k}, A_{(n+2)k}, ... A_{(n+m)k}\}$ correspond to the coefficients at recording and target points, respectively. A set of simulated Fourier coefficients B_s can be defined in the same way. To simulate A_s and B_s, covariance matrix C_k is evaluated for frequencies up to $\omega_{1+K/2} = \pi / \Delta t$. For admissible spatial correlation and spectral density functions, C_k is positively definite and can be expressed as the product of a non-singular lower triangular matrix, L_k, and its transpose by means of Cholesky decomposition:

$$C_k = L_k L_k^T \qquad (9.6)$$

Where the limited-duration segment of the ground motion can be modelled as a Gaussian process, two sets of independent standard normal random variables, $U_k = \{U_{1k}, U_{2k}, ... U_{Nk}\}$ and $V_k = \{V_{1k}, V_{2k}, ... V_{Nk}\}$, can be simulated for each frequency. Sets A_s and B_s are then generated from

$$A_s = L_k U_k, \quad B_s = L_k V_k \qquad (9.7)$$

One can easily show that A_s and B_s have the proper covariance structure. Based on the subsets of simulation values at the recording points, $A_{s\alpha}$ and $B_{s\alpha}$, the linear prediction estimators at the target points are given by,

$$A_{s\beta}^* = C_{\alpha\beta}^T C_{\alpha\alpha}^{-1} A_{s\alpha}, \quad B_{s\beta}^* = C_{\alpha\beta}^T C_{\alpha\alpha}^{-1} B_{s\alpha} \qquad (9.8)$$

The conditional simulation now involves generating for each frequency ω_k, sets of Fourier coefficients $A_{sc,k} = \{A_{(n+1)k}, A_{(n+2)k}, ..., A_{(n+m)k}\}$ and $B_{sc,k} = \{B_{(n+1)k}, B_{(n+2)k}, ..., B_{(n+m)k}\}$ at target points x_β, according to the conditional simulation algorithm (Journel and Huijbregts, 1978).

$$A_{sc,k} = \{A_\beta^* + A_{s\beta} - A_{s\beta}^*\}_{(k)}, \quad B_{sc,k} = \{B_\beta^* + B_{s\beta} - B_{s\beta}^*\}_{(k)} \qquad (9.9)$$

where A_β^* and B_β^* are the linear prediction estimators based on the observed Fourier coefficients at the recording points A_α and B_α:

$$A_\beta^* = C_{\alpha\beta}^T C_{\alpha\alpha}^{-1} A_\alpha \quad B_\beta^* = C_{\alpha\beta}^T C_{\alpha\alpha}^{-1} B_\alpha \qquad (9.10)$$

The reminding Fourier coefficients at frequency ω_k ($k = 2 + K/2, ..., K$) are then obtained using the symmetry conditions given by Equation (9.3). Once coefficients have been generated for the entire frequency range, an inverse FFT is applied to yield a set of ground motion time histories at the target points.

9.3.3 Procedure for Conditional Simulation of Seismic Ground Motion

The conditional simulation of seismic ground motions consists of the following steps:

1) For each frequency ω_k, $k = 1, 2, ..., 1 + K/2 = \pi / \Delta t$ in a limited-duration segment of seismic ground motion: obtain from the recorded motions the Fourier coefficients, A_α and B_α, by means of a direct FFT; assemble the

covariance matrix C_k expressed by Equation (9.5); simulate sets of coefficients A_s and B_s at target and recording points according to Equation (9.7); compute the linear prediction estimators A_β^* and B_β^* based on the known coefficients A_α and B_α using Equation (9.10); compute the linear prediction estimators $A_{s\beta}^*$ and $B_{s\beta}^*$ based on the simulations at recording points $A_{s\alpha}$ and $B_{s\alpha}$ using Equation (9.8); and generate the conditional simulations $A_{sc,k}$ and $B_{sc,k}$, using the algorithm given by Equation (9.9).

2) Generate Fourier coefficients for the entire frequency range using the symmetry conditions expressed by Equation (9.3).

3) Use an inverse FFT to construct the time-history segments at the target points.

4) To account for non-stationary, steps 1 to 3 can be repeated for different input spectral density and frequency-dependent correlation function to generate stationary segments of ground motion. These segments are then assembled together by means of a linear interpolation algorithm (Vanmarcke and Fenton, 1991).

5) The simulation motions may be post-processed by introducing partially predictable wave propagation delays (Boissieres, 1992).

9.4 MULTIPLE-SUPPORT SEISMIC ANALYSIS OF LONG-SPAN SUSPENSION BRIDGES

After having properly simulated seismic ground motions, the method for predicting seismic responses of a long-span suspension bridge under the non-uniform seismic ground motion becomes crucial, although there are many methods for seismic response analysis of structures under the uniform seismic ground motion (Der Kiureghian and Neuenhofer, 1992; Loh and Ku, 1995). Léger *et al.* (1990) investigated the relative performance of two simple analytical methods for multiple-support seismic analysis of large structures under the non-uniform seismic excitations. One is the relative motion method (RMM), which divides the structure response into a dynamic response component and a pseudo-static response component. The other is the large mass method (LMM), which attributes fictitious large mass values at each driven nodal DOF to obtain the total response of the structure.

9.4.1 Equations of Motion for Multiple-support Seismic Analysis

The equations of motion of a long-span suspension bridge with multiple supports subjected to non-uniform seismic ground motions are established in the absolute coordinate. The displacement vectors in the equations of motion are divided into two parts: (1) the displacement vector of the superstructure $\{Y_i\}$; and (2) the displacement vector of the supports $\{Y_s\}$. As a result, the equations of motion of the bridge can be expressed in partitioned form (Léger *et al.*, 1990):

$$\begin{bmatrix} M_{ii} & M_{is} \\ M_{si} & M_{ss} \end{bmatrix} \begin{Bmatrix} \ddot{Y}_i \\ \ddot{Y}_s \end{Bmatrix} + \begin{bmatrix} C_{ii} & C_{is} \\ C_{si} & C_{ss} \end{bmatrix} \begin{Bmatrix} \dot{Y}_i \\ \dot{Y}_s \end{Bmatrix} + \begin{bmatrix} K_{ii} & K_{is} \\ K_{si} & K_{ss} \end{bmatrix} \begin{Bmatrix} Y_i \\ Y_s \end{Bmatrix} = \begin{Bmatrix} 0 \\ 0 \end{Bmatrix} \tag{9.11}$$

where $[M_{ii}]$, $[C_{ii}]$ and $[K_{ii}]$ are the mass, damping and stiffness matrices of the bridge superstructure, respectively; $[M_{ss}]$, $[C_{ss}]$ and $[K_{ss}]$ represent the mass, damping and stiffness matrices of the bridge supports, respectively.

9.4.2 Relative Motion Method

The RMM (Clough and Penzien, 1975) is based on the principle of superposition. The bridge response to multiple-support excitation is divided into a pseudo-static response component, which is a static response included in the bridge structure by different support motions, and a dynamic response relative to the fixed multiple-support system subjected to the ground excitation. The total bridge response can be written as

$$\begin{Bmatrix} Y_i \\ Y_s \end{Bmatrix} = \begin{Bmatrix} Y_i^d \\ 0 \end{Bmatrix} + \begin{Bmatrix} Y_i^s \\ Y_s \end{Bmatrix} \tag{9.12}$$

where $\{Y_i^d\}$ is the relative dynamic response of the superstructure; and $\{Y_i^s\}$ is the pseudo-static structural response which satisfies

$$\begin{bmatrix} K_{ii} & K_{is} \\ K_{si} & K_{ss} \end{bmatrix} \begin{Bmatrix} Y_i^s \\ Y_s \end{Bmatrix} = \begin{Bmatrix} 0 \\ 0 \end{Bmatrix} \tag{9.13}$$

where

$$\{Y_i^s\} = [R]\{Y_s\} \tag{9.14}$$

and $[R]$ is a matrix of influence coefficients obtained from the solution of

$$[K_{ii}][R] = -[K_{is}] \tag{9.15}$$

The coefficients of $[R]$ physically correspond to the displacement of the superstructure due to a unit displacement applied alternately to each support.

To obtain the dynamic response components $\{Y_i^d\}$, the equations of motion of the bridge subjected to a specified support acceleration vector $\{\ddot{Y}_s\}$ can be written as

$$[M_{ii}]\{\ddot{Y}_i^d\} + [C_{ii}]\{\dot{Y}_i^d\} + [K_{ii}]\{Y_i^d\} = -([M_{ii}][R] + [M_{is}])\{\ddot{Y}_s\} \tag{9.16}$$

where $[M_{is}]$ are the off-diagonal coefficients in the mass matrix which introduce coupling between the supports and the superstructure. Once the pseudo-static and dynamic displacement response components have been obtained, the total earthquake-induced bridge responses are computed from the superposition of the dynamic and pseudo-static responses.

9.4.3 Large Mass Method

The LMM is the dynamic equivalent of the penalty method used in static analyses. In order to specify accelerations at the supports, let us consider the inertia term of Equation (9.11) to which large mass values M_{ll} have been added at the driven supports in such a way that inertia forces dominate the response. To reproduce the specified accelerations, the inertia forces developed at the supports should be considered as an external driving force, thus enforcing the identity at the supports. As a result, Equation (9.11) can be written as

$$\begin{bmatrix} M_{ii} & M_{is} \\ M_{si} & (M_{ss}+M_{ll}) \end{bmatrix}\begin{Bmatrix} \ddot{Y}_i \\ \ddot{Y}_s \end{Bmatrix} + \begin{bmatrix} C_{ii} & C_{is} \\ C_{si} & C_{ss} \end{bmatrix}\begin{Bmatrix} \dot{Y}_i \\ \dot{Y}_s \end{Bmatrix} + \begin{bmatrix} K_{ii} & K_{is} \\ K_{si} & K_{ss} \end{bmatrix}\begin{Bmatrix} Y_i \\ Y_s \end{Bmatrix} = \begin{bmatrix} 0 \\ (M_{ss}+M_{ll}) \end{bmatrix}\{\ddot{Y}_s\} \quad (9.17)$$

By this approach, the absolute response can be obtained directly if rigid body modes are included in the response analysis procedure. The earthquake induced internal forces can also be obtained directly, and correspond to the sum of the pseudo-static and dynamic components of the RMM.

9.4.4 Remarks on Two Methods

The RMM and the LMM are two simple analytical methods to perform multi-support seismic analysis of long-span suspension bridges. It is noted that for long-span suspension bridges, the RMM might be very time consuming because it requires an additional static solution at each time step used for the integration of the dynamic equilibrium equations. Also, it requires the integration of the input ground acceleration to provide the ground displacement time history required for the static solution. The RMM cannot be directly extended to study the nonlinear seismic response of long-span suspension bridges. On the other hand, the LMM is not derived from a rigorous mathematical formulation. The large mass should be chosen as large as possible to obtain accurate support motions but should not be so large as to cause excessive round-off errors or overflows. Numerical sensitivity analyses to large mass values are thus recommended. The LMM can be applied directly in a step-by-step integration procedure to compute the nonlinear seismic responses of long-span suspension bridges.

9.5 MONITORING OF SEISMIC EFFECTS ON THE TSING MA BRIDGE

9.5.1 The Tsing Ma Bridge and Earthquake Record

The Tsing Ma Bridge in Hong Kong is a long-span suspension bridge with multiple supports as shown in Figure 9.1. The bridge has a total length of 2,132 m with a main span of 1,377 m between the Tsing Yi tower in the east and the Ma Wan tower in the west. The bridge has a total of 11 supports: (1) the Tsing Yi main anchorage; (2) the Tsing Yi abutment; (3) the Tsing Yi pier T1; (4) the Tsing Yi pier T2; (5) the Tsing Yi pier T3; (6) the Tsing Yi tower; (7) the Ma Wan tower; (8) the Ma Wan pier M2; (9) the Ma Wan pier M1; (10) the Ma Wan anchorage; and (11) the Ma Wan abutment. The two reinforced concrete towers have the same height of about 206 m measured from the base level to the tower saddle. The distances between 11 supports can be seen in Figure 9.1.

Because Hong Kong is located at a low to moderate seismic zone, only one accelerometer was installed on the wall of the Ma Wan main anchorage to monitor its movement due to seismic ground motion. Nevertheless, there are no useable time history records of the anchorage due to seismic ground motions in the past years. As a result, the WASHMS-based evaluation of the seismic effect on the Tsing Ma Bridge could not be performed completely at this moment. In this section, the acceleration time history recorded by the accelerometer installed at the site of the Ting Kau Bridge, which is very close to the Tsing Ma Bridge, during the Taiwan Strait earthquake occurring on 26 December 2006 at Hong Kong Time 20:28 is used as a demonstration of the conditional simulation of seismic ground motions and the evaluation of seismic effect on the bridge.

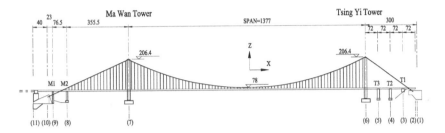

Figure 9.1 Configuration of Tsing Ma Bridge and locations of supports

9.5.2 Conditional Simulation of Ground Motion

Since the accelerometer installed at the site of the Ting Kau Bridge is far away from the seismic centre of the Taiwan Strait earthquake, the frequency components of the acceleration time history recorded by the accelerometer were contaminated by various types of noise. In this regard, a band-pass filter is employed to remove noise components from the acceleration time history as much as possible. The filtered ground acceleration time history is shown in Figure 9.2, in which the time interval is 0.02 sec and the total length is 16.8 sec with 840 time points. This ground acceleration time history is also amplified with a peak acceleration of 0.15g for a 475 year return period.

The acceleration time history is then applied to the Tsing Yi anchorage (support (1)) with the ground vibration perpendicular to the bridge alignment and the direction of wave propagation along the bridge alignment. The seismic wave velocity is set as 1000 m/s according to the site condition of the Tsing Ma Bridge. The following frequency-dependent spatial correlation function of phase-aligned ground accelerations (Harichandran and Vanmarcke, 1986) is adopted.

$$\rho_{\omega_k}\left(\gamma_{ij}\right) = 0.736\exp\left\{-5.063\left|\gamma_{ij}\right|/\theta_\omega\right\} + 0.264\exp\left\{-0.744\left|\gamma_{ij}\right|/\theta_\omega\right\} \tag{9.18}$$

$$\theta_\omega = 5210\left\{1+\left(\frac{\omega}{2.17\pi}\right)^{2.78}\right\}^{-1/2} \tag{9.19}$$

where θ_ω is the frequency-dependent scale of fluctuation (Vanmarcke, 1983).

The conditional simulation method proposed by Vanmarcke *et al.* (1993) is used here to generate a set of ground acceleration time histories for the other 10 supports of the Tsing Ma Bridge. In this regard, the recorded ground acceleration time history shown in Figure 9.2 is divided into four segments with the time durations of 3.02s, 4.04s, 5.08s and 4.66s, respectively. Each segment of data is considered stationary. For each segment, the Fourier coefficients and the spectral density function can be obtained by means of a direct FFT according to Equation (9.2). The covariance matrix can be calculated according to Equation (9.5) based on the spectral density function and the frequency-dependent spatial correlation function. The Fourier coefficients of the ground acceleration time histories at the other 10 supports of the bridge for the entire frequency range can be obtained by using Equations (9.7) to (9.10) and the symmetric condition expressed by Equation (9.3). An inverse FFT is finally used to construct the four segments of acceleration time-histories at the target supports. Afterwards, the four segments can be assembled together through a simple linear interpolation algorithm (Vanmarcke and Fenton, 1991). Consider two adjacent segments of the time history, the first segment starting at time $(n_f - K_e + 1)\Delta t$ and ending at $n_f \Delta t$ and the second segment at time $(n_f + 1)\Delta t$ and ending at $(n_f + K_e)\Delta t$, where K_e denotes the length of each stationary segment and n_f is the time point corresponding to the end of the window. Calling the first segment $Y^i(t_j)$ and the second segment $Y^{i+1}(t_j)$, the final non-stationary time series $Y(t_j)$ can be obtained as follows

$$Y(t_j) = \begin{cases} Y^i(t_j) & \text{if} \quad j < n_f - n_v \\ w(n_f - j)Y^i(t_j) + (1 - w(n_f - j))Y^{i+1}(t_j) & \text{if} \quad n_f - n_v \le j \le n_f - n_v \\ Y^{i+1}(t_j) & \text{if} \quad j > n_f + n_v \end{cases} \quad (9.20)$$

where $w(j)$ is a discrete weighting function defined over the transition region $[-n_v, n_v]$, and it varies linearly such that $w(n_v) = 1$ and $w(-n_v) = 0$. Therefore, $Y(t_{n_f})$ is just equal to the average of $Y^i(t_{n_f})$ and $Y^{i+1}(t_{n_f})$, and has increasing contributions from the appropriate segments as the transition boundaries are approached. In the performance of this non-stationary extension, the size of the transition region n_v is chosen so that it covers a few oscillations. Although Equation (9.20) guarantees continuity of $Y^i(t_j)$ even when $n_v = 0$, such a small value will result in a discontinuous derivative while a large n_v will lead to a continuous derivative.

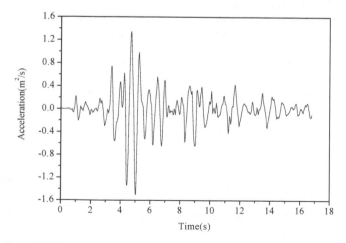

Figure 9.2 The ground acceleration time history at the site of Ting Kau Bridge

Finally, the simulated ground acceleration time histories at the 10 supports are post-processed by introducing partially predictable wave propagation phase delays (Boissieres, 1992). The ground acceleration time histories obtained at the 10 supports of the Tsing Ma Bridge are graphed in Figure 9.3.

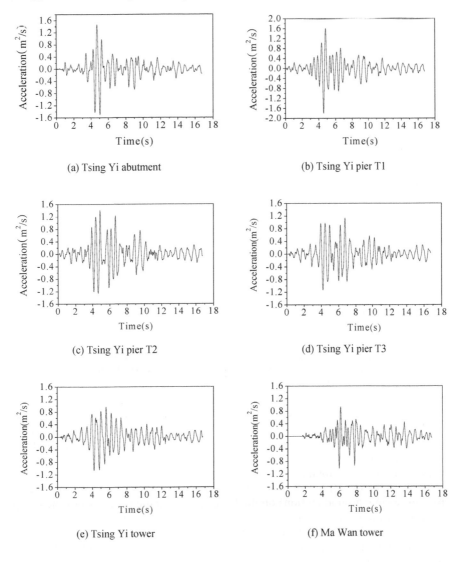

(a) Tsing Yi abutment

(b) Tsing Yi pier T1

(c) Tsing Yi pier T2

(d) Tsing Yi pier T3

(e) Tsing Yi tower

(f) Ma Wan tower

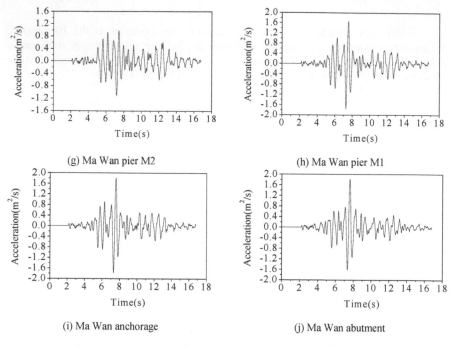

(g) Ma Wan pier M2

(h) Ma Wan pier M1

(i) Ma Wan anchorage

(j) Ma Wan abutment

Figure 9.3 The simulated ground acceleration at supports (2) to (11)

9.5.3 Seismic Response of the Tsing Ma Bridge

The SHM-oriented FE model of the Tsing Ma Bridge, shown in Figure 9.4, is used for seismic response analysis of the bridge. Details of the model can be found in Chapter 4 of this book. Since the bridge model is very large in size and is established in the ABAQUS commercial package, the LMM, which is embedded in the computer software, is used to determine the seismic response of the bridge under non-uniform seismic ground motions. As mentioned above, the big masses should be chosen as large as possible to obtain accurate base motions but should not be so large as to cause excessive round-off errors or overflows. To provide six digits of numerical accuracy, ABAQUS computer package chooses each big mass equal to 10^6 times the total mass of the bridge and each big rotational-inertia equal to 10^6 times the total moment of inertia of the bridge. To assess the effect of non-uniform seismic ground motions on the bridge, the seismic responses of the bridge under a uniform seismic ground motion are also computed, in which the measured ground acceleration time history at support (1) is applied to all 10 other supports.

Although the responses of the entire bridge can be obtained from the computer simulation, only five quantities are shown here: (a) the lateral displacement (in the y-direction) at the top of one leg of the Ma Wan tower; (b) the lateral displacement at the top of one leg of the Tsing Yi tower; (c) the bending moment (about the x-axis) at the bottom of one leg of the Ma Wan tower; (d) the bending moment (about the x-axis) at the bottom of one leg of the Tsing Yi tower; and (e) the lateral displacement of the bridge deck at the middle main span of the bridge. The lateral displacement of the Ma Wan tower and the Tsing Yi tower are shown in Figure 9.5 and Figure 9.6, respectively. The absolute peak displacements of the Ma Wan tower and the Tsing Yi tower under the non-uniform seismic ground motion are 0.0045 m and 0.0056 m, respectively. Their counterparts under the uniform seismic ground motion are 0.0067 m and 0.0061 m, respectively. It can be seen that the lateral displacement of the Ma Wan tower under the non-uniform seismic ground motion is smaller than that under the uniform seismic ground motion. Nevertheless, there is no considerable difference between the lateral displacement of the Tsing Yi tower under the non-uniform and uniform seismic ground motions.

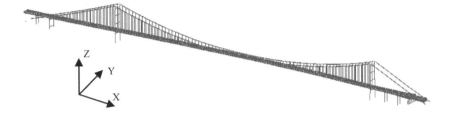

Figure 9.4 3-D finite element model of Tsing Ma Bridge

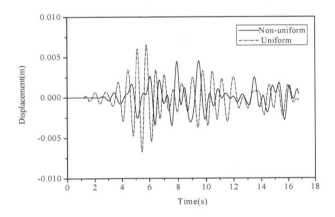

Figure 9.5 The lateral displacement at the top of the Ma Wan tower

The bending moment responses of the Ma Wan tower and the Tsing Yi tower are shown in Figure 9.7 and Figure 9.8, respectively. The absolute peak bending moments of the towers are 197,673 kN·m and 388,806 kN·m under the non-uniform seismic ground motion and 351,565 kN·m and 393,536 kN·m under the uniform seismic ground motion. Again, it can be seen that the bending moment of the Ma Wan tower under the non-uniform seismic ground motion is smaller than that under the uniform seismic ground motion, and there is no considerable difference in the bending moment of the Tsing Yi tower under the non-uniform and uniform seismic ground motions.

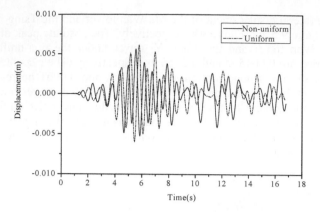

Figure 9.6 The lateral displacement at the top of the Tsing Yi tower

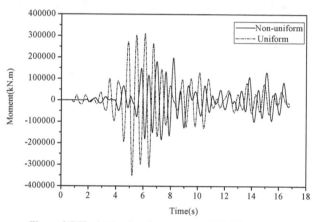

Figure 9.7 The bottom bending moment of Ma Wan tower

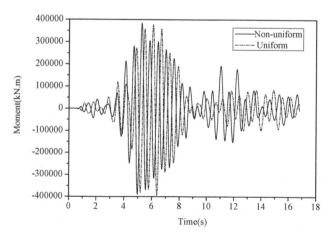

Figure 9.8 The bottom bending moment of Tsing Yi tower

Figure 9.9 shows the lateral displacement responses of the bridge deck at the middle main span under both non-uniform and uniform seismic ground motions. The corresponding absolute peak values are 0.0206 m under the non-uniform seismic ground motion and 0.0187 m under the uniform seismic ground motion. The non-uniform seismic ground motion caused slightly larger lateral displacement of the bridge deck at its middle than the uniform seismic ground motion.

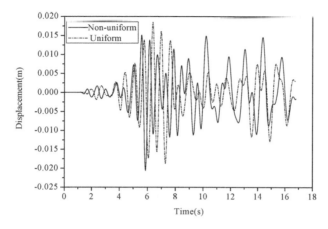

Figure 9.9 The lateral displacement of the bridge deck at its middle

9.6 SUMMARY

The major sensors and sensing systems used for seismic monitoring of long-span suspension bridges have been described in this chapter. The conditional simulation of properly-correlated earthquake ground motions at an arbitrary set of closely-spaced points, which are compatible with known or prescribed motions at other locations, has been presented. The two relatively simple methods, the RMM and the LMM, have been described concisely for predicting seismic responses of a long-span suspension bridge under the non-uniform seismic ground motion. Finally, the Tsing Ma suspension bridge has been taken as an example to illustrate the monitoring of seismic ground motions and seismic effects on the bridge. The results show that consideration of the wave passage, loss of coherence and different local soil conditions is important for simulating the spatial variability of ground motions. The seismic response results, including both bending moment and displacement responses, also show that the spatial variability of ground motions must be taken into consideration for the seismic response analysis of long-span suspension bridges with multiple supports.

9.7 REFERENCES

Abdelghaffar, A. M., 1982, Suspension bridge vibration: continuum formulation. *Journal of the Engineering Mechanics Division, ASCE*, **108(6)**, pp. 1215-1232.

Abdelghaffar, A. M. and Rubin, L.I., 1983, Lateral earthquake response of suspension bridges. *Journal of Structural Engineering, ASCE*, **109(3)**, pp. 664-675.

Boissieres, H.P., 1992, *Estimation of the Correlation Structure of Random Fields*, Ph.D. Dissertation, (Princeton: Princeton University).

Clough, R.W. and Penzien, J., 1975, *Dynamics of Structures*, (New York: McGraw-Hill).

Deodatis, G., 1996, Non-stationary stochastic vector processes: seismic ground motion applications. *Engineering Mechanics*, **11(3)**, pp. 149-168.

Der Kiureghian, A. and Neuenhofer, A., 1992, Response spectrum method for multi-support seismic excitations. *Earthquake Engineering and Structural Dynamics,* **21**, pp. 713-740.

Harichandran, R. and Vanmarcke, E., 1986, Stochastic variation of earthquake ground motion in space and time. *Journal of Engineering Mechanics*, **112(2)**, pp. 154-174.

Heredia-Zavoni, E., 1993, *On the Response of Multi-support MDOF Systems to Spatially Varying Earthquake Ground Motion*, Ph.D. Dissertation, (Princeton: Princeton University).

Heredia-Zavoni, E. and Santa-Cruz, S., 2000, Conditional simulation of a class of nonstationary space-time random fields. *Journal of Engineering Mechanics*, ASCE, **126(4)**, pp. 398-404.

Journel, A. and Huijbregts, C.H., 1978, *Mining Geostatistics*, (London: Academic Press).

Léger, P., Idé, I. M. and Paultre, P., 1990, Multiple-support seismic analysis of large structures. *Computers & Structures*, **36(6)**, pp. 1153-1158.

Loh, C. H. and Ku, B.D., 1995, An efficient analysis of structural response for multiple-support seismic excitations. *Engineering Structures*, **17(1)**, pp. 15-26.

Lueker, M., Marr, J., Ellis, C., Winsted, V. and Akula, S.R., 2010, *Bridge Scour Monitoring Technologies: Development of Evaluation and Selection Protocols for Application on River Bridges in Minnesota*, Report MN/RC 2010-14, (Minnesota: Minnesota Department of Transportation).

Vanmarcke, E.H., 1983, *Random Fields: Analysis and Synthesis*, (Cambridge, Massachusetts: MIT Press).

Vanmarcke, E.H. and Fenton, G.A., 1991, Conditioned simulation of local-fields of earthquake ground motion. *Structural Safety*, **10(1-3)**, pp. 247-264.

Vanmarcke, E.H., Herediazavoni, E. and Fenton, G.A., 1993, Conditional simulation of spatially correlated earthquake ground motion. *Journal of Engineering Mechanics, ASCE*, **119(11)**, pp. 2333-2352.

Zerva, A., 1999, Spatial variability of seismic motions recorded over extended ground surface areas. *Wave Motion in Earthquake Engineering*, edited by Kausel, E. and Manolis, G., (Cambridge, Massachusetts: MIT Press).

Zerva, A., 2008, *Spatial Variation of Seismic Ground Motion: Modeling and Engineering Applications*, (Boca Raton, Florida: CRC Press).

9.8 NOTATIONS

A_{ik}, B_{ik} Fourier coefficients to a segment of the ground motion.

A_s, B_s	Simulated Fourier coefficients.
$A_{sc,k}$, $B_{sc,k}$	Fourier coefficients according to the conditional simulation algorithm.
A_β^*, B_β^*	Linear prediction estimators based on the observed Fourier coefficients.
A_α, B_α	Fourier coefficients from the recorded motions.
A_{su}, B_{su}	Linear prediction estimators at the simulated target points.
C_{ij}, C_k	Coefficient covariance.
$G(\omega)$	One-sided "point" spectral density function.
K_e	Length of each stationary segment.
K	Number of time points.
L_k,	Cholesky decomposition of coefficient covariance.
$[M_{ii}]$, $[C_{ii}]$, $[K_{ii}]$	Mass, damping and stiffness matrices of the bridge superstructure.
$[M_{ss}]$, $[C_{ss}]$, $[K_{ss}]$	Mass, damping and stiffness matrices of the bridge supports.
M_{ll}	Mass values added at the driven supports considering inertia forces.
N, m, n	Number of total points, measurement points, target points.
n_j	The jth time point.
n_f	Time point corresponding to the end of window.
n_v	Size of a transition region.
$[R]$	Matrix of influence coefficients.
t_f	Length of a process.
U_k, V_k	Simulated standard normal random variables.
$w(j)$	Discrete weighting function over a defined transition region.
$Y_i(t)$	A segment of the ground motion.
$\{Y_i\}$, $\{Y_s\}$	Displacement vector of the superstructure and the supports.
$\{Y_i^s\}$	Pseudo-static structural response.
$\{Y_i^d\}$	Relative dynamic response of the superstructure.
θ_ω	Frequency-dependent scale of fluctuation.
ρ_{ω_k}	Frequency-dependent spatial correlation function.
γ_{ij}	Relative position vector of point x_i and x_j.
ω_k	Frequency points.

CHAPTER 10

Monitoring of Other Effects

10.1 PREVIEW

The previous chapters have discussed the monitoring of common loading effects on long-span suspension bridges. Other factors, such as corrosion, ship collision, and scour, may also affect safety, durability, and serviceability of bridges. The basic concepts and monitoring of these factors will be briefly introduced in this chapter to provide complete coverage of the subject.

As the Tsing Ma Bridge has a low risk of experiencing these effects (corrosion, ship collision, and scour), its SHM system does not include the corresponding sensors. For example, design of the bridge has quality control requirements relating to both the concrete constitutions and mixture. Epoxy coated reinforcements and increased cover have been used for concrete components. A maintenance team comprising about 40 workers began painting, checking, and replacing corroding rivets immediately after completion of the bridge. Regarding ship collision, impact forces resulting from ships of 220,000 dead weight tonnes were considered in the design. Moreover, an artificial island was adopted for vessel protection of the Ma Wan tower, as shown in Figure 10.3.

10.2 CORROSION MONITORING

Corrosion is actually not a loading factor, but corrosion of reinforced bars causes deterioration and reduces service life of concrete structures, especially of the structures located in the coastal marine environment. Corrosion generally starts from the ingress of chloride ions and carbon dioxide to the steel surface, which results in expansive stresses and causes cracks and spalling of the concrete cover. This in turn leads to faster ingress of water, oxygen and chlorides, accelerating corrosion of the reinforced bars.

10.2.1 Measurement Methods

Several methods have been developed for detecting and evaluating the effects of steel corrosion in concrete. These methods include visual inspection, non-destructive testing, and electrochemical techniques. Visual inspection is subjective and can only detect significant corrosion. The electrochemical methods require concrete cores to be brought to laboratories for testing. Core sampling is destructive and requires repair of the concrete. In addition, evaluation of the results relies on the subjective.

Methods for corrosion measurement include (Song and Saraswathy, 2007):

1) Open circuit potential measurements
2) Surface potential measurements
3) Concrete resistivity measurement
4) Linear polarization resistance measurement
5) Tafel extrapolation technique
6) Galvanostatic pulse transient method
7) Electrochemical impedance spectroscopy
8) Harmonic analysis
9) Electrochemical noise analysis
10) Embeddable corrosion monitoring sensors
11) Cover thickness measurements
12) Ultrasonic pulse velocity technique
13) X-ray and Gamma radiography measurement
14) Infrared thermograph electrochemical method
15) Visual inspection.

10.2.2 Corrosion Sensors

Embeddable corrosion sensors have been developed and installed in concrete structures. They can monitor different factors affecting steel corrosion, such as chloride content (Huston and Fuhr, 1998; Leung *et al.*, 2009), pH value (Habel *et al.*, 2007; Melhorn *et al.*, 2007), and humidity (Melhorn *et al.*, 2007).

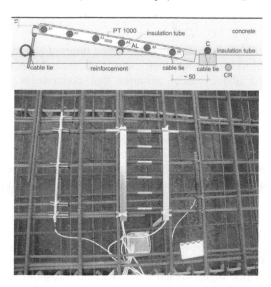

Figure 10.1 The Anode-Ladder sensor for corrosion monitoring
(Courtesy of S+R Sensortec GmbH, Germany)

One popular product is the Anode Ladder System developed in Germany (Raupach and Schießl, 2001), which can measure the chloride-induced corrosion of steel at different levels in the cover of concrete structures (see Figure 10.1). The system actually measures the macrocell current between a steel anode and a stainless steel cathode embedded in concrete. When the critical chloride content is reached or the pH value decreases due to carbonation, the steel surface (anode) is not protected against corrosion. Consequently an electron flow between the anode and cathode is generated and can be measured. The Anode Ladder System uses several anodes placed at different depths and measures the onset of corrosion of the anodes one by one, allowing measurements of the time-dependent corrosion behaviours. For monitoring corrosion of existing structures, an expansion ring anode can be installed into a drill hole, but it is not suitable for long-term use.

The Embedded Corrosion Instrument, a product of Virginia Technologies Incorporated (http://www.vatechnologies.com), monitors the corrosion environment in steel-reinforced concrete over three periods: in construction, during the curing of concrete, and on the long-term use of the structure. The system monitors five key factors in corrosion – linear polarization resistance, open circuit potential, resistivity, chloride ion concentration, and temperature. This provides comprehensive and detailed data.

10.2.3 Theoretical Models of Chloride Penetration

The corrosion induced damage developed in a typical reinforced concrete bridge exposed to chlorides can be divided two stages: the corrosion initiation stage and propagation stage.

The chloride ingress into uncracked concrete can be determined by using Fick's Second Law of Diffusion (Boulfiza *et al.*, 2003)

$$C(x,t) = C_0 \left[1 - erf\left(0.5x / \sqrt{Dt} \right) \right] \tag{10.1}$$

where $C(x,t)$ is the chloride content at depth x and time t, C_0 is the surface chloride content, D is the chloride diffusion coefficient in uncracked concrete, and erf is the error function. The above model assumes that concrete is homogenous and diffusion is the main mechanism of chloride ingress into concrete. The surface chloride content and the chloride diffusion coefficient can be obtained from available field data.

In cracked concrete, chloride ingress through cracks may lead to premature reinforcement corrosion and concrete deterioration (Rodriguez and Hooton, 2003). A simplified smeared approach has been recommended by Japanese researchers to estimate the effect of cracks on chloride ingress, in which an averaged diffusion coefficient is used in Fick's law as (Boulfiza *et al.*, 2003)

$$D_{av} = D + D_{cr} w / s \tag{10.2}$$

where D_{av} is the averaged diffusion coefficient, w is the crack width, s is the crack spacing, and D_{cr} is the diffusion coefficient inside the crack, which may be assumed to be 5×10^{-10} m^2/s.

When the chloride ingress reaches the chloride threshold, the steel corrosion is expected to initiate. Then the corrosion initiation time can be estimated.

10.3 SHIP COLLISION EFFECTS

10.3.1 Introduction

Ship collision is one of the most frequent causes of bridge failures (Harik *et al.*, 1990; Shiraishi and Cranston, 1992; Wardhana and Hadipriono, 2003).

Van Manen and Frandsen (1998) reviewed 150 incidents of ship collision with bridges from 1960 to 1998. The ship collision with the greatest loss of life occurred in Russia in 1983 and killed 176 people (Knott and Prucz, 2000). Other well-known tragedies include the collapse of the Sunshine Skyway Bridge in 1980 with 35 lives lost due to a freighter impacting on a bridge pier, the Interstate 40 Bridge in Webbers Falls, Oklahoma being struck by a towboat on 26 May 2002 which resulted in 14 fatalities and 5 injuries, and the Big Bayou Canot Bridge near Alabama which collapsed after a barge collision and resulted in derailment of a passing train, killing 47 and injuring 103, on 22 September 1993. A recent catastrophe occurred in the Xijiang River in Guangdong Province, China in the early morning of 15 June 2007 (see Figure 10.2). A cargo vessel travelling along the river ploughed into a section of the 1,600-metre-long Jiujiang cable-stayed bridge that spans the river. Two spans collapsed and four vehicles carrying seven people and two road workers plunged into the river in the accident. On 27 March 2008, a cargo vessel hit the Jintang Bridge under construction in Zhejiang Province, China. A 60-metre-long girder fell onto the ship and killed four crew members (see Figure 10.2).

(a) Jiujiang cable-stayed bridge, Guangdong, China   (b) Jintang bridge, Zhejiang, China

Figure 10.2 Collapsed bridges due to ship collision

10.3.2 Preventive Measures

Prevention or reduction of the loss due to the ship collision can be achieved by the following three methods: prevention of collision, detection of collision, and warning of collision.

10.3.2.1 Prevention of Ship-collision

Various types of protective structures for bridges include fender systems, pile supported systems, dolphin protection, artificial island, and floating protection systems (Larsen, 1993). In addition, bridge piers and navigation can be equipped with signs, lights, horns, and buoys to reduce the likelihood of ship collision. The artificial island for vessel protection of the Tsing Ma Bridge is shown in Figure 10.3.

On the other hand, nowadays international ships with gross tonnage of 300 tonnes or more and all passenger ships are required to be equipped with an automatic identification system, which is principally for identification and locating vessels. The system provides a means for ships to electronically exchange data including identification, position, course, and speed with other nearby ships and shore stations. This is the primary method of collision avoidance.

Figure 10.3 The artificial island of the Tsing Ma Bridge

10.3.2.2 Detection of Ship-collision

Early detection of ship-collision events allows a rapid warning to be issued to the motorists and relevant sectors to reduce the potential loss. The detection methods include ship channel surveillance with laser beam, radar system, closed circuit television cameras, radio beam guidance system, bridge continuity sensors, pier vibration sensors, and roadway delineation with reflectors or rail lights (FHWA, 1983).

In practice, ship-bridge collision events have been rarely monitored with equipment. The Vincent Thomas Suspension Bridge, with a total span 1,850 metres long, was struck by a large cargo ship in 2006. Dynamic responses during the collision were recorded by the continuous monitoring system that had been installed before the accident (Yun *et al.*, 2008). Both global system identification and local system identification approaches were employed to assess the bridge condition before and after the incident. In June 2005, shortly after the commissioning of the monitoring system, the Jiangyin Suspension Bridge with a main span of 1,385 m experienced a ship-collision accident. The top of the piling rig in a heavy pile-driving boat collided with the bridge deck and the structural responses were collected by various sensors including accelerometers, displacement transducers, fibre optic strain sensors, load pins and elasto-magnetic sensors of the bridge structure. Hilbert-Huang Transform was employed to analyse the collision data and assess the condition of the bridges after the accidents (Guo *et al.*, 2009).

10.3.2.3 Warning Occurrence of Ship-collision

It has been discovered that many fatalities in ship-bridge collisions have resulted from uninformed motorists (or trains) driving across the collapsed spans. Developing a reliable monitoring and warning system is imperative to detect catastrophic failures of the bridges and warn motorists of such events, with the ultimate aim to reduce further loss of life.

The warning measures are to install signs, flashers, traffic signals, and citizen band radio in appropriate positions on the bridge to alert motorists accordingly. After collapse of the Queen Isabella Memorial Bridge in 2001, the Texas Department of Transportation installed a collision detection system on the bridge to warn motorists of the occurrence of collapse (Mercier *et al.*, 2005). The system consists of a fibre optic cable that carries a current under the bridge deck. Collapse of a span will break the current, activate the flashing lights on the bridge, warn motorists, close gates, and send an alarm signal to the Department of Transportation.

10.3.3 Design and Analysis of Ship-collision

According to the AASHTO *Guide Specification and Commentary for Vessel Collision Design of Highway Bridges* (AASHTO, 1991), all bridge components in a navigable waterway crossing, located in design water depths not less than 600 mm, must be designed for vessel impact. The vessel collision loads should be determined on the basis of the bridge importance and characteristics of the bridge, vessels, and waterway.

Bridge analysis subjected to ship collision typically involves either the static pier analysis under the equivalent static design loading, or use of general-purpose contact nonlinear finite element codes. The former approach is usually used in the bridge design practice and is incapable of taking account of dynamic responses (Consolazio and Cowan, 2005). The latter approach usually involves very fine meshes and small time steps with consideration of material nonlinearity and damage criterion, which is computationally expensive and time-consuming.

To design and evaluate the bridges that might be collided with by off-course ships, determination of load imparted to the bridges is important. In AASHTO *Load and Resistance Factor Design* (AASHTO, 2005), the head-on ship collision impact force on a pier is taken as an equivalent static force:

$$P_s = 1.2 \times 10^5 V \sqrt{DWT} \qquad\qquad (10.3)$$

where P_s is the equivalent static vessel impact force (N), DWT is the dead weight tonnage of vessel (Mg), and V is the vessel impact velocity (m/s).

In the AASHTO code (AASHTO, 1991), the design impact force is based on the kinetic energy of the ship that is calculated as

$$KE = \frac{C_H}{29.2} WV^2 \qquad\qquad (10.4)$$

where KE is the kinetic energy (kip-ft), W is the vessel weight (ton), V is the impact speed (ft/sec), and C_H is the hydrodynamic mass coefficient to account for the additional inertia forces caused by the mass of the water surrounding and moving with the ship. The static equivalent impact load is computed as

$$P_B = \begin{cases} 4112\alpha_B R_B & \alpha_B < 0.34\text{ft} \\ (1349 + 110\alpha_B)R_B & \alpha_B \geq 0.34\text{ft} \end{cases} \qquad\qquad (10.5)$$

where P_B is in kips, α_B is the vessel crush depth (ft) that is empirically predicted as

$$\alpha_B = \left(\sqrt{1 + \frac{KE}{5672}} - 1 \right)\left(\frac{10.2}{R_B} \right) \qquad\qquad (10.6)$$

$R_B = B_B/35$, and B_B is the vessel width (ft).

In Eurocode 1 (2006), the impact action is represented by two mutually exclusive forces, a frontal force and a later force including a friction component. In the absence of a dynamic analysis, indicative forces that depend on the characteristics of ship type are recommended for inland waterways and sea waterways. To facilitate advanced impact analysis, the dynamic design impact forces are also recommended for inland waterways and sea waterways. For example, for inland waterways, it is calculated as:

$$F_{dyn} = \begin{cases} 10.95\sqrt{E_{def}} & E_{def} \leq 0.21\text{MN} \cdot \text{m} \\ 5.0\sqrt{1 + 0.128 E_{def}} & E_{def} > 0.21\text{MN} \cdot \text{m} \end{cases} \qquad\qquad (10.7)$$

where F_{dyn} is the dynamic design impact force (MN), E_{def} is the deformation energy (MN·m), $E_{def} = E_a (1 - \cos\theta)$, in which E_a is the kinetic energy, and θ is the impact angle. For the elastic deformation ($E_{def} \leq 0.21$ MN·m or $F_{dyn} = \leq 5$ MN), the impact force is modelled as a half-sine-wave pulse. For the plastic deformation ($E_{def} > 0.21$ MN·m or $F_{dyn} > 5$ MN), the impact force is modelled as a trapezoidal pulse.

However, the actual impact force is very complicated and time dependent. It depends on the type, size, and speed of the vessel, the collision angle, and the geometry of the bridge. Only a small number of experimental tests have been conducted. For example, Meier-Dornberg (1983) carried out static and dynamic loading tests on scaled models, on which the AASHTO provisions are based. In the laboratory tests, the impact is usually simulated by a falling impact pendulum hammer and the impact energy is determined by the weight and height of the dropping hammer. The use of reduced scale models is a major limitation of laboratory based tests.

The full-scale ship-bridge collision experiment has rarely been conducted due to limited resources. To fill the gap, the Florida Department of Transportation performed a full-scale experimental barge impact testing of the old Saint George Island Causeway Bridge in Apalachicola Florida, before it was replaced and demolished in 2004 (Consolazio *et al.*, 2002; 2003; 2006). The experimental testing aimed to quantify and characterize impact loads that were imparted to bridge piers. Nonlinear dynamic FE analyses were conducted to compare the numerical results and measurement data. It was found that the static design loads specified in the AASHTO for moderate and high energy impacts were substantially larger than the corresponding impact loads measured in the experiments, whereas the static design loads for low energy impacts were smaller than the measurements.

As a full-scale ship-bridge collision experiment is not usually available, numerical simulation analyses with appropriate verifications can obtain detailed results about the impact force, load path, soil-pile interaction, local damage characteristics, and responses. As nonlinear dynamic analyses, especially contact-impact analyses of large-scale models, are very time-consuming, one has to trade-off the model size and accuracy of the results, for example, using multi-scale models (Yu *et al.*, 2009). Constitutive models including damage properties of the materials, and contact models between the ship and bridge are key factors affecting the simulation results. LY-DYNA (1999) is one of the well-known commercial software packages that are widely used in the impact engineering community.

A major drawback of the numerical simulation is that the simulation lacks verification as the real measurement data for ship collision are very rare. It is imperative to develop a monitoring system to prevent, detect, warn, and evaluate ship-bridge collision events. Through real monitoring, the impact force and local and global responses of the bridges can be measured and evaluated, which allows a better understanding of collision. However, to the best knowledge of the authors, there has been no such monitoring system so far. One reason is that ship collision is a rare event for many bridges and the collision location and direction are not known beforehand. In addition, the impact force could be too huge to be measured directly with common transducers.

10.4 SCOUR MONITORING

10.4.1 Introduction

Scour is the erosion of flowing water, washing away the sediments from around the piers and abutments of bridges, which leaves the bridges in an unsafe condition or in danger of collapse. There are generally three types of scours: local scour, contraction scour, and degradational scour (Deng and Cai, 2010). Local scour involves removal of sediments from around the base of the piers or abutments. The basic mechanism causing local scour at piers is the formation of vortices at their base which result from pileup of the water on the upstream surface of the obstruction and subsequent acceleration of the flow around the pier. Contraction scour is the removal of sediments from all or most of the channel bed of the bridge, associated with higher speed of the water as it is caused by narrowing of the channel than the natural river channel. Degradational scour is a natural process that typically includes bed degradation and lateral channel movement which are not covered by local scour and contraction scour, but may remove large amounts of sediments over time.

There are more than 20,000 highway bridges among 590,000 in USA that are rated "scour critical" (Hunt, 2009). Due to the limited time and funding, it is not feasible to repair or replace all of the scour-critical bridges immediately. Therefore, it is important to develop continuous real-time scour measuring and monitoring techniques for deciding which bridge needs rapid repair or replacement.

Hydraulic Engineering Circular No. 23 (Lagasse *et al.*, 2001) recommended three types of scour monitoring: fixed instrumentation, portable instrumentation, and visual monitoring. Fixed instrumentations can be placed on the bridge structure. The recommended fixed instruments include sonar, magnetic sliding collars, float-out devices, and sounding rods. Portable instrumentations can be manually carried and used from one bridge to another. Portable instruments are more cost-effective in monitoring a bridge or multiple bridges than fixed instruments. Visual inspection can be performed at standard regular intervals (usually two years). Portable instrumentation and visual inspection do not offer a continuous monitoring over the bridges.

10.4.2 Scour Monitoring Equipment

There are several types of common instruments used for scour measuring and monitoring as described in the following. A scour monitoring system at a bridge can be comprised of one or more of these types of devices.

1. *Sonar devices* (Lueker *et al.*, 2010). The sonar device is a conventional instrument used underwater for navigation, communication and detection. A sonar transducer sends a sound wave to the streambed and receives the reflected sound wave. The distance from the streambed to the transducer is then calculated from the travel time of the sound wave. This device is particularly suitable to bridges in deep water.

2. *Magnetic sliding collar* (Lagasse *et al.*, 1997). The magnetic sliding collar is a mechanical device for measuring maximum scour depth. The sensor monitors the downward movement of a magnetic sliding collar around a stainless steel pipe as the streambed soil scours away. There is one auto probe inside the pipe for data measuring and recording.

3. *Time domain reflectometry* (Yu and Yu, 2009). Time domain reflectometry is a remote sensing electrical measurement technique which is used to send down a fast rising step pulse or impulse from a rod buried vertically in the streambed. Reflections will reflect to the source when pulse encounters a change interface between material layers with different dielectric properties, i.e. the air-water interface and water-sediment interface. By monitoring the round-trip travel time of a pulse in continuous real-time, the distance to the respective interfaces can be calculated.

4. *Fibre Bragg grating (FBG) sensors* (Lin *et al.*, 2005). FBG sensors have excellent long-term stability and a high reliability in strain and temperature measurements in harsh environments. When the running water flows towards a series of FBG sensors that are mounted on a cantilever beam, deformation strain

will be generated by the bending moment. The measured maximum strain will be used for detection of the local scour depth directly. In addition, FBG scour monitoring sensors can measure both the process of scouring and the variation in the water level.

5. *Mechanical sounding rods* (Lueker *et al.*, 2010). The mechanical sounding rod instruments are manual or automated gravity based physical probes. The major part of the device is a solid steel rod that rests on the streambed and moves with the streambed when scours occur.

6. *Acoustic Doppler velocimeter* (Sarker, 1998). The acoustic Doppler velocimeter is a sensor system based on the acoustic Doppler principle and has been used to study the flow characteristics around scoured bridge piers. The sensor emits acoustic pulses into water which are scattered by the particles present in the flow. The Doppler frequency shift of the reflection is proportional to the particle velocity. Change in velocity profiles around the pier due to scouring can be measured.

7. *Piezoelectric films.* A significant property of a piezoelectric film is that a force in the film will produce a voltage in proportion to the force. When a piezoelectric sensor is buried, it does not move and does not output a signal. As scour occurs, the sensor is moved by the flow and outputs a small current. Thus, it can measure aggradation and degradation of the surrounding soil.

8. *Float-out devices* (Hunt, 2009). This device is buried in the channel bed at a predetermined depth. When the scour reaches that depth, the sensor floats to the stream surface and begins transmitting a radio signal that is detected by a receiver on the bridge.

9. *Thermal sensors* (Washer *et al.*, 2009). This approach monitors the temperature along the height of piers (or piles) embedded in the soil. It was observed that the temperature variation in the water differs from that in the soil. When a scour hole occurs around the pier, the water-soil interface changes, and the change can be detected by the thermal variations.

10.4.3 Preventive Measures

Scour mitigation at bridge sites has received much attention. Scour prevention measures can be generally categorized into two groups: armouring countermeasures and flow altering countermeasures. Lagasse *et al.* (2007) and Barkdoll *et al.* (2007) respectively reviewed different scour countermeasures for bridge piers and abutments.

The basic idea of armouring countermeasures is the addition of one layer on or inside the channel bed, which can provides protection to the channel bed. The most commonly used armouring countermeasure is riprap. Experiments (Lauchlan and Melville, 2001) showed that a deep placement level of the riprap layer could provide good protection against local scour.

Flow altering countermeasures use guidebanks, parallel walls, collars, and other devices for changing the hydraulic properties of flows and therefore reducing the scour effect at bridge piers and abutments.

The selection of different countermeasures is dependent on the application and the nature of the problem. Various countermeasures can be combined to achieve a better scour mitigation effect. Lagasse *et al.* (2001) compared different countermeasures and provided design guidelines for them.

10.5 REFERENCES

AASHTO, 1991, *Guide Specification and Commentary for Vessel Collision Design of Highway Bridges*, (Washington, D.C.: American Association of State Highway and Transportation Officials).

AASHTO, 2005, *LRFD Bridge Design Specification and Commentary*, (Washington, D.C.: American Association of State Highway and Transportation Officials).

Barkdoll, B.D., Ettema, R. and Melville, B.W., 2007, *Countermeasures to Protect Bridge Abutments From Scour*. National Cooperative Highway Research Program (NCHRP) Report No. 587, (Washington, D.C.: Transportation Research Board).

Boulfiza, M., Sakai, K., Banthia, N. and Yoshida, H., 2003, Prediction of chloride ions ingress in uncracked and cracked concrete. *ACI Materials Journal*, **100(1)**, pp. 38-48.

Consolazio, G.R., Cook, R.A., Benjamin Lehr, G. and Bollmann, H.T., 2002, *Barge Impact Testing of the St. George Island Causeway Bridge, Phase I: Feasibility Study*. Structural Research Report No. BC-354-RPWO-23, (Gainesville, Florida: University of Florida).

Consolazio, G.R., Cook, R.A., Biggs, A.E., Cowan, D.R. and Bollmann, H.T., 2003, *Barge Impact Testing of the St. George Island Causeway Bridge, Phase II: Design of Instrumentation Systems*. Structural Research Report No. BC-354-RPWO-56, (Gainesville, Florida: University of Florida).

Consolazio, G.R., Cook, R.A., McVay, M.C., Cowan, D.R. and Biggs, A.E., 2006, *Barge Impact Testing of the St. George Island Causeway Bridge, Phase III: Physical Testing and Data Interpretation*. Structural Research Report No. BC-354-RPWO-76, (Gainesville, Florida: University of Florida).

Consolazio, G.R. and Cowan, D.R., 2005, Numerically efficient dynamic analysis of barge collisions with bridge piers. *Journal of Structural Engineering, ASCE*, **131(8)**, pp. 1256-1266.

Deng, L. and Cai, C.S., 2010, Bridge sour: prediction, modeling, monitoring, and countermeasures – review. *Practice Periodical on Structural Design and Construction, ASCE*, **15(2)**, pp. 125-134.

Eurocode 1, 2006, *Actions on Structures, Part 1-7: General Actions, Accidental Actions*, EN 1991-1-7, (Brussels, Belgium: European Committee for Standardization).

FHWA, 1983, *Pier Protection and Warning Systems for Bridges Subject to Ship Collisions*, Technical Advisory T5140.19, (Washington, D.C.: Federal Highway Administration).

Guo, Y.L., Ni, Y.Q. and Zhou, H.F., 2009, Condition assessment of post-ship-collision bridges using HHT analysis. In *Proceedings of the 5th International Workshop on Advanced Smart Materials and Smart Structures Technology*, Boston, Massachusetts, USA, pp. 57-64.

Habel, W.R., Hofmann, D., Schallert, M., Dantan, N., Krebber, K. and Nother, N., 2007, Fibre optic sensors for long-term SHM in civil engineering applications. *The 3rd International Conference on Structural Health Monitoring of Intelligent Infrastructure*, Vancouver, Canada, edited by Mufti, A., [CD-ROM].

Harik, I.E., Shaaban, A.M., Gesund, H., Valli, G.Y.S. and Wang, S.T., 1990, United States bridge failures, 1951-1988. *Journal of Performance of Constructed Facilities, ASCE*, **4**, pp. 272-277.

Hunt, B.E., 2009, *Monitoring Scour Critical Bridges*, National Cooperative Highway Research Program (NCHRP) Synthesis No. 396, (Washington D.C., USA: Transportation Research Board).

Huston, D.R. and Fuhr, P.L., 1998, Distributed and chemical fiber optic sensing and installation in bridges. *International Workshop on Fiber Optic Sensors for Construction Materials and Bridges*, Newark, N.J., edited by Ansari, F., (Lancaster PA: Technomic Publishing Company Inc.), pp. 79-88.

Knott, M. and Prucz, Z., 2000, Vessel collision design of bridges. *Bridge Engineering Handbook*, edited by Chen, W.F. and Duan, L., (Boca Raton: CRC Press).

Lagasse, P.F., Clopper, P.E., Zevenbergen, L.W. and Girard, L.G., 2007, *Countermeasures To Protect Bridge Piers From Scour*. National Cooperative Highway Research Program (NCHRP) Report No. 593, (Washington D.C.: Transportation Research Board).

Lagasse, P.F., Richardson, E.V., Schall, J.D. and Price, G.R., 1997, *Instrumentation for Measuring Scour at Bridge Piers and Abutments*. National Cooperative Highway Research Program (NCHRP) Report No. 396, (Washington D.C.: Transportation Research Board).

Lagasse, P.F., Zevenbergen, L.W., Schall, J.D. and Clopper, P.E., 2001, *Bridge Scour and Stream Instability Countermeasures: Experience, Selection, and Design Guidance*. Hydraulic Engineering Circular 23, (Washington D.C.: Federal Highway Administration).

Larsen, O.D., 1993, *Ship Collision with Bridges: The Interaction between Vessel Traffic and Bridge Structures*, (Zurich, Switzerland: International Association for Bridge and Structural Engineering).

Lauchlan, C.S. and Melville, B.W., 2001, Riprap protection at bridge piers. *Journal of Hydraulic Engineering, ASCE*, **127(5)**, pp. 412-418.

Leung, C.K.Y., Wan, K.T. and Chen, L., 2009, Practical application of an optical fiber sensor for corrosion in concrete structures. *The 4th International Conference on Structural Health Monitoring of Intelligent Infrastructure*, Zurich, Switzerland, edited by Meier, U., Havranek, B. and Motavalli, M., Paper No. 352.

Lin, Y.B., Chen, J.C., Chang K.C., Chern, J.C. and Lai, J.S., 2005, Real-time monitoring of local scour by using fiber Bragg grating sensors. *Smart Materials and Structures*, **14**, pp. 664-670.

LS-DYNA, 1999, *LS-DYNA User's Manual*, (Livermore, California: Livermore Software Technology Corporation).

Matthew, L., Jeff, M., Chris, E., Vincent, W. and Shankar, R.A., 2010, *Bridge Scour Monitoring Technologies: Development of Evaluation and Selection Protocols for Application on River Bridges in Minnesota*. Report No. MN/RC 2010-14, (Minnesota, USA: Minnesota Department of Transportation).

Meier-Dornberg, K.E., 1983, *Ship Collisions, Safety Zones and Loading Assumptions for Structures in Inland Waterways*. Report No. 496, (Germany: Association of German Engieers), pp. 1-9.

Melhorn, K., Flachsbarth, J., Kowalsky, W. and Johannes, H.H., 2007, Novel sensors for long-term monitoring of pH and humidity in concrete. In *Proceedings of the 6th International Workshop on Structural Health Monitoring*, Stanford, California, edited by Chang, F.K. and Guemes, J.G., (Stanford, California: Stanford University), pp. 387-394.

Mercier, J.J., Alvarez, E., Marfil, J., Bloschock, M.J. and Medlock, R.D., 2005, Bridge collapse detection system in Texas. In *Proceedings of the 6th International Bridge Engineering Conference*, (Boston: Transportation Research Board), pp. 403-408.

Raupach, M. and Schießl, P., 2001, Macrocell sensor systems for monitoring of the corrosion risk of the reinforcement in concrete structures. *NDT & E International*, **34(6)**, pp. 435-442.

Rodriguez, O.G. and Hooton, R.D., 2003, Influence of cracks on chloride ingress into concrete. *ACI Materials Journal*, **100(2)**, pp. 120-126.

Sarker, M.A., 1998, Flow measurement around scoured bridge piers using Acoustic-Doppler Velocimeter (ADV). *Flow Measurement and Instrumentation*, **9(4)**, pp. 217-227.

Shiraishi, N. and Cranston, W.B., 1992, Bridge safety. *Engineering Safety*, edited by Blockley, D., (London: McGraw-Hill), pp. 292-312.

Song, H.W. and Saraswathy, V., 2007, Corrosion monitoring of reinforced concrete structures – a review. *International Journal of Electrochemical Science*, **2**, pp. 1-28.

Van Manen, S.E. and Frandsen, A.G., 1998, Ship collision with bridges: review of accidents. *Ship Collision Analysis*, edited by Gluver, H. and Olsen, D., (Rotterdam: Balkema), pp. 3-11.

Wardhana, K. and Hadipriono, F.C., 2003, Analysis of recent bridge failures in the United States. *Journal of Performance of Constructed Facilities, ASCE*, **17(3)**, pp. 144-150.

Washer, G.A., Rosenblad, B. and Morris, S.E., 2009, *Remote Heatlh Monitoring for Asset Management*, Report No. OR09-019, (Missouri: Missouri Department of Transportation).

Yu, M., Zhou, X.Q., Zha, X.X. and Xia, Y., 2009, Numerical simulation of dynamic responses of a long-span cable-stayed bridge subjected to ship collision. *International Workshop on Structures Response to Impact and Blast*, Haifa, Israel, Paper No. 6-4.

Yu, X.B. and Yu, X., 2009, Time domain reflectometry automatic bridge scour measurement system: principles and potentials. *Structural Health Monitoring*, **8(6)**, pp. 463-476.

Yun, H., Nayeri, R., Tasbihgoo, F., Wahbeh, M., Caffrey, J., Wolfe, R., Nigbor, R., Masri, S.F., Abdel-Ghaffar, A. and Sheng, L.H., 2008, Monitoring the collision of a cargo ship with the Vincent Thomas bridge. *Structural Control and Health Monitoring*, **15**, pp. 183-206.

10.6 NOTATIONS

B_B	Vessel width.
$C(x,t)$	Chloride content at depth x and time t.
C_0	Surface chloride content.
C_H	Hydrodynamic mass coefficient to account for the additional inertia forces caused by the mass of the water surrounding and moving with the ship.
D	Chloride diffusion coefficient in uncracked concrete.
D_{av}	Averaged diffusion coefficient.
D_{cr}	Diffusion coefficient inside the crack.
DWT	Dead weight tonnage of vessel.
E_a	Kinetic energy.
E_{def}	Deformation energy.
erf	Error function.
F_{dyn}	Dynamic design impact force.
KE	Kinetic energy.
P_B	Static equivalent impact load.
P_s	Equivalent static vessel impact force.
s	Crack spacing.
V	Vessel impact velocity.
W	Vessel weight.
w	Crack width.
α_B	Vessel crush depth.
θ	Impact angle.

CHAPTER 11

Structural Damage Detection

11.1 PREVIEW

Civil structures are exposed to natural and man-made hazards, such as typhoons, earthquakes, flood, fire, explosion, and collisions, which may cause structural damages or collapse. Failure of civil structures could be catastrophic not only in terms of losses of life and economy, but also subsequent social and psychological impacts. Detecting the early damage enables maintenance of the structure prior to its complete failure and, consequently saves lives and assets. Damage detection has been a challenging task in SHM systems, particularly for large civil structures subject to multiple loadings and harsh environment. Although many problems in damage detection of long-span suspension bridges have not been solved, the basic concepts and current status of damage detection will be introduced in this chapter to provide complete coverage of the subject.

Structural health has been compared to medical health of human beings (Chang and Liu, 2003). Pulse and blood pressure give an overall indication of the health of the body. Once the body signs show an anomaly, some inspection methods such as radiography may be used to determine the cause of the anomaly. This process is analogous to SHM, and it is hoped that the presence of damage in the structure can be detected from global damage detection methods and the nature of the damage can be identified from local methods. Vibration based damage detection methods are usually regarded as the global methods and non-destructive testing (NDT) methods are the local approaches. These two methods will be described in this chapter, followed by statistical approaches with consideration of uncertainties involved. The detected structural damage will be further assessed for the structural rating, maintenance and repair, which are related in Chapter 12.

11.2 GENERAL CONCEPTS OF STRUCTURAL DAMAGE DETECTION

Although recent improvements in design methodologies and advanced construction technologies have increased the reliability and safety of structures, it is still not possible to build structures with no probability of failure due to a wide variety of unforeseen conditions and circumstances. On the other hand, existing structures have deteriorated due to environmental corrosion and long term fatigue after many years in service. For example, the American Society of Civil Engineers estimated that between one-third and one-half of infrastructure in the United States are structurally deficient, and the investment needs about US$1.6 trillion in a five year period (ASCE, 2005).

Due to economic boom, a huge number of large-scale and complex civil structures such as long-span bridges, high-rise buildings, and large-space structures, have been constructed in China during the past twenty years. Similar to other countries' lessons during the rapid developing period, enormous cost and effort will be required for maintenance of deficient structures (Chang *et al.*, 2009). Accurate and reliable damage detection methods are required for cost-effective maintenance of civil structures.

Commonly used damage detection methods for long-span suspesnion bridges are visual inspection and NDT methods such as acoustic or ultrasonic methods, magnetic field methods, radiographs, eddy current methods, and thermal field methods. These NDT techniques require that the vicinity of the damage is known *a priori* and that the portion of the structure being inspected is readily accessible. Subject to these limitations, these NDT methods can only detect damage on or near the surface of the structure. The need for additional global damage detection methods that can be applied to complex structures has led to the development of vibration methods that examine changes in the vibration characteristics of the structure in the past three decades.

Rytter (1993) divided damage detection methods into four levels:

Level 1: Determination that damage is present in the structure;
Level 2: Determination of the geometric location of the damage;
Level 3: Quantification of the severity of the damage; and
Level 4: Prediction of the remaining service life of the structure.

Most of the damage detection methods developed to date limit themselves to Level 1 to Level 3. Level 4 methods require the knowledge associated with the disciplines in structural design, fracture mechanics, and structural reliability, and are still very limited (Yao and Natke, 1994; Park *et al.*, 1997; Stubbs *et al.*, 2000; Xia and Brownjohn, 2003).

11.3 NON-DESTRUCTIVE TESTING METHODS

Many NDT methods have been developed to detect any damage/change to the material and to evaluate its condition. The process is also frequently called non-destructive evaluation (NDE) or non-destructive inspection (NDI). There are a few books published in this area, such as Hellier (2001) and Shull (2002). *The American Society for Nondestructive Testing* has published a series of handbooks on various NDT methods: leak testing, liquid penetrant testing, infrared and thermal testing, radiographic testing, electromagnetic testing, acoustic emission testing, ultrasonic testing, magnetic testing, and visual testing. Interested readers may refer to these handbooks for details. *American Society for Testing and Materials* (ASTM) has standardized some NDT methods. It is noted that various methods and techniques, due to their particular nature, may lean themselves well to certain applications while being of little value in other applications. Therefore choosing the appropriate methods and techniques is important for NDT. In addition, some methods and techniques may need to be combined to obtain a more comprehensive condition assessment.

The following sections will summarise some commonly used NDT techniques suitable for civil structures, including the ultrasonic pulse velocity method, impact-echo method, acoustic emission method, radiography method, and eddy current method. Basic principles, equipment, applications, advantages and limitations of each technique will be described.

11.3.1 Ultrasonic Pulse Velocity Method

The ultrasonic pulse velocity method has been used to assess the quality of concrete for many years. The method is based on the fact that the velocity of a pulse of compressive waves through a medium is a function of the elastic properties and density of the medium. Its applications include estimating the concrete strength (Anderson and Seals, 1981), determining the homogeneity of concrete, monitoring the setting and hardening process of cement, detecting cracking and deterioration (Knab *et al.*, 1983), and determining the dynamic modulus of elasticity.

The instrument consists of a pulse generator for producing a wave pulse into the concrete and a receiver for sensing the pulse arrival and measuring the travel time of the pulse. Transducers with frequencies of 25 to 100 kHz are usually used for testing concrete. High-frequency transducers (above 100 kHz) may be used for small size specimens with relatively short path lengths, whereas low-frequency transducers (below 25 kHz) may be used for large specimens with relatively longer path lengths. There are three possible configurations in which the transducers may be arranged: direct transmission, semi-direct transmission, and indirect transmission (Naik *et al.*, 2004), as shown in Figure 11.1.

The pulse velocity method is an excellent means for investigating the uniformity of concrete. The test procedure is simple and easy to use. The testing procedures have been standardized as ASTM C597 (ASTM, 2009). Besides the concrete properties, other factors such as transducer contact, temperature of concrete, path length, and reinforcing steels also affect the pulse velocity. Therefore, the pulse velocity method should be used with care so that the pulse velocity is affected by the properties of the concrete only and other adverse factors can be eliminated.

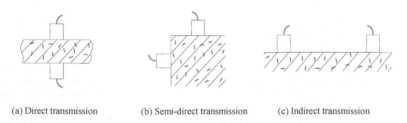

(a) Direct transmission   (b) Semi-direct transmission   (c) Indirect transmission

Figure 11.1 Pulse velocity measurement arrangement

11.3.2 Impact-echo / Impulse-response Methods

The impact-echo method uses a mechanical impact to generate a high-energy stress pulse. The stress pulse propagates into the object along spherical wave fronts as P- and S-waves, which are reflected by internal interfaces (for example, flaws) or external boundaries. The reflected waves, or echoes, can be measured by a receiver and used to detect the depth of the flaws. The principle of the impact-echo method is illustrated in Figure 11.2.

As the energy propagates into the object in all directions and reflections may arrive from many directions, the impact-echo methods are primarily used for testing piles, whose boundary confines most of the energy within the pile. This is generally referred to as "low strain integrity testing" and standardized by ASTM D5882 (ASTM, 2007b). Other than piles, the method has also been applied to detect cracks, voids, and delaminations in concrete slabs or slablike structures. The Standard ASTM C1383 (ASTM, 2004) specifies the use of the impact-echo method to measure the thickness of slablike concrete members.

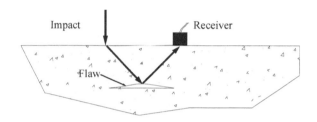

Figure 11.2 Principle of the impact-echo method

An impact-echo test system is composed of three components: an impact source, a receiving transducer, and a data acquisition system with appropriate software for signal analysis and data management. The selection of the impact source is a critical aspect of a successful impact-echo test system. The force-time history of an impact may be approximated as a half-cycle sine curve. The duration of the impact, contact time, determines the frequency content of the stress pulse generated by the impact. A shorter contact time indicates higher frequency components contained. Therefore, smaller defects or shallow defects can be detected. Geophones (velocity transducers) or accelerometers can be used as the receiver.

A variant of the impact method is known as the impulse-response method, transient response method, or impedance testing method (Davis and Hertlein, 1991). Different from the impact-echo method, the time history of the impact force is recorded in the impulse-response method by an instrumented hammer. Given the time history of the impact force and the structural response, the transfer function can be calculated. The transfer function represents the characteristics of a structure, including material properties, geometry, support conditions, and the existence of defects. It is noted that velocity is usually measured and the corresponding transfer function is known as mobility. The impulse-response method is primarily applied for testing integrity of piles. Other successful applications have also been reported, for example, Davis *et al.* (1997) evaluated the integrity of a concrete tank using the impulse-response method and other NDT methods.

11.3.3 Acoustic Emission Method

Acoustic emission refers to a phenomenon that stress waves are generated when a material undergoes a rapid release of energy, for example, cracking. These waves can be picked up by sensors attached to the surface of the material and used to evaluate the health condition of the material. The acoustic emission method has been widely used for highway structural assessment, particularly monitoring cracking and crack development, debonding, and corrosion (Rens *et al.*, 1997; Yoon *et al.*, 2000). The method is also effective for monitoring the wire fracture in stay cables, main cables, and prestressed tendons (Elliot, 1996; Li and Ou, 2008; Vogel *et al.*, 2006). A recent review on acoustic emission monitoring of bridges is presented by Nair and Cai (2010).

    Selection of appropriate transducers is based on the purpose and sensitivity required for the investigation. Resonant sensors are preferable as they are highly sensitive to typical acoustic emission sources. In bridge monitoring, unidirectional sensors and sensors more sensitive to in-plane wave modes may be beneficial (Nair and Cai, 2010).

    The acoustic emission method is applicable for local, global, and continuous monitoring purposes without interrupting traffic over the bridges. It can detect and locate flaws but cannot determine the size of flaws. Quantitative analysis of damage is still difficult in practical bridge applications. In addition, extraneous noise may provide uncertain results in real bridges. For example, slight movement of bolted joints can also generate acoustic signals and may cause false damage detection (Chang and Liu, 2003).

11.3.4 Radiographic Method

Radiography is typically composed of a radiation source and an image collector. Radiation goes directly through a test specimen and exposes the film on the other side of the specimen. Different materials may attenuate the radiation differently, for example, steel attenuates X-rays and gamma rays much more than concrete does. The differences in attenuation can therefore yield a photographic image of the internal structure of the specimen. X-rays and gamma rays are typically used as the radiation source.

    Radiographic methods are used primarily for examining welded products and castings for defects. A field X-ray work on concrete was developed in France for evaluation of prestressed concrete bridges (Dufay, 1985). The equipment has been used for examining the quality of grouting and of concrete and the condition of prestressing cables.

    A more powerful technique called computed tomography could be used to produce a three-dimensional representation of the internal structure of an object such as dimensions, shape, internal defects, and density. The object is essentially radiographed at various orientations and then a computer is used to construct the three-dimensional image.

Conventional radiographic techniques can provide rapid and accurate information on the internal characteristics that is not available via other NDT methods. However, the equipment is generally heavy and the power consumption is large. Power sources 250 kV – 4 MV are often needed to penetrate the thick and dense materials used in civil infrastructure applications (Chang and Liu, 2003). Portability of the equipment and accessibility to the object are two major problems for field implementations.

11.3.5 Eddy Current Method

The eddy current testing uses electromagnetic induction to detect flaws in conductive materials. In this method, a circular probe coil carrying current is placed in proximity to the test specimen. The alternating current in the coil generates a changing magnetic field which induces eddy currents in the test specimen. The presence of a flaw will cause a change in the eddy current, which can be sensed by the probe coil. The eddy current testing is primarily used for surface or subsurface crack detection. It is also used for detecting corrosion in thin materials, and measuring the thickness of paints and other coatings.

Eddy current instruments have a large variety of configurations depending on the applications. A basic eddy current testing instrument consists of an alternating current source, a coil of wire connected to this source, and a voltmeter/ammeter to measure the voltage/current change across the coil. An appropriate coil is the most important part to achieve accurate signals from the probe.

The eddy current testing can detect very small cracks in or near the surface of the material, and is capable of inspecting complex shapes and sizes. The limitations of the method include: 1) only conductive materials can be inspected; 2) the surface of the material must be accessible to the probe; 3) the depth of penetration into the material is limited as the eddy current density decreases with the depth, and 4) flaws such as delaminations parallel to the probe are undetectable.

11.3.6 Infrared Thermographic Method

Infrared radiation has a wavelength longer than visible light, or greater than 700 nanometers. Any object whose temperature is above $0°K$ (273.15°C) radiates infrared energy, which is not visible to the human eye but can be detected by an infrared camera. The infrared thermography is a testing technique measuring the temperature or temperature differences of an object. It can be simply used for inspecting electronic components or mechanical systems where a defect usually causes an increase in temperature.

In bridge inspection applications, the temperature measurements may be taken by day or night as long as heat transfer between the bridge and environment is taking place. Solid concrete is a reasonably good conductor of heat and convection within the concrete can be considered negligible. If the concrete has voids, the conduction paths will be disrupted. The disruptions in the flow of thermal energy lead to temperature differences on the surface, which can be detected by the infrared camera. Infrared thermography has been found an economical and accurate method for the determination of pavement and bridge deck conditions (Zachar and Naik, 1992). The procedures are standardized in ASTM D4788 (ASTM, 2007a).

Various parameters affect the surface temperature measurements: solar radiation, cloud, ambient temperature, wind speed, and surface moisture (Weil, 2004). Therefore, thermographic testing should be carried out on days with no solid cloud cover, with the wind speed below 15 mph, and with the surface dry (Kunz and Eales, 1985).

A complete thermographic data collection and analysis system includes the infrared sensor head, infrared scanning system, data collection system, and image recording and retrieving devices. The whole system can be installed in a specially equipped van in bridge deck assessment applications (Zachar and Naik, 1992).

The infrared thermography is an area-testing technique rather than point-testing as are some other NDT methods. It is more efficient than other invasive methods when testing large areas. One limitation of the technique is that the depth or thickness of a void cannot be determined.

11.4 VIBRATION-BASED DAMAGE DETECTION METHODS

11.4.1 Overview of Vibration-based Damage Detection Methods

Different from the NDT methods, vibration based damage detection methods are regarded as global methods. These methods have been developed on the premise that commonly measured vibration quantities, such as response time-histories and global vibration characteristics, are functions of the physical properties of the structure (mass, damping, boundary conditions and stiffness). Therefore, the changes in the physical properties, such as reductions in stiffness resulting from the onset of cracks or loosening of a connection, will cause detectable changes in these global quantities (Doebling *et al.*, 1996). Identifying the damage from the changes in the vibration properties is the main task of the vibration based damage detection methods.

Vibration based damage detection methods have been developed in the oil industry for offshore platforms during the 1970s and 1980s. The aerospace community began to study the method during the late 1970s and early 1980s in conjunction with the development of the space shuttle. The civil engineering community has studied vibration based damage detection of bridge structures since the early 1980s. Nevertheless, the most successful application of vibration based damage detection technology to date has been for monitoring rotating machinery (Farrar *et al.*, 2001). Doebling *et al.* (1996) conducted a comprehensive review of vibration-based damage detection methods. After that Sohn *et al.* (2003) reviewed the literature between 1996 and 2001.

With regard to the algorithm used, damage detection methods can be classified into direct correlation methods (non-model based) and model updating methods (model-based). The former compares dynamic parameter changes before and after occurrence of damage directly, while the latter adjusts structural parameters iteratively. Via model updating methods, both damage location (Level

2) and severity (Level 3) can be identified by examining the stiffness and mass changes. Direct correlation methods, however, usually cannot quantify the damage severity. The model updating techniques have been introduced in Chapter 4 and will not be described here.

According to the dynamic parameters adopted, damage detection methods can be categorized into time domain, frequency domain, and time-frequency domain methods. The time domain methods use response time-histories, mostly accelerations. These methods are generally based on the error equation of the inertia force, restoring force, and damping force (Agbabian *et al.*, 1991; Hemez, 1993; Zimmerman and Kaouk, 1994). Some researchers also extracted statistical features from the time series data as the damage indicator, for example, high-order moments (Zhang *et al.*, 2008; Xu *et al.*, 2009), time domain periodogram (Zimin and Zimmerman, 2009), and so forth. The time-frequency domain methods are based on time-frequency analysis tools such as Wavelet transform (Daubechies, 1992) and Hilbert-Huang transform (Huang *et al.*, 1998). Development and applications of these tools to damage detection can be found in a number of studies in the past ten years (Hou *et al.*, 2000; Han *et al.*, 2005; Zhu and Law, 2006; Li *et al.*, 2007). The majority of vibration based methods fall into the frequency domain, that is, using frequency response function (FRF), natural frequency, damping, mode shape, mode shape curvature, modal flexibility, and modal strain energy, which will be described in the following sections.

It should be noted that although there are a few nonlinear damage detection methods such as D'Souza and Epureanu (2005) and Li and Law (2010), only linear methods are introduced here. Consequently, the structural responses in both undamaged and damaged states can be obtained using linear equations of motion.

11.4.2 Frequency Changes

Because the natural frequency is the most fundamental vibration parameter, methods directly measuring the shifts in natural frequency (or eigenvalue) have been widely used. Salawu (1997) reviewed damage detection methods with frequency shifts and the problems faced.

Cawley and Adams (1979) may have been be the first researchers to give a formulation for damage detection from frequency changes before and after damage is introduced. For an undamped system, the eigenvalue equation is

$$\left(-\lambda_i [M] + [K]\right) \{\phi_i\} = \{0\} \tag{11.1}$$

where $[M]$ and $[K]$ are the mass and stiffness matrices, respectively, λ_i is the ith eigenvalue and $\{\phi_i\}$ is the associated mode shape vector. Assuming the effect of a damage causes a change in the stiffness matrix, $[\Delta K]$, while the mass is unchanged, Equation (11.1) then becomes

$$\left[-(\lambda_i + \Delta\lambda_i)[M] + ([K] + [\Delta K])\right]\left(\{\phi_i\} + \{\Delta\phi_i\}\right) = \{0\} \tag{11.2}$$

where $\Delta\lambda_i$ and $\{\Delta\phi_i\}$ are respectively the changes in eigenvalue and mode shape vector due to the damage. Left-multiplying $\{\phi_i\}^T$ and substituting Equation (11.1) into (11.2), the following equation can be derived by disregarding the high order terms:

$$\Delta \lambda_i = \frac{\{\phi_i\}^T [\Delta K] \{\phi_i\}}{\{\phi_i\}^T [M] \{\phi_i\}} \tag{11.3}$$

where $\{\}^T$ represents the transposed vector. Then the ratio between eigenvalue shifts for mode i and j, $\Delta\lambda_i/\Delta\lambda_j$, can be obtained from the undamaged state of the structure. An error index is defined as the difference between the analytical $\Delta\lambda_i/\Delta\lambda_j$ and the measured one, due to possible damage at position r, that is,

$$e_{rij} = \begin{cases} \dfrac{\left(\Delta\lambda_{ri}/\Delta\lambda_{rj}\right)^A}{\left(\Delta\lambda_{ri}/\Delta\lambda_{rj}\right)^E} - 1, & if \quad \left(\Delta\lambda_{ri}/\Delta\lambda_{rj}\right)^A \ge \left(\Delta\lambda_{ri}/\Delta\lambda_{rj}\right)^E \\[4mm] \dfrac{\left(\Delta\lambda_i/\Delta\lambda_j\right)^E}{\left(\Delta\lambda_{ri}/\Delta\lambda_{rj}\right)^A} - 1, & if \quad \left(\Delta\lambda_{ri}/\Delta\lambda_{rj}\right)^A < \left(\Delta\lambda_i/\Delta\lambda_j\right)^E \end{cases} \tag{11.4}$$

where subscripts "E" and "A" represent the experimental and analytical data, respectively. The total error, with the assumption of the damage at position r, is the sum of the errors in all the mode pairs. The lowest error indicates the location of the actual damage.

Stubbs *et al.* (1990) and Stubbs and Osegueda (1990a, 1990b) presented a sensitivity method for damage identification that is based on the work by Cawley and Adams (1979). Hearn and Testa (1991) developed a similar damage detection method at the elemental level. Friswell *et al.* (1994) developed Cawley and Adams's work. Other works using frequency changes include Rizos *et al.* (1990), Narkis (1994), Choy *et al.* (1995), and Xu *et al.* (2004).

The effects of temperature on structural natural frequencies were discussed by Adams and Coppendale (1976). This factor was considered in Adams *et al.* (1991) to detect various damages in a space-truss structure using natural frequency changes.

The advantages of damage detection methods based on frequency changes are that the frequency can be measured using very few sensors and has relatively higher precision than other parameters. However, frequencies are not spatially specific and are not very sensitive to damages.

11.4.3 Mode Shape Changes

In this method, the MAC (Modal Assurance Criterion) and its variations are usually used. The MAC value gives an indication of the identical shape of two sets of mode shapes. It is necessary to identify the correlated mode pairs of structural model and experiments, or mode pairs before and after damage. The COMAC (COordinate MAC) is related to the DOF of the structure rather than to the mode numbers. They are expressed as

$$MAC\left(\{\phi_i\},\{\phi_j\}\right) = \frac{\left|\{\phi_i\}^T \{\phi_j\}\right|^2}{\left(\{\phi_i\}^T \{\phi_i\}\right)\left(\{\phi_j\}^T \{\phi_j\}\right)} \tag{11.5}$$

$$COMAC\left(\left[\phi^u\right],\left[\phi^d\right],q\right)=\frac{\left(\sum_{i=1}\left|\left(\phi_i^u\right)_q\left(\phi_i^d\right)_q\right|\right)^2}{\left(\sum_{i=1}\left(\phi_i^u\right)_q^2\right)\left(\sum_{i=1}\left(\phi_i^d\right)_q^2\right)} \tag{11.6}$$

where q is one DOF, and superscripts "u" and "d" represent the undamaged and damaged states, respectively. For a good mode correlation or coordinate correlation the value should be near to 1.

The COMAC has been shown to be capable of damage location (Kim *et al.*, 1992). Ko *et al.* (1994) also pointed out that the COMAC could be used to locate the damage but MAC could not.

Salawu and Williams (1995) conducted a full-scale test on a multi-span RC highway bridge before and after structural repairs. The MAC and COMAC were used to indicate the presence and location of repairs. The authors suggested that MAC threshold value of 0.8 and at least 5% change in frequency would imply the presence of damage with confidence.

Using COMAC it is very simple to detect damage or locate error in a structural model, but it has less physical basis, as compared with other methods. It is advisable to use this method in cases where the structure is tested and modelled in a free-free configuration (Maia *et al.*, 1997). Direct comparison of the pre- and post-damage mode shapes has been found ineffective in identifying the damaged region unless damage is severe (Farrar and Jauregui, 1996).

11.4.4 Modal Damping Changes

As extraction of structural modal damping ratios is usually not as accurate as frequencies and mode shapes, modal damping ratio is not a commonly used damage indicator. Brownjohn and Steele (1979) may be one of the first researchers detecting damage with damping ratios. Salane and Baldwin (1990) tested a composite bridge model and a composite highway bridge. They found that modal damping ratios in the laboratory model decreased after the flange was cut, whereas those in the bridge increased initially and subsequently decreased. Ndambi (2002) investigated the principle of structure modal damping ratio and used it as a damage index for pre-stressed concrete beam structures. Keye (2006) detected delamination damages in a carbon fibre reinforced polymer composite panel using damping ratios. With the Rayleigh damping assumption, the damping ratio changes between the undamaged and damaged models were calculated and the maximum correlation with the measured damping ratio changes indicated the possible damage.

11.4.5 FRF Changes

Ibrahim (1993) used an error indicator computing the Euclidean norm of the FRF vectors measured at discrete frequencies as:

$$e\left(H_{ij}\right) = \frac{\left\|H_{ij}^A - H_{ij}^E\right\|}{\left\|H_{ij}^A\right\|} \tag{11.7}$$

where H is the FRF matrix, and superscripts "A" and "E" represent analytical and experimental items, respectively. Pascual *et al.* (1996) proposed a correlation index to measure the closeness between the measured and analytical FRFs, resemblance to the MAC technique, using the following criterion:

$$FDAC\left(\omega_A, \omega_E, j\right) = \frac{\left|\left\{H_j^A\left(\omega_A\right)\right\}^T \left\{H_j^E\left(\omega_E\right)\right\}\right|^2}{\left(\left\{H_j^A\left(\omega_A\right)\right\}^T \left\{H_j^A\left(\omega_A\right)\right\}\right)\left(\left\{H_j^E\left(\omega_E\right)\right\}^T \left\{H_j^E\left(\omega_E\right)\right\}\right)} \tag{11.8}$$

Samman *et al.* (1991) investigated the change in FRF signals of a scaled highway bridge model caused by cracks in the girders. A pattern recognition method was introduced by utilizing the integer slope and curvature values of FRF wave forms, rather than peak magnitudes. Only one FRF reading per girder was required to detect relatively minor cracks and locate cracks.

Wang and Liou (1991) presented a new method to identify joint parameters by using the two sets of measured FRFs of a substructure with and without the effect of joints. Some strategies were applied to overcome the measurement noise problem that might result in false identification. Numerical simulation and experiments verified the accuracy of the proposed technique.

Biswas *et al.* (1994) developed the modified chain-code method for the rapid detection of a small fault in a structure. Slope and curvature based signatures were derived from the averaged composite FRF signature. The comparison of the intact signatures versus those of the cracked signatures can detect cracks as small as 4 mm in hammer vibration tests of a bridge model, which consists of three steel girders supporting a concrete deck. The method was robust even when noisy data were present.

Xia *et al.* (2007) adopted two damage indicators based on FRFs to detect possible damage to shear connectors in a slab-girder bridge model. Their method is based on the fact that damage to the shear connectors leads to the slab separating from the girders to a certain extent, and so the nearby points on the slab respond differently from those on the girders. One damage index is similar to Equation (11.8), in which the vertical FRFs of the girders and those of the slab were employed. The other damage index is the Euclidean norm of difference of the FRFs. Both indicators could be used to identify damage in shear connectors accurately and consistently. One advantage of the method is that it is a reference-free method, i.e., the undamaged data is not required. In addition, it is not affected by the environmental variation as the slab and girders were measured simultaneously and under identical environmental conditions.

In the FRF based methods, however, the input information is required to form the FRFs. This is quite difficult for large civil engineering structures.

11.4.6 Mode Shape Curvature Changes

An alternative to obtaining spatial damage information is using mode shape derivatives, such as curvature, instead of mode shapes. The logic is based on the fact that a decrease in the flexural stiffness causes an increase in curvature. The mode shape curvatures are obtained by using a central difference approximation as

$$\phi_{q,i}^{"} = \frac{\phi_{q+1,i} - 2\phi_{q,i} + \phi_{q-1,i}}{h^2} \qquad (11.9)$$

where $\phi_{q,i}$ is the ith modal displacement at measurement point q and h is the length of elements.

Pandey *et al.* (1991) demonstrated that the absolute change in mode shape curvatures could be a good indicator of damage for beam structures. Chance *et al.* (1994) detected artificial cracks in a beam and concluded that curvatures were locally sensitive to the fault but mode shapes were not.

It is noted that the accuracy of this method is subject to numerical estimation difficulties resulting from the need for differentiation. Moreover, it is error sensitive because a small noise in the mode shapes may lead to a very different result.

11.4.7 Modal Strain Energy Changes

The strain energy of a Bernoulli-Euler beam is given by Stubbs *et al.* (1995):

$$U = \frac{1}{2} \int_0^L EI \left(\frac{\partial^2 w}{\partial x^2} \right)^2 dx \qquad (11.10)$$

where EI is the flexural rigidity of the beam, and L is the total length of the beam. For a particular i^{th} mode shape, $\phi_i(x)$, the jth member contribution to the mode is

$$U_{ij} = \frac{1}{2} \int_{a_j}^{a_{j+1}} (EI)_j \left(\frac{\partial^2 \phi_i}{\partial x^2} \right)^2 dx \qquad (11.11)$$

where a_j and a_{j+1} are nodal coordinates of the beam element. The fractional energy of the element can be used as a damage indicator. Cornwell *et al.* (1997; 1999) extended it for plate-like structures characterized by two-dimensional curvatures.

Shi *et al.* (1998) defined the modal strain energy (MSE) of the jth element and i^{th} mode as

$$MSE_{ij} = \{\phi_i\}^T \left[K^e \right]_j \{\phi_i\} \qquad (11.12)$$

The change ratio of MSE before and after damages was used as an indicator of the damage location.

Although damage detection methods based on mode shape changes and their derivatives can provide spatial information regarding the location of structural damage, they have several limitations in application. First, a dense array of measurement points is required for an accurate estimate of mode shapes and mode shape curvatures. Second, the mode shape has larger statistical variation than modal frequencies. Third, the mode shape curvature methods are not readily applicable for structures with complex configurations. Finally, it is necessary to select a mode shape, yet it is *a priori* unknown which mode suffers from a significant change due to a particular damage.

11.4.8 Flexibility Changes

With mass normalized mode shapes, the modal flexibility matrix can be obtained as

$$[F]=[\Phi][\Lambda]^{-1}[\Phi]^{T} = \sum_{i=1}^{n}\omega_{i}^{-2}\{\phi_{i}\}\{\phi_{i}\}^{T} \tag{11.13}$$

where $[F]$ is the flexibility matrix, $[\Phi]$ is the mode shape matrix, $[\Lambda] = diag(\omega_{i}^{2})$, and ω_{i} is the ith circular frequency. From Equation (11.13), it can be seen that the mode contribution to the flexibility matrix decreases as the frequency increases. Therefore, a good estimate of the flexibility matrix can be made from a few lower modes rather than the higher frequency modes, which are difficult to measure in practice.

Raghavendrachar and Aktan (1992) demonstrated that the flexibility was more sensitive to local damages than frequencies and mode shapes. Pandey and Biswas (1994) proposed the method based on flexibility changes in damage detection of a free-free steel beam. This method was applied to detect artificial cuts in the I-40 bridge over the Rio Grande in New Mexico, together with several other methods by Farrar and Jauregui (1996). The results demonstrated that this method was not successful in practice.

11.4.9 Comparison Studies

Wang and Zhang (1987) investigated the sensitivity of the vibration characteristics to faults in structures in order to choose sensitive characteristics as the observing properties. The results showed that modal parameters (damped frequencies, natural frequencies and mode shapes) were not sensitive to the structural damage, but the FRFs were when the excitation frequency is near the natural frequencies.

Five damage detection algorithms, including the 1-D strain energy method (Stubbs *et al.*, 1995), mode shape curvature method (Pandey *et al.*, 1991), flexibility change method (Pandey and Biswas, 1994), a method combining mode shape curvature and flexibility change (Zhang and Aktan, 1995), and stiffness change method (Zimmerman and Kaouk, 1994), were respectively applied to detect artificial cuts in the I-40 bridge over the Rio Grande in New Mexico (Farrar and Jauregui, 1996). In general, all methods could identify the damage location correctly for the severest damage, one cut from the mid-web completely through the bottom flange. The methods were inconsistent and did not clearly identify the damage location when they were applied to the three less severe damage cases. Detection results generally showed that the strain energy method performed best.

Cornwell *et al.* (1998) applied two non-model-based methods to detect damage. One is based on the changes in flexibility and the other on the changes in the strain energy. They tested on an aluminium free-free I-beam and a clamped aluminium plate before and after the cut. The results with the two methods showed that changes for serious damage cases could be located successfully but smaller levels of damage could not.

Zhao and DeWolf (1999) reported that modal flexibilities were more sensitive to damage than either natural frequencies or mode shapes.

11.4.10 Challenges in Vibration Based Damage Detection Methods

Although the vibration based damage detection methods have achieved some degree of success in aerospace and mechanical engineering communities, their application in civil structures is still limited due to several inherent weaknesses.

One issue is the dependence on prior analytical models and/or prior test data. Many vibration based damage detection methods assume that the initial FE model of a structure can represent the intact structure. But its accuracy is doubtful because there are many uncertainties in civil structures, e.g. existence of non-structural members, boundary conditions, nonlinear response, and damping. The solution to those problems is to update the FE model properly. This needs the measurement data in the intact state, which are usually not available in practice.

In addition to FE modelling error, measurement noise is also severe for civil structures. Civil structures with very large size and inertia are typically excited under ambient vibration. The frequency components are generally low (below 100 Hz or even a few Hz) and the bandwidth is relatively narrow compared with aerospace structures. This means the vibration based methods are not sensitive to the local damage to structures, which have a more significant influence on high modes. Unless the damage becomes very severe or the sensors are correctly distributed around them, the damage is difficult to detect reliably using limited sensors.

Some kinds of damage in civil structures are a long, gradual, and slow process. Vibration properties do not have an abrupt change in general. In addition, civil structures are located under varying environmental conditions, such as temperature and operational loadings. These environmental conditions affect structural vibration properties, on many occasions, more significantly than structural damage does. Although the temperature effect can be eliminated from the relation between the temperature and the vibration properties, there are a few difficulties in practice: 1) establishment of this relation requires a long period of field measurement (e.g., for years); 2) the global vibration properties are associated with the temperature distribution of the entire structure, not only the air temperature and/or surface temperatures (Xia *et al.*, 2011). This requires the temperature distribution of the whole structure or typical sections; 3) the temperature effect is combined with the operational loading effect. Therefore, environmental variation is a critical and complicated factor for reliable damage detection in civil structures.

Finally, civil structures are unique in terms of geometry, boundary conditions, configuration, and loadings. Therefore the methods may be applicable to one structure only but not to others.

11.5 DAMAGE DETECTION METHODS WITH CONSIDERATION OF UNCERTAINTIES

11.5.1 Uncertainties in Damage Detection

Damage detection in civil structures involves a significant amount of uncertainties. Consequently, the damage detection results are, in effect, the compensation for the uncertainties. If the uncertainties are large, the compensation for the damage detection will distort the actual results and lead to two kinds of false damage identification (Farrar *et al.*, 1998): (1) false-positive damage identification (identifying the intact element as damaged), and (2) false-negative damage identification (failure to identify the damaged elements). The second category of false damage detection can have serious life-safety implications. False-positive readings can also erode confidence in the damage detection process. It is imperative to analyse the source of the uncertainties, quantify the uncertainties and their effects, and evaluate the reliability of the damage identification results.

The uncertainties in damage identification are mainly attributed to:

(1) inaccuracy in the FE model discretization;
(2) uncertainties in geometry and boundary conditions;
(3) variations in material properties;
(4) environmental variability (such as temperature, wind and traffic)
(5) errors associated with measured signals;
(6) errors in post-processing techniques;
(7) improper methods employed in damage identification.

According to their sources, these uncertainties can be classified into three groups, namely methodology errors, modelling errors and measurement noises. Methodology errors are generated by the limitation of the method itself in or damage identification. Modelling errors are related to the uncertainties in modelling the actual structure, mainly including discretisation errors, configuration errors, and mechanical parameter errors. Measurement noises mainly come from procedures and equipment related to the vibration testing. They can be classified into system errors and random errors. Random errors include measurement noise from environmental sources, calibration error in the sensors, and calibration error in the actuators. System errors include imprecise placement and orientation of the sensors and misalignment of the force actuator (Peterson *et al.*, 1996), errors such as the added mass or stiffness due to sensors, and errors that result from the signal processing or identification techniques.

Typical error distributions adopted to simulate the characteristics of the random data are the uniform and normal probability density functions (PDFs) (Sanayei *et al.*, 1992). The latter case is more common for the uncertainties of the FE modelling and the measurement data. Based on the prior PDFs of the uncertainties, different techniques have been developed to estimate the PDFs of the damages or updated structural parameters, which will be described in the following sections. Collins *et al.* (1974) may be the first approach among them, in which a minimum variance method was employed.

11.5.2 Perturbation Approach

When the uncertainties are considered as normally distributed random variables with zero means and given covariance, the measured quantities (structural parameters and modal properties) are regarded as the true values plus the random noises as

$$\alpha_j = \alpha_j^{0} + \alpha_j^{0} X_{\alpha j} = \alpha_j^{0}(1 + X_{\alpha j}) \tag{11.14}$$

$$\lambda_i = \lambda_i^{0}(1 + X_{\lambda i}) \tag{11.15}$$

$$\{\phi_i\} = \{\phi_i\}^{0}(1 + X_{\phi i}) \tag{11.16}$$

for $i = 1, 2, \ldots, nm$ and $j = 1, 2, \ldots, ne$, where α_j is structural parameter, superscript "0" represents the corresponding true value, $X_{\alpha j}$, $X_{\lambda i}$, and $X_{\phi i}$ are the corresponding proportional uncertainties with zero means.

With the perturbation method (Liu, 1995; Xia *et al.*, 2002; Xia and Hao, 2003), parameters in the governing equation are expanded as a Taylor series in terms of the uncertainties. For example, in sensitivity based model updating, Equation (4.31) can be rewritten as

$$\{e\} = [S]\{\Delta\alpha\} \tag{11.17}$$

where $\{e\}$ is an error vector containing the differences between the eigenvalues and mode shapes at the measured DOFs of the structure before and after updating. Iteration "k" is disregarded here for brevity. Equation (11.17) is expanded as a second order Taylor series in terms of X_i (including $X_{\alpha j}$, $X_{\lambda i}$, and $X_{\phi i}$),

$$\{e\} = \{e\}^{0} + \sum_{i=1}^{N} \frac{\partial\{e\}^{0}}{\partial X_i} X_i + \frac{1}{2}\sum_{i=1}^{N}\sum_{j=1}^{N} \frac{\partial^2\{e\}^{0}}{\partial X_i \partial X_j} X_i X_j \tag{11.18}$$

$$[S] = [S]^{0} + \sum_{i=1}^{N} \frac{\partial[S]^{0}}{\partial X_i} X_i + \frac{1}{2}\sum_{i=1}^{N}\sum_{j=1}^{N} \frac{\partial^2[S]^{0}}{\partial X_i \partial X_j} X_i X_j \tag{11.19}$$

$$\{\Delta\alpha\} = \{\Delta\alpha\}^{0} + \sum_{i=1}^{N} \frac{\partial\{\Delta\alpha\}^{0}}{\partial X_i} X_i + \frac{1}{2}\sum_{i=1}^{N}\sum_{j=1}^{N} \frac{\partial^2\{\Delta\alpha\}^{0}}{\partial X_i \partial X_j} X_i X_j \tag{11.20}$$

where N is the total number of the uncertainties. Substituting the above equations into (11.17) and comparing the terms of 1, X_i, and $X_i X_j$, then the unknown quantities can be solved one by one as

$$\{\Delta\alpha\}^{0} = \left([S]^{0}\right)^{+} \{e\}^{0} \tag{11.21}$$

$$\frac{\partial\{\Delta\alpha\}}{\partial X_i} = \left([S]^{0}\right)^{+}\left(\frac{\partial\{e\}}{\partial X_i} - \frac{\partial[S]}{\partial X_i}\{\Delta\alpha\}^{0}\right) \tag{11.22}$$

$$\frac{\partial^2\{\Delta\alpha\}}{\partial X_i \partial X_j} = \left([S]^{0}\right)^{+}\left(\frac{\partial^2\{e\}}{\partial X_i \partial X_j} - \frac{\partial^2[S]}{\partial X_i \partial X_j}\{\Delta\alpha\}^{0} - 2\frac{\partial[S]}{\partial X_i}\frac{\partial\{\Delta\alpha\}}{\partial X_j}\right) \tag{11.23}$$

where "$+$" refers to the pseudo-inverse. As the modal sensitivity matrix $[S]$ is often ill-conditioned, direct solution to Equations (11.21) to (11.23) may yield poor estimates. Hua *et al.* (2008) employed regularisation techniques to handle the problem.

From Equation (11.20) the mean values of $\{\Delta\alpha\}$ are, noting that $E(X_i)=0$,

$$E\left(\{\Delta\alpha\}\right) = E\left(\{\Delta\alpha\}^0\right) + \frac{1}{2}\sum_{i=1}^{N}\frac{\partial^2\{\Delta\alpha\}}{\partial X_i^2}COV\left(X_i, X_i\right) \tag{11.24}$$

where $E(\cdot)$ and $COV(\cdot)$ refer to the mean value and covariance matrix, respectively. It is noted that when the correlation between the updating parameters and the measurement is disregarded, the second-order derivatives in the above equation are not necessary. This has been verified by Khodaparast *et al.* (2008). The covariance matrix of $\{\Delta\alpha\}$ is

$$\left[COV\left(\Delta\alpha, \Delta\alpha\right)\right] = \left[\frac{\partial\{\Delta\alpha\}}{\partial\{X\}}\right]\left[COV\left(X, X\right)\right]\left[\frac{\partial\{\Delta\alpha\}}{\partial\{X\}}\right]^T \tag{11.25}$$

During the above procedures, calculation of the derivatives of $[S]$ to noise vector $\{X\}$ requires the calculation of the eigenvalue derivatives and eigenvector derivatives. These can be obtained using Nelson's method (Nelson, 1976).

Subsequently the statistics of the updated parameters, $\tilde{\alpha}_i$, can be obtained as

$$E(\tilde{\alpha}_i) = E(\alpha_i) + E(\Delta\alpha_i) = \alpha_i^0 + E(\Delta\alpha_i) \tag{11.26}$$

$$COV(\tilde{\alpha}_i, \tilde{\alpha}_j) = COV\left(\alpha_i + \Delta\alpha_i, \alpha_j + \Delta\alpha_j\right)$$

$$= COV\left(\alpha_i, \alpha_j\right) + COV\left(\alpha_i, \Delta\alpha_j\right) + COV\left(\Delta\alpha_i, \alpha_j\right) + COV\left(\Delta\alpha_i, \Delta\alpha_j\right) \tag{11.27}$$

From Equations (11.14) and (11.20), it has

$$COV\left(\alpha_i, \alpha_j\right) = \alpha_i^0\alpha_j^0 COV\left(X_{\alpha i}, X_{\alpha j}\right) \tag{11.28}$$

$$COV\left(\alpha_i, \Delta\alpha_j\right) = E\left[\left(\alpha_i - \alpha_i^0\right)\left(\Delta\alpha_j - \Delta\alpha_j^0\right)\right] = E\left[\left(\alpha_i^0 X_{\alpha i}\right)\left(\sum_{k=1}^{N}\frac{\partial\Delta\alpha_j}{\partial X_k}X_k\right)\right]$$

$$= \alpha_i^0\sum_{k=1}^{N}\frac{\partial\Delta\alpha_j}{\partial X_k}E\left(X_{\alpha i}X_k\right) = \alpha_i^0\sum_{k=1}^{N}\frac{\partial\Delta\alpha_j}{\partial X_k}COV\left(X_{\alpha i}, X_k\right) \tag{11.29}$$

and similarly

$$COV\left(\Delta\alpha_i, \alpha_j\right) = COV\left(\alpha_j, \Delta\alpha_i\right) = \alpha_j^0\sum_{k=1}^{N}\frac{\partial\Delta\alpha_i}{\partial X_k}COV\left(X_{\alpha j}, X_k\right) \tag{11.30}$$

From Equations (11.27) ~ (11.30), the standard deviation of $\tilde{\alpha}_i$, $\sigma(\tilde{\alpha}_i)$, can be derived as the square root of the corresponding diagonal elements of the covariance matrix. With the assumption of normal distribution, the statistical distributions of the stiffness parameters in the updated state can be derived.

Without losing generality, the PDFs of a structural parameter in the undamaged and damaged states are illustrated in Figure 11.3. The interval of the healthy stiffness parameter, $\Omega(\alpha_i, \mu)$, is defined such that the probability of α_i contained within the interval is μ, i.e.

$$P\left(x_\alpha \in \Omega(\alpha_i, \mu)\right) = P(L_\Omega \leq x_\alpha < \infty) = \mu \tag{11.31}$$

where L_Ω is the lower bound of the interval $\Omega(\alpha_i, \mu)$. The probability of damage existence is then defined as that of $\tilde{\alpha}_i$ not within the healthy interval at the confidence level of μ:

$$PDE = P(-\infty < x_{\tilde{\alpha}} \leq L_\Omega) \tag{11.32}$$

The probability of damage existence is a value between 0 ~ 1, which depends on the PDFs and the confidence interval. It is apparent that if the probability of damage existence of an element is close to 1 then most likely the element is damaged; and on the other hand, if the probability of damage existence is close to 0, the damage of the element is very unlikely. Papadopoulos and Garcia (1998) proposed a few different definitions of the probability of damage existence.

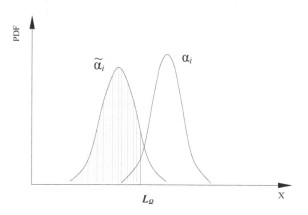

Figure 11.3 Probability density functions of the structural parameters

11.5.3 Bayesian Approach

Bayesian updating has been developed by Beck and his co-workers (Beck and Katafygiotis, 1998; Katafygiotis and Beck, 1998; Beck and Au, 2002; Yuen and Katafygiotis, 2005).

Let H_j, represented with a prior probability $P(H_j)$, denote a hypothesis for a damage event that can contain any number of substructures as damaged. Using the Bayes theorem, the posterior probability $P(H_j|\psi)$, given a set of observed modal parameters ψ, can be represented as:

$$P\left(H_j|\psi\right) = \frac{P\left(\psi|H_j\right)}{P\left(\psi\right)} P\left(H_j\right) \tag{11.33}$$

The most likely damaged substructures are the ones included in the hypothesis H_{max} that has the largest posterior probability (Sohn and Law, 1997), i.e.

$$P\left(H_{max}|\psi\right) = \max_{\forall H_j}\left(P\left(H_j|\psi\right)\right) \tag{11.34}$$

The distribution of the measurement noise and modelling error are explicitly considered within the Bayesian probabilistic framework. To avoid permuting all possible damage events H_j, a branch-and-bound method was devised to search the results. Sohn and Law (2000) applied this algorithm to detect damage in a bridge column. The proposed probabilistic damage detection method was able to locate the damaged region, whereas two deterministic methods were not.

11.5.4 Statistical Pattern Recognition

Pattern recognition is the assignment of some sort of output values to a given input value, according to some specific algorithms. Pattern recognition can be divided into two categories (Watanabe, 1985): 1) supervised classification in which the input pattern is identified as a member of a predefined class, 2) unsupervised classification in which the pattern is assigned to a hitherto unknown class. In the context of damage detection, the unsupervised classification can be applied to the case without damaged data and consequently is limited to Level 1 or Level 2 damage classification, which identifies the presence of damage only. When data are available from both the undamaged and damaged structure, the supervised classification approach can be employed in higher level damage identification such as damage quantification (Sohn *et al.*, 2003).

Sohn *et al.* (2003) reviewed some applications of statistical pattern recognition to structural damage detection. The supervised learning approaches include Response Surface analysis, Fisher Discrimination, Neural Networks, Genetic Algorithms, and Support Vector Machines. The unsupervised learning approaches include Control Chart Analysis, Outlier Detection, Neural Networks, Hypothesis Testing, and Principal Component Analysis.

11.5.5 Monte Carlo Simulation

The Monte Carlo simulation method is a commonly used numerical method for uncertainty analysis. The basic concept of the method is: a large number of samples are generated given a probability distribution of the stochastic parameters; for each set of parameters, a deterministic analysis is performed to obtain the corresponding solution; statistics are eventually estimated from these individual solutions. The technique is usually employed when the analytical solution is unfeasible or impossible to obtain.

As it requires a large number of simulations to obtain accurate and reliable statistics, the Monte Carlo simulation method is computationally intensive. Its accuracy is closely related to the size of the samples. Some variance reduction techniques can be applied to improve the variance of simulation results without increasing the sample size (Kottegoda and Rosso, 1997).

In this context, Agbabian *et al.* (1988) employed the Monte Carlo simulation method to identify the statistical properties of stiffness coefficients in a linear system. They computed the time histories of the applied excitation as well as the accelerations, velocities, and displacements of the system. The calculated data were then corrupted with a set of Gaussian noise. By separately applying the model updating procedure to different time segments, stiffness coefficients were identified. Statistics such as mean, variance, and PDFs were subsequently obtained. This work has later been extended to statistical identification of a nonlinear system approximated by an equivalent linear one (Smyth *et al.*, 2000). Banan *et al.* (1994), Sanayei and Saletnik (1996), Yeo *et al.* (2000), and Zhou *et al.* (2003) adopted similar approaches for studying the effect of measurement noise on identification results.

The researchers from Los Alamos National Laboratory employed the Monte Carlo simulation method to estimate the statistical confidence intervals on modal parameters identified from measured vibration data (Farrar *et al.*, 1998), and then study the effect of the measurement noise on damage detection (Doebling *et al.*, 1997). The Monte Carlo simulation method was also used to verify other approximate techniques, for example, Xia *et al.* (2002), Xia and Hao (2003), and Hua *et al.* (2008) employed it to verify the perturbation method described in Section 11.5.2.

11.6 SUMMARY

The application of the vibration based damage detection methods to civil structures has been limited to numerical analyses, laboratory model studies, and some small to medium size practical structures. For long-span bridges, vibration data are usually employed for examining the global properties of the structures rather than detecting damage. NDT techniques including visual inspection are applied to long-span bridges for inspection purposes. (Vibration data are also often used for measuring tension force of stay cables, while they are usually not regarded for damage detection but inspection). The inspection results can then be used for bridge rating and making maintenance strategies, which will be studied in the next chapter. However, as explained previously, the NDT techniques require that the location of the damage is known, which means the NDT methods are very time-consuming and real damage may be undetected. Development of novel and reliable damage detection methods for long-span bridges is imperative and needs great efforts. Consequently a laboratory-based test-bed will be established as a benchmark problem to apply different sensors, test various damage detection algorithms, identify their strengths and weaknesses, and develop more feasible algorithms for real long-span suspension bridges. The test-bed will be detailed in Chapter 13.

11.7 REFERENCES

Adams, R.D., Brownjohn, J.M.W. and Cawley, P., 1991, The detection of defects in GRP lattice structures by vibration measurements. *NDT & E International*, **24**, pp. 123-134.

Adams, R.D. and Coppendale, J., 1976, Measurement of the elastic moduli of structural adhesives by a resonant bar technique. *Journal of Mechanical Engineering Science*, **18**, pp. 149-158.

Agbabian, M.S., Masri, S.F. and Miller, R.K., 1991, System identification approach to detection of structural changes. *Journal of Engineering Mechanics, ASCE*, **117(2)**, pp. 370-390.

Agbabian, M.S., Masri, S.F., Miller, R.K. and Caughey, T.K., 1988, A system identification approach to the detection of changes in structural parameters. In *Proceedings of Workshop on Structural Safety Evaluation Based on System Identification Approach*, edited by Natke, H.G. and Yao, J.T.P., (Wiesbaden: Friedrick Vieweg & Son), pp. 341-356.

Anderson, D.A. and Seals, R.K., 1981, Pulse velocity as a predictor of 28 and 90 day strength. *ACI Journal Proceedings*, **78(9)**, pp. 116-122.

ASCE, 2005, *http://www.asce.org/reportcard/2005/index2005.cfm.*

ASTM, 2004, *Standard C1383: Standard Test Method for Measuring the P-Wave Speed and the Thickness of Concrete Plates Using the Impact-Echo Method*, (West Conshohocken, Pennsylvania: ASTM International).

ASTM, 2007a, *Standard D4788: Standard Test Method for Detecting Delaminations in Bridge Decks Using Infrared Thermography*, (West Conshohocken, Pennsylvania: ASTM International).

ASTM, 2007b, *Standard D5882: Test Method for Low Strain Integrity Testing of Piles*, (West Conshohocken, Pennsylvania: ASTM International).

ASTM, 2009, *Standard C597: Standard Test Method for Pulse Velocity through Concrete*, (West Conshohocken, Pennsylvania: ASTM International).

Banan, M.R., Banan, M.R. and Hjelmstad, K.D., 1994, Parameter estimation of structures from static response I: computational aspects. *Journal of Structural Engineering, ASCE*, **120**, pp. 3243-3258.

Beck, J.L. and Au, S.K., 2002, Baysian updating of structural models and reliability using Markov chain Monte Carlo simulation. *Journal of Engineering Mechanics, ASCE*, **128(4)**, pp. 380-391.

Beck, J.L. and Katafygiotis, L.S., 1998, Updating models and their uncertainties I: Bayesian statistical framework. *Journal of Engineering Mechanics, ASCE*, **124**, pp. 455-461.

Biswas, M., Pandey, A.K. and Bluni, S., 1994, Modified chain-code computer vision techniques for interrogation of vibration signatures for structural fault detection. *Journal of Sound and Vibration*, **175(1)**, pp. 89-104.

Brownjohn, J.M.W. and Steele, G.H., 1979, *Non-destructive Testing Using Measurements of Structural Dampings*, Thesis, (Bristol, UK: University of Bristol).

Cawley, P. and Adams, R.D., 1979, The locations of defects in structures from measurements of natural frequencies. *Journal of Strain Analysis*, **14(2)**, pp. 49-57.

Chance, J., Tomlinson, G.R. and Worden, K., 1994, A simplified approach to the numerical and experimental modeling of the dynamics of a cracked beam. In *Proceedings of the 12th International Modal Analysis Conference*, edited by DeMichele D.J., (Bethel, Connecticut: Society for Experimental Mechanics), pp. 778-785.

Chang, P.C. and Liu, S.C, 2003, Recent research in nondestructive evaluation of civil infrastructures. *Journal of Materials in Civil Engineering, ASCE*, **15(3)**, pp. 298-304.

Chang, S.P., Yee, J.Y. and Lee, J., 2009, Necessity of the bridge health monitoring system to mitigate natural and man-made disasters. *Structure and Infrastructure Engineering*, **5(3)**, pp. 173-197.

Choy, F.K., Liang, R. and Xu, P., 1995, Fault identification of beams on elastic foundation. *Computers and Geotechnics*, **17**, pp. 157-176.

Collins, J.D., Hart, G.C., Hasselman, T.K. and Kennedt, B., 1974, Statistical identification of structures. *AIAA Journal*, **12(2)**, pp. 185-190.

Cornwell, P.J., Doebling, S.W. and Farrar, C.R., 1997, Application of the strain energy damage detection method to plate-like structures. In *Proceedings of the 15th International Modal Analysis Conference*, Orlando, FL, edited by Wicks, A.L., (Bethel, Connecticut: Society for Experimental Mechanics), pp. 1312-1318.

Cornwell, P.J., Doebling, S.W. and Farrar, C.R., 1999, Application of the strain energy damage detection method to plate-like structures. *Journal of Sound and Vibration*, **224(2)**, pp. 359-374.

Cornwell, P.J., Kam, M., Carlson, B., Hoerst, B., Doebling, S.W. and Farrar, C.R., 1998, Comparative study of vibration-based damage ID algorithms. In *Proceedings of the 16th International Modal Analysis Conference*, Santa Barbara, CA, edited by Wicks, A.L. and DeMichele D.J., (Bethel, Connecticut: Society for Experimental Mechanics), pp. 1710-1716.

Daubechies, I., 1992, *Ten Lectures on Wavelets*, (Philadelphia, Pennsylvania: Society for Industrial and Applied Mathematics).

Davis, A.G., Evans, J.G. and Hertlein, B.H., 1997, Nondestructive evaluation of concrete radioactive waste tanks. *Journal of Performance of Constructed Facilities, ASCE*, **11(4)**, pp. 161-167.

Davis, A.G. and Hertlein, B.H., 1991, Development of nondestructive small-strain methods for testing deep foundations: a review. *Transportation Research Record*, **1331**, pp. 15-20.

Doebling, S.W., Farrar, C.R. and Goodman, R.S., 1997, Effects of measurement statistics on the detection of damage in the Alamosa Canyon bridge. In *Proceedings of the 15th International Modal Analysis Conference*, Orlando, Florida, edited by Wicks, A.L., (Bethel, Connecticut: Society for Experimental Mechanics), pp. 919-929.

Doebling, S.W, Farrar, C R, Prime, M.B and Shevitz, D.W, 1996, *Damage Identification and Health Monitoring of Structural and Mechanical Systems from Changes in their Vibration Characteristics: A Literature Review*. Report LA-13070-MS, (Los Alamos: Los Alamos National Laboratory).

D'Souza, K. and Epureanu, B., 2005, Damage detection in nonlinear systems using system augmentation and generalized minimum rank perturbation theory. *Smart Materials and Structures*, **14**, pp. 989-1000.

Dufay, J.C., 1985, Scorpion: premier systeme de radioscopie televisee hauteenergie pour le controle non destructif des ouvrages d'art en beton precontraint. *Bull Liaison Labo. Ponts et Chaussees*, **139**, pp. 77-83.

Elliot, J.F., 1996, Monitoring of prestressed structures. *Civil Engineering, ASCE*, **66(7)**, pp. 61-63.

Farrar, C.R., Doebling, S.W. and Cornwell, P.J., 1998, A comparison study of modal parameter confidence intervals computed using the Monte Carlo and Bootstrap techniques. In *Proceedings of the 16th International Modal Analysis Conference*, Santa Barbara, California, edited by Wicks, A.L. and DeMichele D.J., (Bethel, Connecticut : Society for Experimental Mechanics), pp. 936-944.

Farrar, C.R., Doebling, S.W. and Nix, D.A., 2001, Vibration-based structural damage identification. *Philosophical Transactions of the Royal Society: Mathematical, Physical & Engineering Sciences*, **359(1778)**, pp. 131-149.

Farrar, C.R. and Jauregui, D.V., 1996, *Damage Detection Algorithms Applied to Experimental and Numerical Modal Data From the I-40 Bridge*. Report LA-13074-MS, (Los Alamos: Los Alamos National Laboratory).

Friswell, M.I., Penny, J.E.T. and Wilson, D.E.L., 1994, Using vibration data and statistical measures to locate damage in structures. *Modal Analysis: The International Journal of Analytical and Experimental Modal Analysis*, **9(4)**, pp. 239-254.

Han, J.G., Ren, W.X. and Sun, Z.S., 2005, Wavelet packet based damage identification of beam structures. *International Journal of Solids and Structures*, **42(26)**, pp. 6610-6627.

Hearn, G. and Testa, R.G., 1991, Modal analysis for damage detection in structures. *Journal of Structural Engineering, ASCE*, **117(10)**, pp. 3042-3063.

Hellier, C.J., 2001, *Handbook of Nondestructive Evaluation*, (New York: McGraw-Hill).

Hemez, F.M., 1993, *Theoretical and Experimental Correlation Between Finite Element Models and Modal Tests in the Context of Large Flexible Space Structures*, Ph.D. Dissertation, (Boulder, Colorado: University of Colorado).

Hou, Z., Noori, M. and Amand, R.St., 2000, Wavelet-based approach for structural damage detection. *Journal of Engineering Mechanics, ASCE*, **126(7)**, pp. 677-683.

Hua, X.G., Ni, Y.Q., Chen, Z.Q. and Ko, J.M., 2008, An improved perturbation method for stochastic finite element model updating. *International Journal for Numerical Methods in Engineering*, **73(13)**, pp. 1845-1864.

Huang, N.E., Shen, Z., Long, S.R., Wu, M.C., Shih, H.H., Zheng, Q., Yen, N.C., Tung, C.C. and Liu, H.H., 1998, The empirical mode decomposition and the Hilbert spectrum for nonlinear and non-stationary time series analysis. *Proceedings of Royal Society London A*, **454**, pp. 903-995.

Ibrahim, S.R., 1993, Correlation and updating methods: finite element dynamic model and vibration test data. In *Proceedings of International Conference on Structural Dynamics Modelling, Test, Analysis and Correlation*, Cranfield, UK, (Glasgow, Scotland: Bell & Bain Ltd), pp. 323-347.

Katafygiotis, L.S. and Beck, J.L., 1998, Updating models and their uncertainties II: model identifiability. *Journal of Engineering Mechanics*, **124**, pp. 463-467.

Keye, S., 2006, Improving the performance of model-based damage detection methods through the use of an updated analytical model. *Aerospace Science and Technology*, **10(3)**, pp. 199-206.

Khodaparast, H.H., Mottershead, J.E. and Friswell, M.I., 2008, Perturbation methods for the estimation of parameter variability in stochastic model updating. *Mechanical Systems and Signal Processing*, **22(8)**, pp. 1751-1773.

Kim, J.H., Jeon, H.S. and Lee, C.W., 1992, Application of the modal assurance criteria for detecting and locating structural faults. In *Proceedings of the 10th International Modal Analysis Conference*, edited by DeMichele D.J., (Bethel, Connecticut : Society for Experimental Mechanics), pp. 536-540.

Knab, L.I., Blessing, G.V. and Clifton, J.R., 1983, Laboratory evaluation of ultrasonics for crack detection in concrete. *ACI Journal Proceedings*, **80(1)**, pp. 17-27.

Ko, J.M., Wong, C.W. and Lam, H.F., 1994, Damage detection in steel framed structures by vibration measurement approach. In *Proceedings of the 12th International Modal Analysis Conference*, edited by DeMichele D.J., (Bethel, Connecticut: Society for Experimental Mechanics), pp. 280-286.

Kottegoda, N.T. and Rosso, R., 1997, *Statistics, Probability, and Reliability for Civil and Environmental Engineers*, (New York: McGraw-Hill).

Kunz, J.T. and Eales, J.W., 1985, Evaluation of bridge deck condition by the use of thermal infrared and ground penetrating radar. In *Proceedings of the 2nd*

Annual International Bridge Conference, Pittsburgh, Pennsylvania, USA, (North Vancouver, British Columbia: Buckland & Taylor Ltd.), pp. 121-127.

Li, D.S. and Ou, J.P., 2008, Acoustic emission monitoring and critical failure identification of bridge cable damage. In *Proceedings of SPIE Nondestructive characterization for composite materials, aerospace engineering, civil infrastructure, and homeland security*, edited by Shull, P.J., Wu, H.F., Diaz, A.A. and Vogel, D.W., (Bellingham, Washington: SPIE), **6934**, pp. 1-5.

Li, H.L., Deng, X.Y. and Dai, H.L., 2007, Structural damage detection using the combination method of EMD and wavelet analysis. *Mechanical Systems and Signal Processing*, **21(1)**, pp. 298-306.

Li, X.Y. and Law, S.S., 2010, Adaptive Tikhonov regularization for damage detection based on nonlinear model updating. *Mechanical Systems and Signal Processing*, **24(6)**, pp. 1646-1664.

Liu, P.L., 1995, Identification and damage detection of trusses using modal data. *Journal of Structural Engineering, ASCE*, **121(4)**, pp. 599-608.

Maia, N.M.M., Silva, J.M.M., He, J., Lieven, N.A.J., Lin, R.M., Skingle, G.W., To, W. and Urgueira, A.P.V., 1997, *Theoretical and Experimental Modal Analysis*, (England: Research Studies Press Ltd).

Naik, T.R., Mailhotra, V.M. and Popovics, J.S., 2004, The ultrasonic pulse velocity method. In *Handbook on Nondestructive Testing of Concrete*, edited by Malhotra, V.M. and Carino, N.J., (Boca Raton, Florida: CRC Press).

Nair, A. and Cai, C.S., 2010, Acoustic emission monitoring of bridges: review and case studies. *Engineering Structures*, **32(6)**, pp. 1704-1714.

Narkis, Y., 1994, Identification of crack location in vibrating simply supported beams. *Journal of Sound and Vibration*, **172(4)**, pp. 549-558.

Ndambi, M.J.M., 2002, *Damage Assessment in Reinforced Concrete Beams by Damping Analysis*, Ph.D. Dissertation, (Brussel, Belgium: Vrije University).

Nelson, R.B., 1976, Simplified calculation of eigenvector derivatives. *AIAA Journal*, **14**, pp. 1201-1205.

Pandey, A.K. and Biswas, M., 1994, Damage detection in structures using changes in flexibility. *Journal of Sound and Vibration*, **169(1)**, pp. 3-17.

Pandey, A.K., Biswas, M. and Samman, M.M., 1991, Damage detection from changes in curvature mode shapes. *Journal of Sound and Vibration*, **145(2)**, pp. 321-332.

Papadopoulos, L. and Garcia, E., 1998, Structural damage identification: a probabilistic approach. *AIAA Journal*, **36(11)**, pp. 2137-2145.

Park, S., Stubbs, N. and Sikorsky, C., 1997, Linkage of nondestructive damage evaluation to structural system reliability. *Smart Systems for Bridges, Structures and Highways*, edited by Stubbs, N., (Washington, D.C.: IEEE Computer Society), pp. 234-245.

Pascual, R., Golinval, J.C. and Razeto, M., 1996, Testing of FRF based model updating methods using a general finite element program. In *Proceedings of the 21st International Seminar on Modal Analysis*, Leuven, Belgium, pp. 1933-1945.

Peterson, L.D., Bullock, S.J. and Doebling, S.W., 1996, The statistical sensitivity of experimental modal frequencies and damping ratios to measurement noise. *Modal Analysis: The International Journal of Analytical and Experimental Modal Analysis*, **11(1)**, pp. 63-75.

Raghavendrachar, M. and Aktan, A.E., 1992, Flexibility by multireference impact testing for bridge diagnostics. *Journal of Structural Engineering, ASCE,* **118(8),** pp. 2186-2203.

Rens, K.L., Wipf, T.J. and Klaiber, F.W., 1997, Review of nondestructive evaluation techniques of civil infrastructure. *Journal of Performance of Constructed Facilities, ASCE,* **11(4),** pp.152-160.

Rizos, P.F., Aspragathos, N. and Dimarogonas, A.D., 1990, Identification of crack location and magnitude in a cantilever beam from the vibration modes. *Journal of Sound and Vibration,* **138(3),** pp. 381-388.

Rytter, A., 1993, *Vibration Based Inspection of Civil Engineering Structures,* Ph.D. Dissertation, (Aalborg, Denmark: Aalborg University).

Salane, H.J. and Baldwin, J.W., 1990, Identification of modal properties of bridges. *Journal of Structural Engineering,* **116(7),** pp. 2008-2021.

Salawu, O.S., 1997, Detection of structural damage through changes in frequency: a review. *Engineering Structures,* **19(9),** pp. 718-723.

Salawu, O.S. and Williams, C., 1995, Bridge assessment using forced-vibration testing. *Journal of Structural Engineering, ASCE,* **121(2),** pp. 161-173.

Samman, M.M., Biswas, M. and Pandey, A.K., 1991, Employing pattern recognition for detecting cracks in a bridge model. *Modal Analysis: The International Journal of Analytical and Experimental Modal Analysis,* **6(1),** pp. 35-44.

Sanayei, M., Onipede, O. and Babu, S.R., 1992, Selection of noisy measurement locations for error reduction in static parameter identification. *AIAA Journal,* **30(9),** pp. 2299-2309.

Sanayei, M. and Saletnik, M.J., 1996, Parameter estimation of structures from static strain measurements II: error sensitivity analysis. *Journal of Structural Engineering, ASCE,* **122,** pp.563-572.

Shi, Z.Y., Law, S.S. and Zhang, L.M., 1998, Structural damage localization from modal strain energy change. *Journal of Sound and Vibration,* **218(5),** pp. 825-844.

Shull, P.J., 2002, *Nondestructive Evaluation: Theory, Techniques, and Applications,* (New York: Marcel Dekker Inc).

Smyth, A.W., Masri, S.F., Caughey, T.K. and Hunter, N.F., 2000, Surveillance of mechanical systems on the basis of vibration signature analysis. *Journal of Applied Mechanics, ASME,* **67,** pp. 540-551.

Sohn, H., Farrar, C.R., Hemez, F.M., Czarnecki, J.J., Shunk, D.D., Stinemates, D.W. and Nadler, B.R., 2003, *A Review of Structural Health Monitoring Literature: 1996-2001.* Report LA-13976-MS, (Los Alamos: Los Alamos National Laboratory).

Sohn, H. and Law, K.H., 1997, A Bayesian probabilistic approach for structure damage detection. *Earthquake Engineering and Structural Dynamics,* **26,** pp. 1259-1281.

Sohn, H. and Law, K.H., 2000, Bayesian probabilistic damage detection of a reinforced-concrete bridge column. *Earthquake Engineering and Structural Dynamics,* **29,** pp. 1131-1152.

Stubbs, N., Broome, T.H. and Osegueda, R., 1990, Nondestructive construction error detection in large space structures. *AIAA Journal,* **28(1),** pp. 146-152.

Stubbs, N., Kim, J.T. and Farrar, C.R., 1995, Field verification of a nondestructive damage localization and sensitivity estimator algorithm. In *Proceedings of the*

13th International Modal Analysis Conference, edited by Wicks, A. L., (Bethel, Connecticut: Society for Experimental Mechanics), pp. 210-218.

Stubbs, N. and Osegueda, R., 1990a, Global non-destructive damage evaluation in solids. *Modal Analysis: The International Journal of Analytical and Experimental Modal Analysis*, **5(2)**, pp. 67-79.

Stubbs, N. and Osegueda, R., 1990b, Global damage detection in solids-experimental verification. *Modal Analysis: The International Journal of Analytical and Experimental Modal Analysis*, **5(2)**, pp. 81-97.

Stubbs, N., Park, S., Sikorsky, C. and Choi, S., 2000, A global non-destructive damage assessment methodology for civil engineering structures. *International Journal of System Science*, **31**, pp. 1361-1373.

Vogel, T., Schechinger, B. and Fricker, S., 2006, Acoustic emission analysis as a monitoring method for prestressed concrete structures. In *Proceedings of the 9th European Conference on NDT*, Berlin, Germany, edited by Farley, M, pp. 281-298.

Wang, J.H. and Liou, C.M., 1991, Experimental identification of mechanical joint parameters. *Journal of vibration and Acoustics, Transaction of the ASME*, **113**, pp. 28-36.

Wang, W.J. and Zhang, A.Z., 1987, Sensitivity analysis in fault diagnosis of structures. In *Proceedings of the 5th International Modal Analysis Conference*, London, edited by DeMichele D.J., (Bethel, Connecticut: Society for Experimental Mechanics), pp. 496-501.

Watanabe, S., 1985, *Pattern Recognition: Human and Mechanical*, (New York: Wiley).

Weil, G.J., 2004, Infrared thermographic techniques. In *Handbook on Nondestructive Testing of Concrete*, edited by Malhotra, V.M. and Carino, N.J., (Boca Raton, Florida: CRC Press).

Xia, P.Q. and Brownjohn, J.M.W., 2003, Residual stiffness assessment of structurally failed reinforced concrete structure by dynamic testing and finite element model updating. *Experimental Mechanics*, **43**, pp. 372-378.

Xia, Y. and Hao, H., 2003, Statistical damage identification of structures with frequency changes. *Journal of Sound and Vibration*, **263(4)**, pp. 853-870.

Xia, Y., Hao, H., Brownjohn, J.M.W. and Xia, P.Q., 2002, Damage identification of structures with uncertain frequency and mode shape data. *Earthquake Engineering and Structural Dynamics*, **31(5)**, pp. 1053-1066.

Xia, Y., Hao, H., and Deeks, A.J., 2007, Dynamic assessment of shear connectors in slab-girder bridges. *Engineering Structures*, **29(7)**, pp. 1475-1486.

Xia, Y., Xu, Y.L., Wei, Z.L., Zhu, H.P. and Zhou, X.Q., 2011, Variation of structural vibration characteristics versus non-uniform temperature distribution. *Engineering Structures*, **33(1)**, pp.146-153.

Xu, Y.L., Zhang, J., Li, J.C. and Xia, Y., 2009, Experimental investigation on statistical moment-based structural damage detection method. *Structural Health Monitoring*, **8(6)**, pp. 555-571.

Xu, Y.L., Zhu, H.P. and Chen, J., 2004, Damage detection of mono-coupled multistory buildings: numerical and experimental investigations. *Structural Engineering and Mechanics*, **18(6)**, pp. 709-729.

Yao, J.T.P. and Natke, H.G., 1994, Damage detection and reliability evaluation of existing structures. *Structural Safety*, **15**, pp. 3-16.

Yeo, I., Shin, S., Lee, H.S. and Chang, S.P., 2000, Statistical damage assessment of framed structures from static responses. *Journal of Engineering Mechanics, ASCE,* **126**, pp. 414-421.

Yoon, D.J., Weiss, W.J. and Shah, S.P., 2000, Assessing damage in corroded reinforced concrete using acoustic emission. *Journal of Engineering Mechanics, ASCE,* **126(3)**, pp. 273-283.

Yuen, K.V. and Katafygiotis, L.S., 2005, Model updating using noisy response measurements without knowledge of the input spectrum. *Earthquake Engineering and Structural Dynamics,* **34**, pp. 167-187.

Zachar, J. and Naik, T.R., 1992, Principles of infrared thermography and application for assessment of the deterioration of the bridge deck at the zoo interchange. In *Proceedings of the Materials Engineering Congress,* Atlanta, Georgia, edited by White, T.D., (New York: American Society of Civil Engineers), pp. 107-115.

Zhang, J., Xu, Y.L., Xia, Y. and Li, J., 2008, A new statistical moment-based structural damage detection method. *Structural Engineering and Mechanics,* **30(4)**, pp. 445-466.

Zhang, Z. and Aktan, A.E., 1995, The damage indices for the constructed facilities. In *Proceedings of the 13ᵗʰ International Modal Analysis Conference,* Nashville, Tennessee, edited by DeMichele D.J., (Bethel, Connecticut: Society for Experimental Mechanics), pp. 1520-1529.

Zhao, J. and DeWolf, J.T., 1999, Sensitivity study for vibrational parameters used in damage detection. *Journal of Structural Engineering, ASCE,* **125(4)**, pp. 410-416.

Zhou, J., Feng, X. and Fan, Y.F., 2003, A probabilistic method for structural damage identification using uncertain data. In *Proceedings of the 1st International Conference on Structural Health Monitoring and Intelligent Infrastructure,* edited by Wu, Z.S. and Abe, M., (Lisse, Netherlands: A.A., Balkema), pp. 487-492.

Zhu, X.Q. and Law, S.S., 2006, Wavelet-based crack identification of bridge beam from operational deflection time history. *International Journal of Solids and Structures,* **43(7-8)**, pp. 2299-2317.

Zimin, V.D. and Zimmerman, D.C., 2009, Structural damage detection using time domain periodogram analysis. *Structural Health Monitoring,* **8(2)**, pp. 125-135.

Zimmerman, D.C. and Kaouk, M., 1994, Structural damage detection using a minimum rank update theory. *Journal of Vibration and Acoustics,* **116**, pp. 222-231.

11.8 NOTATIONS

a_j, a_{j+1}	Nodal coordinates of a beam element.
$COV(\cdot)$	Covariance matrix.
$E(\cdot)$	Mean values.
EI	Flexural rigidity of a beam.
$e(H_{ij})$	An error index defined as the Euclidean norm of the frequency response function.

e_{rij}	An error index defined as the difference in the change of the analytical and measured eigenvalues due to possible damage at position r.
$\{e\}$	An error vector containing the differences before and after updating.
$[F]$	Modal flexibility matrix.
$H(\omega)$	Frequency response function at ω.
$H^A(\omega)$, $H^E(\omega)$	Analytical and experimental frequency response function.
H_j	A hypothesis for a damage event.
h	Length of an element.
$[K]$, $[M]$	Stiffness matrix, mass matrix.
$[K^e]$	Elemental stiffness matrix.
L	Total length of a beam.
L_Ω	Lower bound of the interval $\Omega(\alpha_i, \mu)$.
MSE_{ij}	Modal strain energy of the jth element and i^{th} mode.
N	Total number of the uncertainties.
ne	Number of elements.
nm	Number of modes.
$P(H_j)$	Prior probability of a damage event.
$P(H_j|\psi)$	Posterior probability based on a set of observed modal parameters ψ.
PDE	Probability of damage existence.
q	The degree of freedom.
$[S]$	Modal sensitivity matrix.
U	Strain energy of a Bernoulli-Euler beam.
$X_{\alpha j}$, $X_{\lambda i}$, $X_{\phi i}$	Proportional uncertainties with zero means of structural parameter, eigenvalue and eigenvector.
$\{X\}$	Noise vector.
α_j	Structural parameters.
ψ	A set of observed modal parameters.
μ	Confidence level of a structural parameter contained within an interval.
$\Omega(\alpha_i, \mu)$	The interval of the healthy stiffness parameter.
$\sigma(\cdot)$	Standard deviation.
λ_i, $[\Lambda]$	The ith eigenvalue, matrix of eigenvalues.
λ_i^u, λ_i^d, $\Delta\lambda_i$,	Eigenvalues in the undamaged and damaged states, eigenvalue changes.
ω	Circular frequency.
$\{\phi_i\}$, $[\Phi]$	The ith eigenvector, matrix of eigenvectors.
$\phi_{q,i}^{\cdot}$	Mode shape curvature.
ϕ_i^u, ϕ_i^d, $\{\Delta\phi_i\}$	Eigenvectors (mode shapes) in the undamaged and damaged states, mode shape changes

CHAPTER 12

Bridge Rating System

12.1 PREVIEW

Long-span suspension bridges begin to deteriorate once they are built and continuously accumulate damage during their service life due to natural hazards and harsh environments such as typhoons, earthquakes, highway road vehicles, trains, temperature and corrosion. To ensure the serviceability and safety of long-span suspension bridges, bridge rating systems are often adopted by bridge management authorities as guidance in determining the time intervals for inspection and the actions to be taken in the event of defects being identified. Regular bridge inspection is one of the most important maintenance programmes to achieve desirable solutions to maintain satisfactory infrastructure performance from the long-term economic point of view. Nevertheless, most of the currently used bridge rating methods for long-span suspension bridges are based on practical experience with some engineering analyses. On the other hand, SHM technologies have gained rapid development in recent years (Mufti, 2001; Wenzel, 2009). SHM systems have been designed and installed in a number of long-span cable-supported bridges to monitor their serviceability and safety, including the Tsing Ma Bridge (Wong *et al.,* 2001a; 2001b). However, the measurement data obtained from the SHM system have not been fully utilized for the current bridge rating. There is insufficient link between the bridge rating method and the SHM system to fulfil the common goals.

In this regard, this chapter presents an SHM based bridge rating method for inspection of long-span suspension bridges. The fuzzy based analytic hierarchy approach (F-AHP) is employed, and the hierarchical structure for synthetic rating of each structural component of a bridge is proposed. The criticality and vulnerability analyses are performed largely based on field measurement data from the SHM systems to offer relatively accurate condition evaluation of the bridges and to reduce uncertainties involved in the existing rating method. The procedures for determining relative weighs and fuzzy synthetic ratings for both criticality and vulnerability are then suggested. The fuzzy synthetic decisions for inspection are made in consideration of the synthetic ratings of all structural components. Finally, the SHM based bridge rating method is applied to the Tsing Ma suspension bridge as a case study.

12.2 LONG-SPAN SUSPENSION BRIDGE RATING METHODS

12.2.1 Current Status of Bridge Rating

Bridge condition rating is often used for short and medium span bridges (Bevc *et al.*, 1999; AASHTO, 2003; FHA, 2006). In condition rating, a bridge is first classified into a few groups. The rating scores are then given to the groups or components in consideration of deterioration extent and intensity, the component importance and urgency. A synthesis rating for the whole bridge is finally worked out on the basis of the rating scores for individual components in terms of good, fair, poor, and bad.

Long-span suspension bridges have many more categories of components than small and medium span bridges. The inspection work is therefore much tougher than short and medium span bridges (Tabatabai, 2005; Dudgeon, 2007). It is reasonable and economical to inspect different components with different time intervals for long-span suspension bridges, which not only saves unnecessary inspection costs but also increases the inspection quality by concentrating resources on where they are most needed. For long-span suspension bridges, criticality or vulnerability rating methods are more suitable. For instance, the Master Maintenance Manual has been developed by the Hong Kong Highways Department (HyD) for the management, operation and maintenance of the Tsing Ma suspension bridge in Hong Kong (HyD, 1996; 2001). The structural components of the bridge have been assessed, according to a bridge rating system, for their proneness to damage or deterioration and their importance in maintaining structural integrity. Criticality and vulnerability ratings are given to each component and are to be used as the reference or guidance in defining the frequencies and time intervals for inspections (Wong, 2006). The criticality or vulnerability rating is different from condition rating in that the former emphasizes the possibility of deterioration while the latter put emphasis on the actual deterioration.

Nevertheless, most of the currently used bridge rating methods for long-span suspension bridges are based on practical experience with some engineering analyses. Efforts have been made by some scholars to pursue more scientific yet practical bridge rating systems. Kushida *et al.* (1993) proposed a membership function to quantify subjective uncertainties included in empirical knowledge on bridge rating. Melhem (1994) presented a fuzzy inference model based on a priority setting obtained through the solution of an eigenvector problem involving a pair-wise comparison matrix of importance. Aktan *et al.* (1996) integrated analytical and experimental research to assess global conditions and evaluate the serviceability and safety of bridges. Liang *et al.* (2001) set up a multi-layer fuzzy method for evaluating damage conditions of existing bridges. Kawamura and Miyamoto (2003) proposed a concrete bridge rating expert system for deteriorated concrete bridges, constructed from multi-layer neural networks. Sasmal *et al.* (2006) developed a systematic procedure and formulations for rating existing bridges using fuzzy mathematics. This section will discuss the F-AHP and the hierarchical structure for synthetic rating of each structural component of a bridge (HKPU, 2008; 2009c).

12.2.2 Formation of a Hierarchical Structure

The analytic hierarchy process (AHP) was developed by Saaty (1980) based on an axiomatic foundation. The AHP attracted the interest of many researchers because of its interesting mathematical properties and easy applicability. The diverse applications of AHP in industrial and financial sectors are due to its simplicity and ability to cope with complex decision making problems. The main steps in the application of AHP to the current problem are as follows: (1) to decompose a general decision problem into hierarchical sub-problems that can be easily comprehended and evaluated; (2) to determine the priorities of the items at each level of the decision hierarchy; and (3) to synthesize the priorities to determine the overall priorities of the decision alternatives. Since a long-span suspension bridge is a very complex system and the decision making takes place in a situation in which the pertinent data and the sequences of possible actions are not precisely known, it is important to adopt fuzzy data to express such situations in decision making on inspection, leading to the so-called F-AHP bridge rating method.

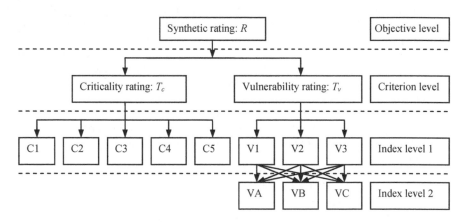

Figure 12.1 Analytic hierarchical structure for each bridge component

For a bridge, the top level of the F-AHP bridge rating system can be assigned as an objective level upon which the best decision for inspection could be made for each structural component. The next level of the hierarchical structure can be defined as a criterion level upon which the criticality rating and the vulnerability rating can be respectively determined for each structural component based on the criticality and vulnerability rating criteria in the next level called the index level. After the hierarchical structure is constructed, one can then determine the relative weights (priorities) of the items at each level of the decision hierarchy based on the mathematical properties of AHP. Finally, one can synthesize the relative weights at all levels to make the best decision on inspection. The hierarchical structure for synthetic rating of each structural component of a bridge is shown in Figure 12.1. The criticality rating criteria of each component are composed of five criticality factors from C1 to C5. The vulnerability rating criteria are set up based on three vulnerability factors V1 to V3. Each of the vulnerability factors is rated in three serial effects of VA, VB, and VC.

12.2.3 Determination of Relative Weights for Each Level

The AHP often uses the eigenvalue solution of comparative matrices to find the best relative weights (relative importance) for different elements in each level. The first step is to carry out pair-wise comparisons of elements in each level of the hierarchical structure. By assuming that the index level for criticality rating consists of A_1, A_2, ..., A_n items, the comparative matrix $[A]$ can be constructed by comparing objective i with objective j to obtain the relative weights $\alpha_{ij} = \omega_i/\omega_j$ ($i, j = 1, 2, ..., n$), as shown in Table 12.1. The relative importance value of ω_i ranges from 1 to 9, as suggested by Saaty (1980). Their definitions are described in Table 12.2.

Table 12.1 Comparative weight matrix.

	A_1	A_2	$\cdots$	A_n
A_1	α_{11}	α_{12}	$\cdots$	α_{1n}
A_2	α_{21}	α_{22}	$\cdots$	α_{2n}
$\cdots$	$\cdots$	$\cdots$	$\cdots$	$\cdots$
A_n	α_{n1}	α_{n2}	$\cdots$	α_{nn}

Table 12.2 Measurement of comparative matrix.

Intensity of relative importance	Definition
1	Equal importance
3	Weakly more importance than the other
5	Strongly more importance than the other
7	Demonstrably more importance than the other
9	Absolutely more importance than the other
Other numbers from 1 to 9	Intermediate values

A decision-maker could provide only the upper triangle of the above comparison matrix. The reciprocals placed in the lower triangle do not need any further analysis because of the following characteristics.

$$\alpha_{ij} > 0 \qquad (12.1)$$

$$\alpha_{ij} = 1/\alpha_{ji} \qquad (12.2)$$

$$\alpha_{ii} = 1 \qquad\qquad (12.3)$$

where $i, j = 1, 2, ..., n$. Each entry of matrix $[A]$ represents a pair-wise judgement. In the consistent reciprocal matrix $[A]$, it can be proved that matrix $[A]$ has rank 1 with non-zero eigenvalue equal to n_A (Saaty, 1980). In most of the practical problems, the pair-wise comparisons are not perfect, and one must find the principal right-eigenvalues that satisfies

$$[A]\{\phi\} = \lambda_{\max}\{\phi\} \qquad\qquad (12.4)$$

where $\{\phi\}$ is the eigenvector with respect to eigenvalue n_A; and $\lambda_{\max} \approx n_A$. The next step is to check the consistency of the comparative matrices in terms of the consistency ratio CR, which is determined by first estimating $\lambda_{\max}$ of matrix $[A]$. A consistency index CI of matrix $[A]$ is defined as

$$CI = (\lambda_{\max} - n_A)/(n_A - 1) \qquad\qquad (12.5)$$

The consistency ratio CR is then calculated by dividing CI with a random index RI as listed in Table 12.3. Each RI is an average random consistency index derived from a sample of size 500 of randomly generated reciprocal matrices. If the previous approach yields a CR greater than 0.10 then a re-examination of the pair-wise judgements is recommended until a CR less than or equal to 0.10 is achieved.

Table 12.3 Random index (RI).

N	1	2	3	4	5	6	7	8	9	10
RI	0	0	0.58	0.9	1.12	1.24	1.32	1.41	1.45	1.49

If the consistency of the comparison matrix is satisfied, the relative weights are calculated based on the normalized eigenvector corresponding to the maximum eigenvalue. The above procedures can be applied to determine the relative weights for vulnerability index level 1, criticality index level 1, and criterion level in Figure 12.1.

12.2.4 Fuzzy Synthetic Ratings for Criticality and Vulnerability

The criticality rating and vulnerability rating for each structural component are based on the criticality factors and vulnerability factors in relation to the criticality criteria and vulnerability criteria, respectively. Although the utilization of the SHM will reduce the uncertainties in the estimation of criticality factors and vulnerability factors, the accuracy of these factors is still not precisely known. Therefore, these factors are treated as fuzzy data in the decision making for inspection and their numbers are set from 0 and 100 in the F-AHP based bridge rating method.

The triangular fuzzy numbers are preferred in this study (Sasmal and Ramanjaneyulu, 2008). A fuzzy number M on $U \in (-\infty, +\infty)$ is defined to be a triangular fuzzy number if its membership function $\mu_m : U \rightarrow [0.1]$ is equal to (Dubois and Prade, 1979)

$$\mu_m(x) = \begin{cases} \dfrac{x-r}{m-r} & x \in [r,m] \\[2mm] \dfrac{x-u}{m-u} & x \in [m,u] \\[2mm] 0 & others \end{cases} \tag{12.6}$$

where $r \le m \le u$; r and u stand for the lower and upper values of the support for the decision of the fuzzy number M, respectively; m is the modal value. For example, when r and m are both equal to 0 or m and u equal to 100, the value of $\mu_m(x)$ is assumed as 1. The triangular fuzzy number is denoted as $M = (r, m, u)$. Let us select the five-point fuzzy rating set $\{G\}$ as

$$\{G\} = \{0 \quad 25 \quad 50 \quad 75 \quad 100\} \tag{12.7}$$

The triangular fuzzy numbers (0, 0, 50), (0, 25, 75), (0, 50, 100), (25, 75, 100), (50, 100, 100) can be generated to improve the decision making of criticality rating and vulnerability rating. Figure 12.2 shows the five triangular fuzzy numbers defined with the corresponding membership functions.

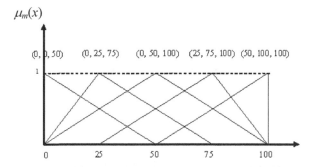

Figure 12.2 Fuzzy membership functions for the five point fuzzy rating set

In the F-AHP based bridge rating method, the criticality rating for each structural component can be determined by the following steps:

a) Calculate the criticality factors (C1, C2, C3, C4, and C5 in Figure 12.1) for each structural component using SHM-based computer simulations and measurement data recorded by the SHM system, as described in Sections 12.3 and 12.4.

b) Work out membership degree matrix $[R_c]$ based on the fuzzy membership functions and the criticality factors.

$$[R_c] = \begin{pmatrix} r_{11} & \cdots & r_{1m} \\ \vdots & \ddots & \vdots \\ r_{n1} & \cdots & r_{nm} \end{pmatrix} \tag{12.8}$$

where n is the number of items in the criticality index level; m is the number of the fuzzy rating values in the fuzzy rating set $\{G_c\}$; r_{ij} denotes the membership degree of the ith item to the jth fuzzy membership function; and subscript c denotes the criticality.

c) The fuzzy synthetic rating vector $\{B_c\}$ for criticality can then be determined by

$$\{B_c\} = \{\omega_c\}^T [R_c] \tag{12.9}$$

d) Finally, the fuzzy synthetic rating T_c for the criticality of the concerned structural component can be obtained as

$$T_c = \{G_c\}\{B_c\}^T \tag{12.10}$$

In the F-AHP based bridge rating method, the vulnerability rating for each structural component can be determined by the following steps:

a) Calculate the vulnerability factors (V1-VA, V1-VB, V1-VC, V2-VA, V2-VB, V2-BC, V3-VA, V3-VB, and V3-VC in Figure 12.1) for each structural component in the vulnerability index level, as described in Sections 12.3 and 12.4.

b) Calculate the vulnerability factors (V1, V2, and V3 in Figure 12.1) for each structural component in the vulnerability layer using the weighted product model:

$$\begin{cases} V1 = VA1^{1/3} \times VB1^{1/3} \times VC1^{1/3} \\ V2 = VA2^{1/3} \times VB2^{1/3} \times VC2^{1/3} \\ V3 = VA3^{1/3} \times VB3^{1/3} \times VC3^{1/3} \end{cases} \tag{12.11}$$

c) Work out the membership degree matrix $[R_v]$ based on the fuzzy membership functions and the vulnerability factors, where the subscript 'v' denotes the vulnerability.

d) The fuzzy synthetic rating vector $\{B_v\}$ for vulnerability can then be determined by

$$\{B_v\} = \{\omega_v\}^T [R_v] \tag{12.12}$$

e) Finally, the fuzzy synthetic rating T_v for the vulnerability of the structural component concerned can be obtained as

$$T_v = \{G_v\}\{B_v\}^T \tag{12.13}$$

12.2.5 Fuzzy Synthetic Decision for Inspection

The fuzzy synthetic rating R at the objective level can be calculated based on the fuzzy synthetic ratings, T_c and T_v, and the relative weights at the criterion level:

$$R = \{\omega_{cv}\}^T \begin{Bmatrix} T_c \\ T_v \end{Bmatrix} \tag{12.14}$$

After the fuzzy synthetic ratings of all the structural components are obtained, the prioritization or optimum for inspection frequency (fuzzy synthetic decision for inspection) can be determined. The larger the value of R, the smaller is the inspection time interval. One example is shown in Table 12.4.

Table 12.4 Fuzzy synthetic decision for inspection.

Fuzzy synthetic rating R	Time interval for inspection
$75 \leq R \leq 100$	6 months
$57.5 \leq R \leq 75$	1 year
$25 \leq R \leq 57.5$	2 years
$0 \leq R \leq 25$	6 years

12.3 CRITICALITY AND VULNERABILITY FACTORS

12.3.1 Criticality Factors

The criticality factors include five items in this study for the Tsing Ma Bridge. Table 12.5 shows the definition, range, and points for each criticality factor. To facilitate the decision making using the F-AHP based bridge rating method, the numbers for each criticality factor range from 0 to 100.

Table 12.5 Criticality factor: definition and values.

CF	Definition	Range	Points
C1	Any alternative load path?	No Yes, affect global structural performance Yes, not affect global structural performance	100 67 0
C2	Design normal combined loads (based on strength utilization factor)	0%~100%	100 67 33
C3	Design fatigue loads (based on fatigue life)	High: < 200 years Normal: between 200 to 300 years Low: > 300 years or not applicable	100 67 0
C4	Known or discovered imperfections but not serious enough to warrant immediate repair	Any, non-repairable Any, repairable None	100 67 0
C5	Failure mechanisms	Catastrophic collapse Partial collapse Structural damage	100 67 33

Criticality factor C1 – it is understandable that the structural component with an alternative load path or redundancy is robust, that is, no serious failure consequence will be induced by limited damage to this structural component. Therefore, the greater the redundancy for a given structural component, the lower is the criticality factor C1 for this component. This criticality factor could be represented by three numerical values of 100, 67, and 0 in this study.

Criticality factor C2 – this factor represents the strength reliability of a structural component, which is determined based on the strength utilization factor (SUF). The SUF is calculated using Equation (12.15). If the SUF of a certain structural component reaches the extreme value, C2 for this component should be 100. Otherwise, C2 should be the ratio of the SUF divided by the extreme value.

$$SUF = \frac{\gamma_L \sigma_L (1.0 + I)}{\varphi R_N - \gamma_D \sigma_D}$$ (12.15)

where R_N is the as-built nominal resistance; φ is the strength reduction factor; γ_L is the partial load factor for live loads; γ_D is the partial load factor for dead loads; σ_L is the stresses due to live loads; σ_D is the stresses due to dead loads; and I is the impact factor for dynamic live loads.

Criticality factor C3 – this item represents the fatigue reliability of a structural component. The relative fatigue criticality could be represented by 100, 67, and 0 in this study.

Criticality factor C4 – this item emphasizes the imperfections of structural components detected by previous inspections. In this regard, the severity of imperfections and the urgency to repair is respectively considered and represented by 100, 67, and 0 in this study.

Criticality factor C5 – this item represents the ultimate loading carrying capacity of structural components under extreme loading events. The structural component of foremost failure will be the most critical component and the corresponding energy demand for failure will be the least. The relative ultimate loading carrying capacity of different structural components could be represented by three numerical values of 100, 67, and 33 in this study.

12.3.2 Vulnerability Factors

The vulnerability factors include three items in this study for the Tsing Ma Bridge, namely, V1 – corrosion, V2 – damage, and V3 – wear. Each of these factors is rated in three serial effects of: (1) degree of exposure, (2) likelihood of detection in superficial inspection, and (3) likely influence on structural integrity. Table 12.6 shows the definition, range and points for each vulnerability factor. Similarly, the numbers for each vulnerability factor are between 0 and 100.

V1 represents the damage due to extra-slowly varying effects, such as alkali-silica reaction, chloride contamination, carbonation of concrete, and corrosion of steel. V2 represents the damage due to rapidly varying effects, such as accidental damage caused by vehicle collisions and ship collisions. V3 represents the damage due to slowly varying effects, such as movement of bearings and joints due to daily temperature variations. All of these three factors consist of three sub-items (VA, VB, and VC) and each sub-item can be represented by the three numerical values of 100, 50, and 0 respectively, according to their ranges.

12.4 CRITICALITY AND VULNERABILITY ANALYSES

Criticality and vulnerability analyses can be carried out to determine criticality factors and vulnerability factors for each structural component of the bridge. In principle, criticality and vulnerability analyses must be conducted based on the measured results from SHM systems and the SHM-oriented methodologies introduced in the previous chapters. Before the analyses, the key structural components of the bridge must be classified.

Table 12.6 Vulnerability factor: definition and values.

VF	Sub-item	Definition	Range	Points
V1: Corrosion	VA	Exposure or degree of protection	Internal or Adequate	0
			Partial or Average	50
			Extreme or None	100
	VB	Likelihood of detection in superficial inspection	Likely	0
			Possible	50
			Unlikely	100
	VC	Likely influence on structural integrity	Likely	0
			Possible	50
			Unlikely	100
V2: Damage	VA	Exposure to damage	None	0
			Medium	50
			High	100
	VB	Likelihood of detection in superficial inspection	Likely	0
			Possible	50
			None	100
	VC	Likely influence on structural integrity	Low	0
			Medium	50
			High	100
V3: Wear	VA	Relative wear rate per annum	Low	0
			Medium	50
			High	100
	VB	Likelihood of detection in routine maintenance	Likely`	0
			Medium	50
			Unlikely	100
	VC	Likely influence on structural integrity	Low	0
			Medium	50
			High	100

12.4.1 Structural Component Classification

The Tsing Ma Bridge is a complex structure comprising tens of thousands of structural components. In this study, the key structural components of the bridge are classified into 15 groups and 55 components for criticality and vulnerability analyses. The 15 groups are basically the key components of the bridge for direct and indirect load-transfer: (1) suspension cables, (2) suspenders, (3) towers, (4) anchorages, (5) piers, (6) outer longitudinal trusses, (7) inner longitudinal trusses, (8) main cross frames, (9) intermediate cross frames, (10) plan bracings, (11) deck, (12) railway beams, (13) bearings, (14) movement joints, and (15) Tsing Yi approach deck. Details of classification in each group are illustrated in Table 12.7.

Table 12.7 Classification of structural components of Tsing Ma Bridge.

Name of Group	Name of Component	Group No.	Component No.	Serial No.
Suspension Cables	Main Cables	1	(a)	1
	Strand Shoes		(b)	2
	Shoe Anchor Rods		(c)	3
	Anchor Bolts		(d)	4
	Cable Clamps & Bands		(e)	5
Suspenders	Hangers	2	(a)	6
	Hanger Connections: Stiffeners		(b)	7
	Hanger Connections: Bearing Plates		(c)	8
Towers	Legs	3	(a)	9
	Portals		(b)	10
	Saddles		(c)	11
Anchorages	Chambers	4	(a)	12
	Prestressing Anchors		(b)	13
	Saddles		(c)	14
Piers: M1, M2, T1, T2, T3	Legs	5	(a)	15
	Cross Beams		(b)	16
Outer Longitudinal Trusses	Top Chord	6	(a)	17
	Diagonal		(b)	18
	Vertical Post		(c)	19
	Bottom Chord		(d)	20
Inner Longitudinal Trusses	Top Chord	7	(a)	21
	Diagonal		(b)	22
	Vertical Post		(c)	23
	Bottom Chord		(d)	24
Main Cross Frames	Top Web	8	(a)	25
	Sloping Web		(b)	26
	Bottom Web		(c)	27
	Bottom Chord		(d)	28
Intermediate Cross Frames	Top Web	9	(a)	29
	Sloping Web		(b)	30
	Bottom Web		(c)	31
	Bottom Chord		(d)	32
Plan Bracings	Upper Deck	10	(a)	33
	Lower Deck		(b)	34
Deck	Troughs	11	(a)	35
	Plates		(b)	36
Railway Beams	T-Sections	12	(a)	37
	Top Flanges		(b)	38
	Connections		(c)	39
Bearings	Rocker Bearings at Ma Wan Tower	13	(a)	40
	PTFE Bearings at Tsing Yi Tower		(b)	41
	PTFE Bearings at Pier T1		(c)	42
	PTFE Bearings at Pier T2		(d)	43
	PTFE Bearings at Pier T3		(e)	44
	PTFE Bearings at Tsing Yi Anchorage		(f)	45

	Rocker Bearings at M2		(g)	46
	PTFE Bearings at M1		(h)	47
	Hinge Bearing at Lantau Anchorage		(i)	48
Movement Joints	Highway Movement Joint	14	(a)	49
	Railway Movement Joint		(b)	50
	Top Chord		(a)	51
Tsing Yi Approach Deck	Diagonal		(b)	52
	Vertical Post	15	(c)	53
	Bottom Chord		(d)	54
	Diagonals (K-Bracings)		(e)	55

12.4.2 Criticality Analysis

To determine the criticality factors for each structural component of the Tsing Ma Bridge, an SHM-oriented FE model must be established based on the approach of one analytical member representing one real member at a stress level (Liu *et al.*, 2009). The SHM-oriented FE model of the bridge (see Figure 12.3) and model updating with the measurement data are detailed in Chapter 4 of this book.

For criticality factor C1, the load path analysis must be carried out based on the SHM-oriented FE model. However, the computational effort for the load path analysis is beyond the current computer capacity because of the huge size of the model (21,946 elements in total). Some methodologies must be found to solve this problem. As a result, the criticality factor C1 is adopted from the currently-used values (Wong *et al.*, 2006), which are based on practical experience with some engineering analysis.

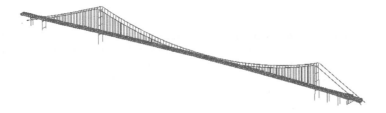

Figure 12.3 SHM-oriented finite element model of the Tsing Ma Bridge

To determine the criticality factor C2 for each structural component of the bridge, the criticality analysis is performed focusing on strength of the bridge. In principle, this must be done according to the measured results from SHM systems and the SHM-oriented methodologies introduced in the previous chapters. However, the current criticality analysis is performed based on normal combined design loads in terms of the strength utilization factor. The strength utilization factor is one of the five criticality factors in the criticality rating of the bridge. To fulfil this task, the SHM-oriented FE model is employed. Seven types of loads (dead loads, super-imposed dead loads, temperature loads, highway loads, railway loads, wind loads, and seismic loads) and three load combinations are considered in the stress analysis (HKPU, 2009a). Except for the dead loads, the superimposed dead loads, and the seismic loads, there are two, 24, eight, and three load cases for the temperature loads, the highway loads, the railways loads, and the wind loads, respectively. In the three load combinations, there are also a total of 52 load cases. For each load case, the stresses in the major structural components are determined, and the stress distributions are obtained for each major structural component. Based on the obtained stress distribution results, the stresses in the structural components at five key bridge deck sections are provided. The strength utilization factors of the key structural components are calculated, from which the critical locations of each key structural components are identified. For the bridge towers made of reinforced concrete, the strength analysis is carried out using the load-moment strength interaction method, and the strength utilization factors of the two tower legs are determined. Finally, the strength utilization factors for the key structural components are used together with other factors for rating the bridge components.

To determine the criticality factor C3 for each structural component of the bridge, the criticality analysis is performed focusing on the fatigue life of the components under traffic loads (HKPU, 2009b). A traffic induced stress analysis method is proposed based on the SHM-oriented FE model and the SHM-oriented methods introduced in the previous chpaters for the determination of stress time histories. Vehicle spectrum analysis is implemented to obtain the actual train spectrum and the actual road vehicle spectrum in terms of the measurement data of traffic loading. Fatigue-critical locations of different bridge components are determined on the basis of stress time histories caused by a standard train. Respective fatigue damage due to train and road vehicles, and fatigue life due to both train and road vehicles at different fatigue-critical components are estimated using the vehicle spectrum method. Finally, the estimated fatigue lives of the key structural components are used together with other factors for rating the bridge components.

Any damage and imperfections detected by the SHM system of the bridge should be included in determining criticality factor C4 for the structural components. As the bridge has been completed and opened to public traffic for only 12 years and the bridge has been carefully maintained, most C4 values are set equal to zero except for those of PTFE (polytetrafluoroethylene) bearings, in which minor repairable defects (due to extrusion and protrusion of PTFE sheets) were discovered within the first five years of service.

The collapse analysis, considering both geometrical nonlinearities and material nonlinearities, should be carried out to determine criticality factor C5 for varying structural components. The structural component of foremost failure will be the most critical component and the corresponding energy demand for failure will be the least. The collapse analysis should be carried out until the whole structure collapses using the SHM-oriented FE model, the static and dynamic loads predicted by the measurement data from the SHM system, and the material properties confirmed by the SHM system. However, the computational effort for the collapse analysis is beyond the current computer capacity because of the huge size of the bridge model. Novel methodologies must be found to solve this problem. As a result, the criticality factor C5 is adopted from the currently-used empirical values (Wong *et al.*, 2006).

The SHM-based criticality analyses are carried out for the 55 components listed in Table 12.7 based on the factors and ratings of criticality as given in Table 12.5. The final results of the criticality rating used in the case study are given in Table 12.8.

Table 12.8 Criticality factors for structural components.

Name of Group	Name of Component	Criticality factors				
		C1	C2	C3	C4	C5
Suspension Cables	Main Cables	100	65	0	0	100
	Strand Shoes	67	67	0	0	100
	Shoe Anchor Rods	67	67	0	0	100
	Anchor Bolts	67	67	0	0	100
	Cable Clamps & Bands	67	67	0	0	33
Suspenders	Hangers	67	16	0	0	33
	Hanger Connections: Stiffeners	67	100	0	0	33
	Hanger Connections: Bearing Plates	67	100	0	0	33
Towers	Legs	100	40	0	0	67
	Portals	100	67	0	0	67
	Saddles	100	67	0	0	67
Anchorages	Chambers	100	33	0	0	100
	Prestressing Anchors	67	100	0	0	67
	Saddles	100	67	0	0	67
Piers: M1, M2, T1, T2, T3	Legs	100	33	0	0	100
	Cross Beams	100	67	0	0	67
Outer Longitudinal Trusses	Top Chord	100	62	0	0	67
	Diagonal	100	75	100	0	67
	Vertical Post	100	20	67	0	67
	Bottom Chord	100	76	100	0	67
Inner Longitudinal Trusses	Top Chord	67	100	67	0	67
	Diagonal	67	53	0	0	67
	Vertical Post	67	32	67	0	67
	Bottom Chord	67	100	67	0	67
Main Cross Frames	Top Web	67	71	0	0	67
	Sloping Web	100	67	67	0	67
	Bottom Web	100	100	0	0	67
	Bottom Chord	100	100	0	0	67
Intermediate Cross Frames	Top Web	67	31	0	0	67
	Sloping Web	67	67	67	0	67
	Bottom Web	67	100	67	0	67
	Bottom Chord	67	100	67	0	67
Plan Bracings	Upper Deck	100	85	0	0	67
	Lower Deck	100	57	0	0	67
Deck	Troughs	67	100	67	0	67
	Plates	67	100	0	0	67
Railway Beams	T-Sections	100	0	67	0	67
	Top Flanges	100	33	0	0	67
	Connections	100	33	0	0	67
Bearings	Rocker Bearings at Ma Wan Tower	100	100	0	0	67
	PTFE Bearings at Tsing Yi Tower	100	100	0	67	67
	PTFE Bearings at Pier T1	100	100	0	67	67
	PTFE Bearings at Pier T2	100	100	0	67	67
	PTFE Bearings at Pier T3	100	100	0	67	67

	PTFE Bearings at Tsing Yi Anchorage	100	100	0	67	67
	Rocker Bearings at M2	100	100	0	0	67
	PTFE Bearings at M1	100	100	0	67	67
	Hinge Bearing at Lantau Anchorage	100	100	0	0	67
Movement Joints	Railway Movement Joint	100	67	100	0	33
	Highway Movement Joint	100	67	100	0	33
Tsing Yi Approach Deck	Top Chord	67	38	0	0	67
	Diagonal	67	36	0	0	67
	Vertical Post	67	28	67	0	67
	Bottom Chord	67	78	67	0	67
	Diagonals (K-Bracings)	67	19	67	0	67

12.4.3 Vulnerability Analysis

V1 represents damage due to extra-slowly varying effect. Sub-item V1-VA can be expressed by the corrosion rate (defined as reciprocal of time demand to reach the critical corrosion value since present). For a larger corrosion rate, a bigger numerical value should be allocated. It is noted that this treatment can take corrosion processes into consideration. Item V1-VB can be based on the experience of inspectors. The numerical values allocated to various structural components should be differentiated considering the accessibility for superficial inspection. The numerical value of criticality factor C1 can be used for sub-item V1-VC. Since there are no corrosion sensors installed in the Tsing Ma Bridge, the current vulnerability factor V1 (Wong *et al.*, 2006) is adopted without change in the case study.

V2 represents damage due to rapidly varying effect. Sub-item V2-VA can be evaluated by an impact analysis using the SHM-oriented FE model, the dynamic loads predicted by the measurement data from the SHM system, and the material properties confirmed by the SHM system. The item V2-VB can be based on the experience of inspectors. The numerical values allocated to various structural components should be differentiated considering the accessibility for superficial inspection. The numerical value of criticality factor C1 can be used for item V2-VC. However, the computational effort for the dynamic impact analyses is beyond the current computer capacity because of the huge size of the bridge model. The currently-used vulnerability factor V2 (Wong *et al.*, 2006) is adopted without change in the case study.

V3 represents damage due to slowly varying effect. Sub-item V3-VA can be evaluated based on the measurement data from the SHM system. From the observations of the past 10 years, the movement joints and bearings are subjected to serious wear. The numerical values of V3-VA allocated to all the joints and bearings could be set to 100 while other components could be given zero. Sub-item V3-VB can be evaluated based on the experience of inspectors. The numerical values allocated to various structural components should be differentiated considering the accessibility for superficial inspection. The numerical value of criticality factor C1 can be used for item V3-VC.

Vulnerability ratings are concerned with the consequences of corrosion, accidental or malicious damage, and wear (see Equation (12.11)). The reason for taking product rating in the rating of each vulnerability factor is mainly because the effects of exposure, likelihood of detection, and influence on structural integrity are in series, i.e., the effects are sequentially inter-related. The final results of vulnerability rating of the bridge are given in Table 12.9.

Table 12.9 Vulnerability factors for structural components.

Name of Group	Name of Component	Vulnerability factors								
		Corrosion (V1)			Damage (V2)			Wear (V3)		
		VA	VB	VC	VA	VB	VC	VA	VB	VC
Suspension Cables	Main Cables	100	100	100	100	100	100	0	100	100
	Strand Shoes	50	100	100	50	100	100	0	100	100
	Shoe Anchor Rods	50	100	100	50	100	100	0	100	100
	Anchor Bolts	50	100	100	50	100	100	0	100	100
	Cable Clamps & Bands	100	100	100	50	50	100	0	50	100
Suspenders	Hangers	100	100	50	100	100	50	0	50	50
	Stiffeners	100	100	50	100	100	50	0	50	0
	Bearing Plates	100	100	50	100	100	50	0	50	0
Towers	Legs	100	100	100	0	50	100	0	50	0
	Portals	100	100	100	0	50	100	0	50	0
	Saddles	100	100	100	0	50	100	0	50	0
Anchorages	Chambers	100	100	100	0	100	100	0	50	0
	Prestressing Anchors	50	100	100	0	100	100	0	100	100
	Saddles	50	100	100	0	100	100	0	100	100
Piers	Legs	100	100	100	0	50	100	0	50	0
	Cross Beams	100	100	100	0	50	100	0	50	0
Outer Longitudinal Trusses	Top Chord	50	100	100	50	100	100	0	50	100
	Diagonal	50	100	100	50	100	100	0	50	100
	Vertical Post	50	100	100	50	100	100	0	50	100
	Bottom Chord	50	100	100	50	100	100	0	50	100
Inner Longitudinal Trusses	Top Chord	50	100	100	50	100	100	0	50	100
	Diagonal	50	100	100	50	100	100	0	50	100
	Vertical Post	50	100	100	50	100	100	0	50	100
	Bottom Chord	50	100	100	50	100	100	0	50	100
Main Cross Frames	Top Web	50	100	100	50	100	100	0	50	100
	Sloping Web	50	100	100	50	100	100	0	50	100
	Bottom Web	50	100	100	50	100	100	0	50	100
	Bottom Chord	50	100	100	50	100	100	0	50	100
Intermediate Cross Frames	Top Web	50	100	100	50	100	100	0	50	100
	Sloping Web	50	100	100	50	100	100	0	50	100
	Bottom Web	50	100	100	50	100	100	0	50	100
	Bottom Chord	50	100	100	50	100	100	0	50	100
Plan Bracings	Upper-Deck	100	100	100	50	50	100	0	50	100
	Lower-Deck	100	100	100	50	50	100	0	50	100
Deck	Troughs	50	100	100	50	100	50	0	50	100
	Plates	50	100	100	50	100	50	0	50	100
Railway Beams	T-Sections	100	100	100	50	50	100	0	50	100
	Top Flanges	100	100	100	50	50	100	0	50	100
	Connections	100	100	100	50	50	100	0	50	100
Bearings	Ma Wan Tower	100	100	100	50	100	100	100	100	100
	Tsing Yi Tower	100	100	50	100	100	100	100	100	100

	Pier T1	100	100	50	100	100	100	100	100	100
	Pier T2	100	100	50	100	100	100	100	100	100
	Pier T3	100	100	50	100	100	100	100	100	100
	Tsing Yi Anchorage	100	100	50	100	100	100	100	100	100
	M2	100	100	100	50	100	100	100	100	100
	M1	100	100	50	100	100	100	100	100	100
	Lantau Anchorage	100	100	50	50	100	100	100	100	100
Movement Joints	Highway Movement Joint	50	100	100	100	100	100	100	100	100
	Railway Movement Joint	50	100	100	100	100	100	100	100	100
Tsing Yi Approach Deck	Top Chord	50	100	100	50	100	100	0	50	100
	Diagonal	50	100	100	50	100	100	0	50	100
	Vertical Post	50	100	100	50	100	100	0	50	100
	Bottom Chord	50	100	100	50	100	100	0	50	100
	Diagonals (K-Bracings)	50	100	100	50	100	100	0	50	100

12.5 BRIDGE RATING SYSTEM FOR THE TSING MA BRIDGE

12.5.1 Relative Weights

To apply the proposed SHM-based F-AHP bridge rating system to the Tsing Ma Bridge, the relative weights must be determined for each level. According to the AHP procedure, the comparative matrix and the relative weights for the criticality index level 1 (CR1) are found and listed in Table 12.10. The counterparts for the vulnerability index level 1 (VR1) are listed in Table 12.11. The consistency of the comparative matrices is also checked and the results are listed in Table 12.12.

Table 12.10 Comparative matrix and relative weights for CR1.

Index layer 1 (L)	C1	C2	C3	C4	C5	Relative weights
C1	1	1/3	1/2	1	1	0.1237
C2	3	1	2	3	3	0.3945
C3	2	1/2	1	2	2	0.2343
C4	1	1/3	1/2	1	1	0.1237
C5	1	1/3	1/2	1	1	0.1237

Table 12.11 Comparative matrix and relative weights for VR1.

Index layer 1 (R)	V1	V2	V3	Relative weights
V1	1	2	2	0.5
V2	1/2	1	1	0.25
V3	1/2	1	1	0.25

Since the comparative matrix is not unique, its effect on the final results must be investigated. Table 12.13 and Table 12.14 tabulate another set of comparative matrix and relative weights for the criticality index level 1 (CR2) and vulnerability index level 1 (VR2), respectively. The consistency of the comparative matrices is also checked and the results are listed in Table 12.15.

Table 12.12 Consistency of comparative matrices for CR1 and VR1.

Layer	λ_{max}	$CI=(\lambda_{max}-n_A)/(n_A-1)$	RI	CR=CI/RI
Index level CR1	3.0	0.0	0.58	0.0
Index level VR1	5.01	0.0025	1.12	0.0022

Table 12.13 Comparative matrix and relative weights for CR2.

Index layer 1 (L)	C1	C2	C3	C4	C5	Relative weight
C1	1	1/4	1/3	1	1	0.0985
C2	4	1	2	4	4	0.4304
C3	3	1/2	1	3	3	0.2741
C4	1	1/4	1/3	1	1	0.0985
C5	1	1/4	1/3	1	1	0.0985

Table 12.14 Comparative matrix and relative weights for VR2.

Index layer 1 (R)	V1	V2	V3	Relative weight
V1	1	3	3	0.6
V2	1/3	1	1	0.2
V3	1/3	1	1	0.2

Table 12.15 Consistency of comparative matrices for CR2 and VR2.

Layer	λ_{max}	$CI=(\lambda_{max}-n_A)/(n_A-1)$	RI	CR=CI/RI
Index level CR2	3.0	0.0	0.58	0.0
Index level VR2	5.02	0.005	1.12	0.0044

Table 12.16 Comparison of relative weights between two cases.

Item	Relative weights		(Case 2 − Case 1)/Case 1×100%
	Case 1	Case 2	
V1	0.5	0.6	20.00%
V2	0.25	0.2	-20.00%
V3	0.25	0.2	-20.00%
C1	0.1237	0.0985	-20.37%
C2	0.3945	0.4304	9.10%
C3	0.2343	0.2741	16.99%
C4	0.1237	0.0985	-20.37%
C5	0.1237	0.0985	-20.37%

If the importance of criticality is regarded to be the same as that of vulnerability, the relative weight vector for the criterion level can be taken as $\{\omega_{cv}\}=\{0.5, 0.5\}^{\mathrm{T}}$ for both cases. The comparison of relative weights between the two cases is shown in Table 12.16. It can be seen that the difference of relative weights reaches as high as 20% and as low as -20.37%.

12.5.2 Time Intervals for Inspection

Based on the two sets of relative weights and according to the SHM-based F-AHP bridge rating system, the decision on the time intervals for inspection can be determined and the results are listed in Table 12.17 for the two weight cases. It can be seen that the time intervals for inspection are the same for most structural components except for six structural components out of 55, of which the time intervals for inspection are longer in weight case 1 than those in weight case 2. The difference of fuzzy synthetic rating between the two cases varies from -0.9% to 5.68%, which is much smaller compared with the range of difference in relative weights. Therefore, one may conclude that the effect of the relative weights from different comparative matrices on the decision on the time intervals for inspection is small. It can also be seen that for the bridge components concerned, the time intervals for inspection are either one year or two years.

Table 12.17 Time intervals for inspection.

Group No.	Serial No.	Weight Case R1		Weight Case R2		(R2−R1)/R1 ×100%
		Fuzzy synthetic rating	Time internal for inspection	Score of fuzzy rating system	Time interval for inspection	
		R1	Year	R2	Year	
Suspension Cables	1	53.3	2	54.4	2	2.06
	2	51.9	2	53.5	2	3.08
	3	51.9	2	53.5	2	3.08
	4	51.9	2	53.5	2	3.08
	5	45.3	2	46.2	2	1.99
Suspenders	6	44.3	2	43.9	2	-0.90
	7	57.6	1	60.2	1	4.51
	8	57.6	1	60.2	1	4.51
Towers	9	50.5	2	51.6	2	2.18
	10	50.1	2	51.5	2	2.79
	11	50.1	2	51.5	2	2.79
Anchorages	12	51.1	2	54.0	2	5.68
	13	55.1	2	57.4	2	4.17
	14	52.2	2	53.2	2	1.92
Piers	15	51.1	2	52.3	2	2.35
	16	50.1	2	51.5	2	2.79
Outer Longitudinal Trusses	17	50.6	2	51.3	2	1.38
	18	60.5	1	62.3	1	2.98
	19	50.1	2	50.5	2	0.80
	20	60.6	1	62.9	1	3.80
Inner Longitudinal	21	56.8	2	59.5	1	4.75
	22	45.1	2	45.8	2	1.55

Trusses	23	47.3	2	47.8	2	1.06
	24	56.8	2	59.5	1	4.75
Main Cross Frames	25	51.4	2	53.0	2	3.11
	26	54.9	2	56.3	2	2.55
	27	59.0	1	61.3	1	3.90
	28	59.0	1	61.3	1	3.90
Intermediate Cross Frames	29	43.0	2	43.2	2	0.47
	30	52.8	2	54.6	2	3.41
	31	57.0	2	59.5	1	4.39
	32	57.0	2	59.5	1	4.39
Plan Bracings	33	54.4	2	56.2	2	3.31
	34	47.4	2	48.2	2	1.69
Deck	35	56.8	2	59.5	1	4.75
	36	55.1	2	57.5	1	4.36
Railway Beams	37	48.0	2	48.5	2	1.04
	38	50.2	2	51.0	2	1.59
	39	50.2	2	51.0	2	1.59
Bearings	40	59.6	1	62.2	1	4.36
	41	59.2	1	61.8	1	4.39
	42	59.2	1	61.8	1	4.39
	43	59.2	1	61.8	1	4.39
	44	59.2	1	61.8	1	4.39
	45	59.2	1	61.8	1	4.39
	46	59.6	1	62.2	1	4.36
	47	59.2	1	61.8	1	4.39
	48	59.6	1	62.2	1	4.36
Movement Joints	49	62.3	1	64.1	1	2.89
	50	62.3	1	64.1	1	2.89
Tsing Yi Approach Deck	51	44.7	2	45.2	2	1.12
	52	44.2	2	44.6	2	0.90
	53	46.4	2	46.8	2	0.86
	54	53.8	2	55.8	2	3.72
	55	47.3	2	47.9	2	1.27

12.5.3 Decision for Inspection

The structural components of the Tsing Ma Bridge have been assessed for their proneness to damage or deterioration and their importance in maintaining structural integrity. Criticality and vulnerability ratings are given to key structural components and are used as references or guidance in defining the inspection frequencies, time intervals and actions to be taken in the event of defects being identified. Currently, there are two main categories of inspection in the Master Maintenance Manual (HyD, 1996; 2001) for long-span cable-supported bridges in Tsing Ma Control Area, namely, Routine Inspection and Special Inspection. The Routine Inspection is further sub-divided into five types of inspection, i.e. Routine Superficial Inspection, Routine Statutory Inspection, Routine Functional Inspection, Routine General Inspection, and Routine Principal Inspection. Table 12.18 tabulates the types, natures, frequencies, and time durations of inspections required by the Master Maintenance Manual.

Table 12.18 shows that of the five types of routine inspections, only Routine General Inspection requires criticality and vulnerability ratings in the planning and scheduling of the frequencies and time durations for execution of inspections.

Table 12.18 Types of inspections stipulated in the Master Maintenance Manual.

Inspection categories	Type of inspection	Nature of inspection	Frequencies and time intervals
Routine inspection	Routine superficial inspection	Visual inspection	1 week
	Routine statutory inspection	Not applicable	Not applicable
	Routine functional inspection	Not applicable	Not applicable
	Routine general inspection	Visual inspection, close visual inspection	Dependent upon the results of criticality ratings and vulnerability ratings
	Routine principal inspection	Visual inspection, close visual inspection, detailed visual inspection, non-destruction inspection, destructive inspection	Six years or as required by the government
Special inspection	Not routine inspections, arise based on extreme events	Close visual inspection	Immediate after gust wind speed exceeding 90 km/h and ambient air temperature rising to within 10% of the design criteria, or required following an identified defect, or required by the government.
		Monitoring inspection	Depends on nature of defect and to be agreed by the government.

12.6 SUMMARY

An SHM-based F-AHP bridge rating system has been proposed for long-span suspension bridges and applied to the Tsing Ma Bridge in this chapter. The effects of different comparative matrices and relative weights on the decision on time intervals for inspection have been investigated. It is shown that the time intervals for inspection are the same as those currently used for most of the structural components except for six structural components out of 55, of which the time intervals for inspection are longer in weight case 1 than those in weight case 2. The effect of the relative weights from different comparative matrices on the decision on the time intervals for inspection is small. For the bridge components concerned, the time intervals for inspection are either one year or two years. The results from the case study indicate that the proposed bridge rating method is feasible and can be used in practice for long-span suspension bridges with the SHM systems. Nevertheless, this is a preliminary study only, in which some criticality and vulnerability factors used are not from the SHM-oriented rating method. A future deep research is required in this topic.

12.7 REFERENCES

AASHTO, 2003, *Guide Manual for Condition Evaluation and Load and Resistance Factor Rating (LRFR) of Highway Bridges*, (Washington, DC: American Association of State Highway and Transportation Officials).

Aktan, A.E., Farhey, D.N., Brown, D.L., Dalal, V., Helmicki, A.J., Hunt, V.J. and Shelley, S.J., 1996, Condition assessment for bridge management. *Journal of Infrastructure Systems*, **2(3)**, pp. 108-117.

Bevc, L, Mahut, B. and Grefstad, K., 1999, *Review of Current Practice for Assessment of Structural Condition and Classification of Defects*, (Ljubljana, Slovenia: Slovenian National Building and Civil Engineering Institute).

Dudgeon, I., Webster, D. and Duff, A., 2007, *Forth Replacement Crossing Study*. Report 1, (Glasgow, Scotland: Jacobs U.K. Limited).

Dubois, D. and Prade, H., 1979, Fuzzy real algebra: some results. *Fuzzy Sets and Systems*, **2(4)**, pp. 327-348.

FHA, 2006, *Bridge Inspector's Reference Manual*, (New Jersey, Washington: Federal Highway Administration).

HKPU, 2008, *Establishment of Bridge Rating System for Tsing Ma Bridge: Criticality and Vulnerability Rating Method*. Report No. 7, (Hong Kong: Department of Civil and Structural Engineering, The Hong Kong Polytechnic University).

HKPU, 2009a, *Establishment of Bridge Rating System for Tsing Ma Bridge: Criticality and Vulnerability Analysis: Strength*. Report No. 8, (Hong Kong: Department of Civil and Structural Engineering, The Hong Kong Polytechnic University).

HKPU, 2009b, *Establishment of Bridge Rating System for Tsing Ma Bridge: Criticality and Vulnerability Analysis: Fatigue*. Report No. 9, (Hong Kong: Department of Civil and Structural Engineering, The Hong Kong Polytechnic University).

HKPU, 2009c, *Establishment of Bridge Rating System for Tsing Ma Bridge: Bridge Rating and System Integration*. Report No. 11, (Hong Kong: Department of Civil and Structural Engineering, The Hong Kong Polytechnic University).

HyD, 1996, *Master Maintenance Manual for Long-Span Cable Supported Bridges, Appendix 14 of the Contract for Management, Operation and*

Maintenance of the Tsing Ma Control Area (TMCA), (Hong Kong: Hong Kong
   Highways Department).
HyD, 2001, *Master Maintenance Manual for Long-Span Cable Supported
   Bridges, Appendix 14 of the Contract for Management, Operation and
   Maintenance of the Tsing Ma Control Area (TMCA),* (Hong Kong: Hong Kong
   Highways Department).
Kawamura, K., Miyamoto, A., 2003, Condition state evaluation of existing
   reinforced concrete bridges using neuro-fuzzy hybrid system. *Computer and
   Structures,* **81(18-19),** pp. 1931-1940.
Kushida, M., Tokuyama, T. and Miyamoto, A., 1993, Quantification of subjective
   uncertainty included in empirical knowledge on bridge rating. In *Proceedings of
   the 2nd International Symposium on Uncertainty Modeling and Analysis,* College
   Park, Maryland, (Washington: IEEE Computer Society), pp. 22-28.
Liang, M.T., Wu, J.H. and Liang, C.H., 2001, Multiple layer fuzzy evaluation for
   existing reinforced concrete bridges. *Journal of Infrastructure System*s, **7(4),** pp.
   144-159.
Liu, T.T., Xu, Y.L., Zhang, W.S., Wong, K.Y., Zhou, H.J. and Chan, K.W.Y.,
   2009, Buffeting-Induced stresses in a long suspension bridge: structural health
   monitoring orientated stress analysis. *Wind and Structures,* **12(6),** pp. 479-504.
Melhem, H., 1994, Fuzzy logic for bridge using an eigenvector of priority settings.
   In *Proceedings of the First International Joint Conference of the North
   American Fuzzy Information Processing Society Biannual Conference, the
   Industrial Fuzzy Control and Intelligent Systems Conference, and the NASA
   Joint Technolo,* San Antonio, Texas, (Washington: IEEE Computer Society), pp.
   279-286.
Mufti, A.A., 2001, *Guidelines for Structural Health Monitoring,* (Winnipeg,
   Canada: Intelligent Sensing for Innovative Structures).
Saaty, T.L., 1980, *The Analytic Hierarchy Process,* (New York: McGraw Hill).
Sasmal, S. and Ramanjaneyulu, K., 2008, Condition evaluation of existing
   reinforced concrete bridges using fuzzy based analytic hierarchy approach.
   Expert Systems with Applications, **35(3),** pp. 1430-1443.
Sasmal, S., Ramanjaneyulu, K., Gopalakrishnan, S. and Lakshmanan, N., 2006,
   Fuzzy logic based condition rating of existing reinforced concrete bridges.
   Journal of Performance of Constructed Facilities, ASCE, **20(3),** pp. 261-273.
Tabatabai, H., 2005, *Inspection and Maintenance of Bridge Stay Cable Systems-a
   Synthesis of Highway Practice,* NCHRP Synthesis 353, (Milwaukee, Wisconsin:
   University of Wisconsin-Milwaukee).
Wenzel, H., 2009, *Health Monitoring of Bridges,* (New York: John Wiley &
   Sons).
Wong, K.Y., 2006, Criticality and vulnerability analyses of Tsing Ma Bridge. In
   *Proceedings of the International Conference on Bridge Engineering –
   Challenges in the 21st Century,* Hong Kong, edited by Yim, K.P., (Hong Kong:
   The Hong Kong Institution of Engineers), pp. 209.
Wong, K.Y., Man, K.L. and Chan, W.Y.K., 2001a, Monitoring of wind load and
   response for cable-supported bridges in Hong Kong. In *Proceedings of SPIE 6th
   International Symposium on NDE for Health Monitoring and Diagnostics,
   Health Monitoring and Management of Civil Infrastucture Systems,* Newport

Beach, California, edited by Kundu, T., (Bellingham, Washington: SPIE), pp. 292-303.

Wong, K.Y., Man, K.L. and Chan, W.Y.K., 2001b, Application of global positioning system to structural health monitoring of cable-supported bridges. In *Proceedings of SPIE 6th International Symposium on NDE for Health Monitoring and Diagnostics, Health Monitoring and Management of Civil Infrastructure Systems*, Newport Beach, California, edited by Kundu, T., (Bellingham, Washington: SPIE), pp. 390-401.

12.8 NOTATIONS

$[A]$	Comparative matrix for judgement.
$\{B_c\}$	Fuzzy synthetic rating vector for criticality.
$\{B_v\}$	Fuzzy synthetic rating vector for vulnerability.
C1,..., C5	Criticality factors.
$[R_c]$	Membership degrees matrix based on the criticality factors.
$[R_v]$	Membership degrees matrix based on the vulnerability factors.
$\{G\}$	Fuzzy rating set.
I	Impact factor for dynamic live loads.
M	Fuzzy number.
m	Modal value, number of the fuzzy rating values in the fuzzy rating set.
n	Size of matrix $[A]$.
n_A	Non-zero eigenvalue of $[A]$.
r, u	Lower and upper values of the support for the decision of the fuzzy number.
r_{ij}	Membership degree of the ith item to the jth fuzzy membership function.
R	Fuzzy synthetic rating at the objective level.
R_N	As-built nominal resistance.
T_c	Fuzzy synthetic rating for the criticality of the concerned structural component.
T_v	Fuzzy synthetic rating for the vulnerability of the concerned structural component.
U	Domain of Fuzzy number.
V1, V2, V3,	Vulnerability factors (Level 1).
VA, VB, VC	Vulnerability factors (Level 2).
α	Relative weight of important value.
φ	Strength reduction factor.
λ	Eigenvalue.
$\{\phi\}$	Eigenvector.
ω	Relative important value.
σ_D	Stresses due to dead loads.
σ_L	Stresses due to live loads.
γ_D	Partial load factor for dead loads.
γ_L	Partial load factor for live loads.
μ_m	Fuzzy membership function.

CHAPTER 13

Establishment of Test-beds

13.1 PREVIEW

Comprehensive structural health monitoring (SHM) systems have been developed and installed in several long suspension bridges in the world aiming to monitor structural health conditions of the bridges in real time. Numerous damage detection methods have been developed in either the frequency domain, time domain, or time-frequency domain, as discussed in Chapter 11. Nevertheless, many key issues remain unsolved as to how to take full advantage of the health monitoring system for effective and reliable damage detection of such important but complex structures. Some of the damage detection methods have been successfully applied in aerospace engineering and mechanical engineering, but successful applications to long-span suspension bridges in civil engineering are still limited. It is desirable to establish a laboratory-based test-bed to allow researchers to simulate different loading cases, to realize rational damage scenarios, to apply different sensors and sensing networks, and to test various damage detection algorithms before they are applied to real long-span suspension bridges.

This chapter first emphasizes the necessity and importance of establishing test-beds for long suspension bridges in laboratories. The design principles of the laboratory-based test-bed are then introduced. The design and setup of a physical model for a long-span suspension bridge with various damage scenarios considered are presented. The geometric measurements and the modal tests are subsequently carried out on the physical bridge model to identify its geometric configuration and dynamic characteristics, respectively. Finally, the FE modelling of the physical bridge model is established using a commercial software package, which is followed by an FE model updating in terms of the measured modal properties. The test-bed comprising the delicate physical model and the updated FE model of a long suspension bridge could serve as a benchmark problem for SHM of long-span suspension bridges.

13.2 NECESSITY OF ESTABLISHING TEST-BEDS

Many innovative long-span suspension bridges built in recent years are exposed to harsh conditions such as heavy traffic loads, strong winds and severe earthquakes. This engenders many challenges to professionals on how to ensure these structures function properly during their long service period and how to prevent them from sudden failure and fatal disaster during strong winds, severe earthquakes, terrorist attacks, and other abnormal events. Recently-developed SHM technology can provide a better solution for the problems concerned (Chang, 1997; Mufti, 2001; Wong *et al.*, 2001a and 2001b). The main objectives of the SHM are to monitor the real-time behaviour of a long suspension bridge, to assess its performance under various service loads, to verify or update the rules used in its design stage, to detect its damage or deterioration, and to guide its maintenance or repair work in terms of a comprehensive sensory system and a sophisticated data processing system implemented with advanced information technology and supported by refined computer algorithms. Nevertheless, many key issues remain unsolved to achieve the utmost goal of SHM technologies (Glaser *et al.*, 2007). One of the key issues is how to make use of SHM systems for effective and reliable damage detection of such important but complex structures.

Numerous damage detection methods have been developed in either the frequency domain, time domain, or time-frequency domain (Doebling *et al.*, 1998; Farrar *et al.*, 2003), as discussed in Chapter 11. They may be vibration-based or static-based, model-based or signal-based, deterministic or statistical, linear or nonlinear, and local or global. Some of these methods have been successfully applied in aerospace engineering and mechanical engineering, but successful applications to long suspension bridges in civil engineering are still limited. This is because a long-span suspension bridge is large in dimension and complex in structural system but damage to the bridge is often local. Therefore it is not practical to predict or detect the structural damage using one single damage index or one damage detection method. A long-span suspension bridge is also subjected to complex operational environments such as multiple loading conditions and varying temperature. The operational environments have significant effects on measurement results, and the measured structural responses are often not attributed to a single loading. Most of the existing damage detection methods are sensitive to changes in the operational environments. Furthermore, many uncertainties are associated with the damage detection of a long-span suspension bridge, which include measurement noise in recorded data and modelling errors in the material and structural properties and geometric features of a bridge. This may be one of the reasons why some damage detection methods work theoretically well without considering uncertainties but are not practical for long-span suspension bridges. To overcome the aforementioned difficulties in SHM of long-span suspension bridges, it is desirable to establish a laboratory-based test-bed to allow researchers to simulate different loading cases, to realize rational damage scenarios, to apply different sensors and sensing networks, and to test various damage detection algorithms before they are applied to real long suspension bridges (Glaser *et al.*, 2007). Similar work has been done for SHM of building structures (Johnson *et al.*, 2004).

13.3 DESIGN PRINCIPLES OF TEST-BED

The Tsing Ma suspension bridge in Hong Kong is the longest suspension bridge in the world carrying both highway and railway, as shown in Figure 13.1. The Tsing Ma Bridge has been equipped with a comprehensive SHM system since 1997 (Wong *et al.*, 2001a and 2001b). This bridge is therefore selected as a reference for the design of a physical bridge model as a test-bed. All of the major structural components in the bridge should be included in the bridge model so that the model can best represent the real long-span suspension bridge and the strain-level measurements can be performed on the major structural components. All of the connections and boundary conditions of the bridge should be reproduced in the physical model so that rational damage scenarios can be best simulated. The physical model could be installed on a shake table which can generate a proper ground motion to the model as one type of external loading. The size of the physical model should be large enough to facilitate the installation of various sensing systems and the measurements of various structural responses.

It would be desirable to design such a bridge model to satisfy the laws of similarity necessary to translate a model into a prototype as approximately done in the aeroelastic model for aerodynamic studies of a long-span suspension bridge (Simiu and Scanlan, 1996) or in the dynamic model for seismic studies of a long-span suspension bridge. However, it has proved to be very difficult to meet the design requirements of the bridge model described above for SHM studies. Since the ratio of prototype length to model length is often very large, if one uses the materials of prototype modulus to build the bridge model, a large amount of additional mass should be added to and distributed over the bridge model and the sizes of major structural components would be very small. This will lead to many difficulties in the manufacture of the scaled model and the damage detection of major structural components. If one uses the materials of prototype density to build the bridge model, one must use the materials of very low modulus such as plexiglass. These materials are, however, sensitive to temperature, and the material properties are also not stable when the scaled model is subjected to large deformation. In consideration of all the aforementioned factors, the compromise in similarity requirements has to be made in this study to design and set up a bridge model as a physical representation of the Tsing Ma Bridge for SHM researches. It is thus decided that the geometric configuration of the bridge model should follow the designated geometric ratio. The shapes and sequence of the first a few vibration modes of the scaled model should follow those of the prototype bridge whereas the materials and cross section properties of major structural components will not follow the laws of similarity. The caution should thus be seriously taken when using and interpreting the test results.

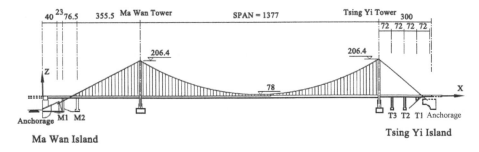

Figure 13.1 Configuration of prototype Tsing Ma Bridge (unit: m)

13.4 PHYSICAL MODEL DESIGN AND SETUP

13.4.1 Brief Description of the Tsing Ma Bridge

The Tsing Ma Bridge has an overall length of 2,132 m and a main span of 1,377 m between the Tsing Yi tower in the east and the Ma Wan tower in the west (see Figure 2.1). The height of the two reinforced concrete towers is 206 m, measured from the base level to the tower saddle. The two main cables of 1.1 m diameter and 36 m apart in the north and south are accommodated by the four saddles located at the top of the tower legs. On the Tsing Yi side, the main cables are extended from the tower saddles to the main anchorage through the spay saddles, forming a 300 metre Tsing Yi side span. On the Ma Wan side, the main cables extended from the Ma Wan tower are held first by the saddles on Pier M2 at a horizontal distance of 355.5 m from the Ma Wan tower and then by the main anchorage through the spay saddles at the Ma Wan abutment. The bridge deck is a hybrid steel structure consisting of Vierendeel cross frames supported on two longitudinal trusses acting compositely with stiffened steel plates. The width of the bridge deck in the main span is 41m. The bridge deck is suspended by suspenders in the main span to allow a sufficiently large navigation channel and in the Ma Wan side span to minimize the number of substructures in the sea. Due to the highway layout requirement, the deck on the Tsing Yi side is supported by three piers rather than suspenders.

The translational movements of the bridge deck at the Ma Wan abutment are restricted in three directions, but the deck is free for rotation about the lateral axis. At the Tsing Yi abutment, only the vertical and lateral movements of the bridge deck are restrained; the longitudinal movement and the rotations about the lateral axis are allowable. At the Ma Wan tower, the bridge deck is connected to the bottom cross beam of the tower through four articulated link bearings (or rockers) and to the tower legs through four lateral bearings (rollers). The articulated link bearings allow the deck to move within the horizontal plane but restrict the movement in the vertical direction. The lateral bearings are to restrain the lateral movement of the deck but to allow movement within the vertical plane. Therefore, the deck is allowed to move along the longitudinal direction of the bridge. At the Tsing Yi tower, there are also four bottom bearings connecting the deck to the lowest cross beam of the tower and four lateral bearings connecting the deck to the tower legs. All the piers allow the bridge deck to move in the longitudinal direction.

The ambient vibration study of the Tsing Ma Bridge shows that the lowest natural frequency of the bridge of 0.069 Hz corresponds to the first lateral mode of vibration. The first vertical mode of vibration is almost anti-symmetric in the main span at a natural frequency of 0.113 Hz, and the first torsional mode of vibration occurs at a natural frequency of 0.267 Hz. The further details regarding the Tsing Ma Bridge and its natural frequencies and mode shapes can be found in Chapter 4 of this book and Xu *et al.* (1997).

13.4.2 Geometric Scale and Shake Table

In consideration of the maximum length available in the Structure Dynamics Laboratory of The Hong Kong Polytechnic University, the geometric scale is selected as 1:150 for the configuration of the bridge (see Figure 13.2). According to this geometric scale, the physical model has a main span 9.18 m with an overall length 14.34 m. The height of the two bridge tower models is 1.37 m measured from the base level to the tower saddle. The two main cables in the bridge model are 0.24 m apart in the north and south. The width of the bridge deck model in the main span is 0.273 m. Since the physical model of very low frequencies will be sensitive to environmental disturbance, yield very small sizes of structural components, and make damage detection very difficult, the lowest natural frequency of the physical bridge model is targeted around 4 Hz with its mode shape corresponding to the lateral vibration of the bridge. Two kinds of materials were chosen in the bridge model design and fabrication: (1) steel is used for the bridge towers, piers, cables, suspenders and anchorages; and (2) aluminium is used for the bridge deck. Based on the aforementioned considerations together with the consideration that different kinds of damage scenarios should be readily simulated, the detailed design of the bridge model was performed and the dimensions of structural members were determined in an iterative way with the help of the modal analysis of a strain-oriented FE model.

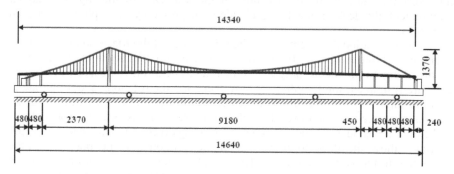

Figure 13.2 Configuration of Tsing Ma Bridge model and shake table (unit:mm)

To generate the necessary ground motion to the bridge model, a shake table was designed and built. The shake table was made of a rectangular hollow-section steel beam of a length of 14.64 m. The cross section of the beam is 400 mm wide, 200 mm high, and 6.3 mm thick to facilitate the installation of the bridge model. As shown in Figure 13.2, the beam of a total weight of 851 kg sits on five sets of roller bearings with a uniform spacing of 3.16 m. Each set of roller bearing consists of one top plate, one bottom plate and two rollers (see Figure 13.3). The contact surfaces between the plates and rollers were polished to reduce the friction force to a very low level. The top plate was welded to the steel beam (table), whereas the bottom plate was fastened to a steel plate embedded in a concrete block on the ground. The five steel plates embedded in their corresponding concrete blocks should be kept in the same horizontal level so as to ensure the table (steel beam) in the horizontal level. If the two rollers in each set of bearing are arranged along the x-axis of the bridge model (see Figure 13.3), an electromagnetic exciter will be installed in the middle of the table and apply a force normal to the table so that the table will generate the ground motion to the bridge model in the lateral direction (y-axis). If the two rollers in each set of bearing are arranged in the y-axis of the bridge model, the electromagnetic exciter will be moved to the right end of the table and apply a force parallel to the table so that the table will generate the ground motion to the bridge model in the longitudinal direction.

Figure 13.3 Roller bearing of shake table

13.4.3 Bridge Towers and Piers

The prototype bridge towers are reinforced concrete structures but the two bridge tower models were made of mild steel to facilitate the manufacture and the late damage detection of the bridge model. Each tower model consisted of two steel legs which are linked by four horizontal cross beams, as shown in Figure 13.4. The cross section of the lower cross beams is larger than the upper one. The two tower models are almost of identical geometric and structural configurations. The net height of the tower is about 1,306 mm from the top surface of the base plate to the flat top of the saddle, and the distance between the two legs in the tower plan is about 266 mm at the base level and reduces to 240 mm at the top of the leg. The thickness of the tower leg in the tower plan is constant at 15 mm but the width of the tower leg reduces from 40 mm at the base to 20 mm at the top. The bottom ends of two legs of each tower are welded on a steel plate which is then fastened to the shake table using four bolts. A slice of steel could be placed between the bottom steel plate of the tower and the shake table at its first place, and then the removal of the steel slice could simulate the settlement of the bridge tower. The verticality of the towers was monitored by the laser makers. The two bridge tower models were installed on the shake table, standing 9.18 m apart from each other in the longitudinal direction and forming a main span of the bridge model.

Figure 13.5 shows the configuration and size of the saddle installed on the top of the bridge tower to hold the main cables. The tower leg and the saddle were actually cut from a piece of steel plate though they are shown in separate figures.

On the top surface of the saddle there is a narrow groove in order to accommodate the main cable.

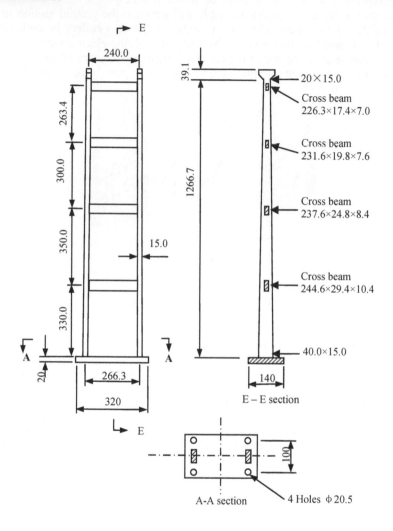

Figure 13.4 Bridge tower model (unit: mm)

All the supporting piers in the prototype Tsing Ma Bridge are made of reinforced concrete but they were all manufactured as free-standing steel frames in the scaled model. Figure 13.6 shows the model of pier M2 and the associated saddle model which is pin-connected to the top of the leg of pier M2 to hold the main cable before it is passed to the anchorage on the Ma Wan side as arranged in the prototype bridge. All of the supporting pier models were installed on the shake table through steel slices in order to simulate their settlement.

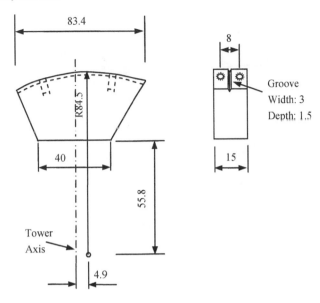

Figure 13.5 Tower saddle model (unit: mm)

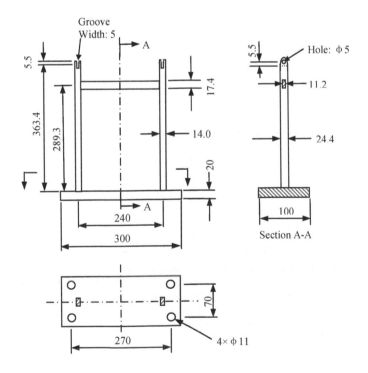

(a) Pier M2 model

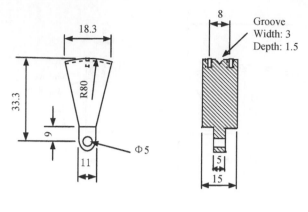

(b)   Pier M2 saddle model

Figure 13.6 Pier M2 and saddle models (unit: mm)

13.4.4 Bridge Deck

The Tsing Ma bridge deck is a hybrid steel structure consisting of a series of cross-frames supported on longitudinal trusses acting compositely with stiffened steel plates. The model of the bridge deck also comprises a series of cross frames together with four longitudinal trusses but they are all made of aluminium so that the members in the cross frames and the longitudinal trusses are not too small to be manufactured and the targeted dynamic characteristics can be achieved. A typical cross-frame model in the main span consists of top and bottom chords, inner struts, outer struts, and upper and lower inclined edge members, as shown in Figure 13.7(a). For the cross frames near the two bridge abutments and those near the tower portal frames, their configurations are different from those in the main span, as plotted in Figure 13.7(b). There are a total of 242 cross frames in the bridge deck model. These cross frames are linked up by two outer longitudinal trusses and two inner longitudinal trusses by welding (see Figure 13.8). The diagonal bracings are arranged between two neighbouring cross frames throughout the entire two outer longitudinal trusses. In the two inner longitudinal trusses, the diagonal bracings are arranged between two neighbouring cross frames only in the side spans and the areas near the bridge towers. To simulate the stiffened steel plates of the prototype bridge deck, aluminium plates of 0.5 mm thick are laid on the lower level of the bridge deck in the middle by special adhesive. The aluminium plates can be detached in part from the bridge deck to simulate different damage scenarios of the steel plates. The longitudinal trusses and the cross frames at some joints can also be disconnected to simulate different damage scenarios of deck joints. For some cross frames, the damage of either chords or struts can be simulated.

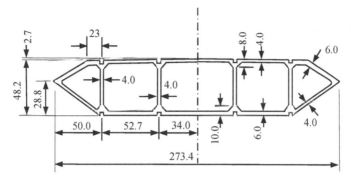

(a) Cross frame in main span

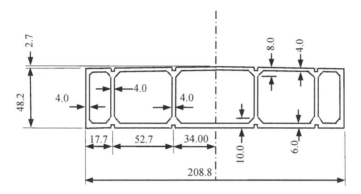

(b) Cross frame near bridge tower

Figure 13.7 Cross frame models (unit: mm)

Figure 13.8 Bridge deck model

Figure 13.9 Sliding bearing models on Pier T2

The bridge deck model is hung at some cross-frames by the suspenders in the main span and the Ma Wan side span at a horizontal interval of 120 mm. Piers T1, T2, T3 and M1 provide vertical supports to the bridge deck while pier M2 provides both vertical supports and lateral restraints to the bridge deck (see Figure 13.9). All the piers allow the bridge deck to move longitudinally. In addition to the five piers, there is an abutment at each end of the bridge deck. The bridge deck is pin-connected to the abutment on the Ma Wan side but it can slide over the abutment on the Tsing Yi side (see Figure 13.10). The two abutments are both made of steel plates and fastened on the shake table. The translational movements of the bridge deck at the Ma Wan abutment are restricted but the deck is free for rotation about the lateral axis. At the Tsing Yi abutment, the vertical and lateral movements of the bridge deck are restrained but the deck longitudinal movement and the rotations about the lateral axis are allowed. The vertical motion of the bridge deck at the Ma Wan tower is restricted by connecting the corresponding part of the bridge deck with the portal beam of the tower through 4 steel pin-links, as shown in Figure 13.11. The vertical motion of the bridge deck at the Tsing Yi tower is restricted by using 4 sliding bearings. The lateral motions of the bridge deck at the two towers are also restricted through sliding bearings. The longitudinal motions of the bridge deck along the bridge axis are, however, permitted. The damage scenarios of either pin-links or sliding bearings can be simulated by the removal or the fixing of some of the links or bearings.

(a) Abutment on Ma Wan side   (b) Abutment on Tsing Yi side

Figure 13.10 Abutments and connections with bridge deck

Figure 13.11 Bearing models on Ma Wan Tower

Figure 13.12 Tsing Yi side span

13.4.5 Bridge Cables and Suspenders

In terms of the global geometrical scale of 1:150, the initial geometric configurations of the main cables of the bridge model are determined in reference to those of the main cables of the prototype bridge. Each main cable model consists of 19 strands of parallel steel wires with a diameter about 0.6 mm, and the resultant cable has an overall diameter of approximately 3 mm. The two main cables of about 240 mm apart are accommodated by the saddles sitting on the top of the two towers. On the Tsing Yi side, the main cables are extended from the tower saddles to the Tsing Yi anchorages, forming a 2,260 mm Tsing Yi side span (see Figure 13.12). On the Ma Wan side, the main cables extending from the tower saddles are first secured by M2 pier saddles at the deck level and then by the anchorages. The two anchorages at each side of the bridge are equipped with two specially-designed clamps securing the main cables and two load cells measuring the cable tension forces. Figure 13.13(a) and Figure 13.13(b) show the arrangement and details of the Ma Wan anchorage and the Tsing Yi anchorage, respectively. The initial tension forces in the main cables at both ends are set about 11 N when only the tower-cable system is considered without suspenders connecting the bridge deck to the main cables. Besides, all the anchorages are capable of being adjusted upward, downward, forward and backward for the simulation of the settlements of the anchorages.

      (a) Anchorages on Ma Wan side   (b) Anchorages on Tsing Yi side

Figure 13.13 Anchorages and load cells

      Different from the main cables, each suspender is just made up of a steel wire with a diameter about 0.5 mm, clamped to the main cables every 120 mm along the bridge longitudinal axis and attached to the side chords of the main cross frames of the bridge deck (see Figure 13.14). The laser makers were used to monitor the verticality of suspenders during the installation. As stated previously, the set-up of the physical bridge model aims to provide a test-bed in which the development and validation of various monitoring schemes with different damage scenarios could be carried out. Each suspender in the model is so designed that it can be easily removed and readily re-installed between the clampers attached to the main cables and the side chords of main cross frames to simulate the possible suspender breakout. The clampers installed on the main cables can be considered as added lumped masses which should be taken into account in the finite element modelling. Besides, the main cable portions laid along the saddle grooves are encapsulated with the plastic tubes lubricated by oil in order to abate the friction between the saddles and the cables as much as possible and on the other hand to simulate the relative slips of the main cables to the saddles/ towers.

      (a) Suspender and main cable   (b) Suspender and cross-frames

Figure 13.14 Connections between suspender, main cable, and deck

After the bridge deck was connected to the main cables through the suspenders, the tension forces in the main cables were adjusted to about 117 N on the Tsing Yi side and about 112 N on the Ma Wan side to meet the geometric requirement of the profiles of both the bridge deck and the main cables. Figure 13.15 shows the completed bridge model. To satisfy the targeted dynamic characteristics, a total of twelve supplementary lumped masses, each of about 174 gram, are added on the two sides of the main cross frames in the middle of the main span on both sides (Figure 13.16(a)); and a total of eight supplementary lumped masses, each of about 171 g, are attached to the lower deck plate in the middle (Figure 13.16(b)). Due to the added masses, the tension forces in the main cables were further adjusted to about 146 N on the Tsing Yi side and about 140 N on the Ma Wan side to keep the profiles of the bridge deck and main cables similar to the targeted ones.

Figure 13.15 Completed Tsing Ma Bridge model

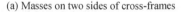

(a) Masses on two sides of cross-frames   (b) Masses on middle of lower bridge deck

Figure 13.16 Arrangement of supplementary masses

Figure 13.17 Setup for profile measurements

13.5 GEOMETRIC MEASUREMENTS AND MODAL TESTS

13.5.1 Geometric Measurements and Results

The geometric profiles of the physical bridge model must comply with those of the prototype bridge in terms of the global geometric scale. To measure the geometric profiles of the bridge model, a small movable platform with two poles was manufactured and installed on the shake table (see Figure 13.17). Two laser displacement meters were installed on the upper parts of two poles to measure the heights of two main cables and two more laser displacement meters were installed on the lower parts of the poles to measure the heights of the bridge deck. The small platform could move almost from one end of the bridge model to another along the steel beam (shake table), by which the vertical profiles of both the main cables and bridge deck could be measured. Before the measurement, the laser displacement meters with signal conditionals should be calibrated. Figure 13.18 displays the vertical profiles of both the physical bridge model and the prototype bridge in terms of the global geometric scale and by taking the upper surface of the steel beam as zero reference. It can be seen that the vertical profile of the physical model matched well with that of the prototype bridge. In Figure 13.18, "Ds" denotes the south side of deck; "Cs" means the south main cable; "Dp" means the prototype bridge deck; and "Cp" means the prototype bridge main cables. The maximum errors of the vertical profile are 2.17% between Ds and the Dp, and 4.37% between Cs and the Cp. The counterparts of the north side were also measured, and the maximum errors are 2.84% between Dn and the Dp, and still 4.37% between Cn and the Cp.

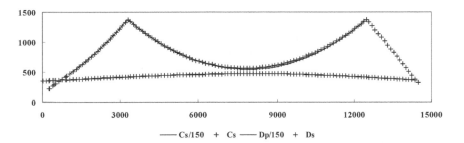

$$\text{——} \ Cs/150 \quad + \quad Cs \quad \text{——} \ Dp/150 \quad + \quad Ds$$

Figure 13.18 Comparison of profiles between the bridge model and prototype bridge (unit: mm)

13.5.2 Modal Tests and Results

Modal tests were carried out to determine the natural frequencies, modal shapes and damping ratios of the bridge model. The detailed information on system identification can be found in Chapter 4. The shapes and sequence of the first few vibration modes of the bridge model should follow those of the prototype bridge. The measured dynamic characteristics of the bridge model will be used for updating the FE model of the bridge model. The updated FE model will be used as a baseline for the subsequent studies of damage detection. In the dynamic modal tests, a series of impulses generated by an instrumented hammer was applied to the physical model at a designated location. A soft hammer tip was used so that the impulses exerted on the physical model included appropriate frequency components of interest and the signals of higher frequency components would not mask those of the fundamental modes. Accelerometers were mounted on the bridge deck and main cables for measuring the bridge model responses. A schematic layout of the accelerometers is presented in Figure 13.19, in which the arrows indicate the measured vibration direction. The number of accelerometers in each measurement is limited so that the additional weights of the accelerometers have little influence on the dynamic properties of the bridge model. In this regard, the entire bridge deck was divided into eighteen segments to be measured one by one. For each segment, two measurements were conducted: one for vertical vibration and the other for lateral vibration. Nine accelerometers were used in each measurement: four on cables, four on decks, and one as a common reference.

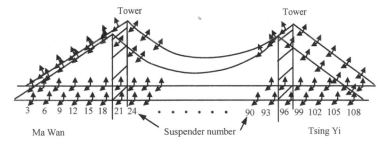

Figure 13.19 Layout of accelerometer locations

The natural frequencies and mode shapes of the bridge model were then identified through spectral analyses of the recorded force and acceleration response time histories. The mode shapes of the bridge model could be classified as lateral modes, vertical modes, and torsional modes as the prototype. The first 12 natural frequencies, f_e, are listed in the third column of Table 13.1. It is shown that the lowest natural frequency of the physical model is 3.907 Hz corresponding to the first lateral vibration mode. Figure 13.20(a) depicts an isometric view of the measured first lateral mode in the form of a half wave which is approximately symmetric in the main span with the in-phase movements of both the main cables and the deck. Displayed in Figure 13.20(b) is the second lateral mode, which appears almost anti-symmetric with one wave in the main span. The first vertical mode of the bridge model is almost anti-symmetric in the main span at a frequency of 5.273 Hz, as plotted in Figure 13.20(c). Due to the existence of suspenders, the motion of the deck is in phase with those of the two main cables. The second mode shape in the vertical plane is symmetric with a half wave in the main span, as shown in Figure 13.20(d). The symmetrical vertical mode at a frequency of 5.664 Hz, appearing after the anti-symmetrical mode, marks a significant dynamic characteristic observed in the prototype bridge. The first torsional mode occurs at a frequency of 8.789 Hz. As displayed in Figure 13.20(e), one side of the bridge deck moves upward while the other side of the deck moves downward in the main span. Furthermore, the vertical motion at the middle of each main cross frame is very small, indicating a pure torsional mode shape of the bridge with a half wave. The identified second torsional vibration mode shape appears in one wave in the main span at a natural frequency of 9.766 Hz, as shown in Figure 13.20(f). The two main cables move in phase with the bridge deck. The aforementioned results demonstrate that the shapes and sequence of the first few vibration modes of the bridge model do follow those of the prototype bridge.

Table 13.1 Measured and computed modal parameters before modal updating.

Direction	Mode No.	f_e (Hz)	f_a (Hz)	Diff. (%)	MAC	Span
Lateral	1	3.907	3.960	1.35	0.98	Main
	2	4.981	4.778	-4.07	0.94	Main
	3	5.371	5.100	-5.04	0.94	Main
	4	7.875	7.654	-2.81	0.99	Ma Wan
	5	7.911	7.868	-0.55	0.97	Main
	6	8.643	8.352	-3.36	0.97	Main
	7	9.081	8.601	-5.28	0.99	Tsing Yi
Vertical	1	5.273	6.199	17.56	0.55	Main
	2	5.664	6.367	12.40	0.54	Main
	3	9.191	8.628	-6.13	0.99	Tsing Yi
Torsional	1	8.789	9.521	8.33	0.92	Main
	2	9.766	10.350	5.98	0.94	Main

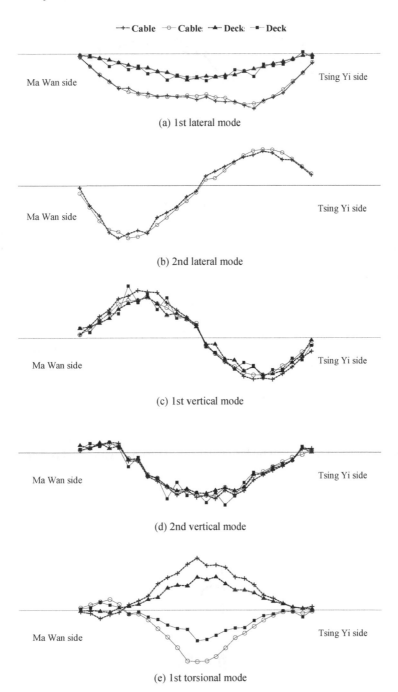

(a) 1st lateral mode

(b) 2nd lateral mode

(c) 1st vertical mode

(d) 2nd vertical mode

(e) 1st torsional mode

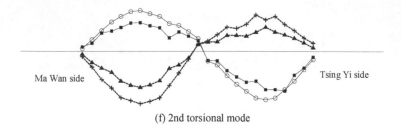

Ma Wan side   Tsing Yi side

(f) 2nd torsional mode

Figure 13.20 The first few measured mode shapes

13.6 ESTABLISHMENT OF FINITE ELEMENT MODEL

This section presents the FE modelling of the physical bridge model by using the commercial software ANSYS (ANSYS, 2004). The detailed FE modelling technique can be found in Chapter 4.

13.6.1 Modelling of Bridge Towers and Piers

Since the cross sections of the tower legs linearly decrease from the base to the top, three-dimensional tapered spatial beam elements (Beam44) with six DOFs at each end were used to model the tower legs. The portal beams between two adjacent tower legs were modelled with prismatic spatial beam elements (Beam4). The bridge towers were thus represented by three-dimensional multilevel portal frames with the bottom of the tower legs fixed at the base, as shown in Figure 13.21. The density, elastic modulus and Poisson's ratio of the beam elements used in the FE modelling are taken as those of the physical model. Piers M1, M2, T1, T2 and T3 were all modelled by beam elements (Beam4) with fixed constraints.

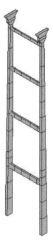

Figure 13.21 Finite element model of main tower

13.6.2 Modelling of Bridge Deck

The bridge deck is a suspended type with two typical structural configurations of 60 mm long segment. Hence, the modelling of the entire bridge deck can be conducted by repeatedly assembling the typical suspended deck segments along the bridge longitudinal direction. Each typical segment was represented by a three-dimensional FE sectional model using beam elements, as shown in Figure 13.22. Besides, elastic shell elements with four nodes, 24 DOFs, and 0.5 mm in thickness were utilized to model the orthotropic plate lying on the bottom chords of the cross frames. The supplementary lumped masses were modelled using mass elements with 6 DOFs.

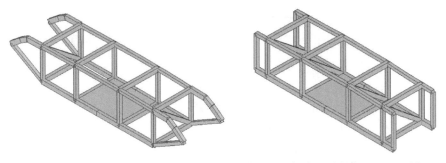

(a) Isometric view of deck segment at main span   (b) Isometric view of deck segment at side span

Figure 13.22 Isometric view of two typical deck segments

13.6.3 Modelling of Cables and Suspenders

Two-node link elements with six DOFs, having the unique feature of uniaxial tension-only, were applied to model the main cables and the suspenders. The static equilibrium profiles of both main cables were calculated based on the static horizontal tensions and the unit weight of both main cables and bridge deck given in the design drawings. The main cable between two adjacent suspender units was modelled by one link element. Each suspender unit was modelled by one link element as well. The geometric nonlinearity in the main cables due to cable tension and the tension in the suspender units due to the weight of the deck were considered in the elements by activating the stress stiffness effect. In order to account for the masses of the clampers, mass elements, denoted as "MO" in Figure 13.23, were added at the joints.

13.6.4 Modelling of Supports and Restraints

The vertical bearings between the deck and the Ma Wan tower, the deck and pier M2 were represented by swing rigid links so that the relative motion along the bridge axis was permissible. The vertical bearings between the deck and the Tsing Yi tower, the deck and piers T1, T2, T3 and M1 were simulated as one-dimensional longitudinal spring-damper elements to allow free longitudinal motion of the bridge. The lateral bearings of the deck to the two towers were modelled using one-dimensional longitudinal spring-damper elements. Only the rotation about the lateral axis was permissible in the Ma Wan abutment whereas only the longitudinal movement and lateral rotation were allowable at the Tsing Yi abutment. The top surfaces of the cable saddles were simply modelled with the joints coupled with the nodes at the tower leg tops. Similarly, the connection between pier M2 and the main cable was simulated with coupling in both the vertical and lateral directions. The ends of the two main cables were fixed in all DOFs. An overview of the FE model of the entire bridge is plotted in Figure 13.24. The natural frequencies obtained from the modal analysis of the FE model, denoted as f_a, are listed in the fourth column of Table 13.1.

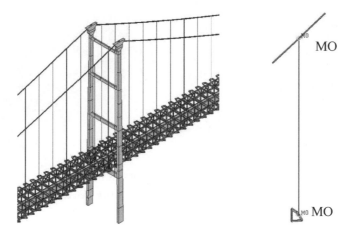

Figure 13.23 Finite element models of main cables and suspenders

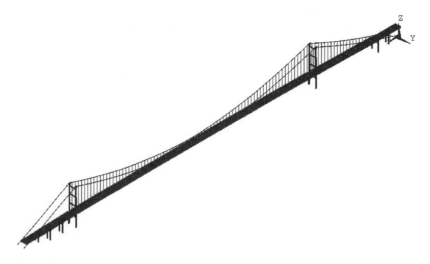

Figure 13.24 Overview of the finite element model of the entire bridge model

13.6.5 Correlation Analysis

To correlate the initial FE analysis results with the experimental modal analysis results, the MAC (Allemang and Brown, 1982) is calculated. The natural frequencies and mode shapes obtained from the initial FE model are compared with those measured from the vibration tests, including seven lateral modes, three vertical bending modes and two torsional modes. The results are compared in Table 13.1. The correlation of the mode shapes expressed by MAC values is good except for the first and second vertical bending modes, for which both MAC values are below 0.6. Besides, the frequency correlation is not so good with the maximum error of 17.56% in the first vertical mode and 12.40% in the second. The initial FE model for the physical model was insufficiently accurate to represent the test-bed. Model updating is needed to improve the FE model.

13.7 FINITE ELEMENT MODEL UPDATING

A great deal of effort has been invested in the research of FE model updating (see Chapter 4) and it generally involves two approaches: the system matrix updating (or direct methods) and the structure parameter updating (or iterative methods) (Mottershead and Friswell, 1993; Zhang *et al.*, 2000; Brownjohn *et al.*, 2001; Jaishi and Ren, 2005, Carvalho *et al.*, 2006). As aforementioned, the geometric stiffnesses of the cables and suspenders associated with their axial forces mainly caused by the weights of the bridge are included in the FE analysis, and therefore the stiffness of the bridge model is related to its mass in the FE analysis and the subsequent model updating, which is unlikely to be served by the system matrix updating method because it cannot handle the coupling of the mass and stiffness matrices. Consequently, the parameter updating method becomes the choice of this study in searching the optimal updated parameters.

Frequencies can be identified more accurately than mode shapes in general. On the other hand, mode shapes contain more spatial information about the dynamic behaviour of the structure. Thus, both frequencies and mode shapes are used as reference in the FE model updating in this study. In general, the model updating procedure contains three steps: (1) select updating parameters; (2) define objective functions and state variables; and (3) implement the optimization.

13.7.1 Parameters for Updating

The selection of the parameters for updating involves two aspects, that is, how many parameters should be selected and which parameters from the many candidates are preferred? This always demands physical and mathematical insights as well as the engineering experience. Physically, the selected parameters should represent some uncertainties of the model and the updated results should preserve the physical meaning of the model. Mathematically, the good conditioning of the problem should be ensured.

Sources of the discrepancies between the experimental and analytical results may lie in the imprecision of the geometrical and material parameters, the boundary conditions, and the connections of the member components. For this reduced-scale bridge model, the geometrical parameters can be accurately measured comparatively and the boundary conditions can be simulated under control. Thus, the geometrical parameters and the boundary conditions in the initial FE model will not be updated. However, the rigid connections among beam elements and between beam elements and shell (plate) elements in particular may not well represent the real connections, as do the spring connections between the deck and towers. It is thus reasonable to consider adjusting the stiffness of the cross frame, plate, longitudinal beams, and diagonal bracings as well as the springs. Besides, the axial forces in the cables and suspenders in the physical model, to some extent, are likely different from those in the FE analysis. In this connection, the sectional stiffness of the suspenders and the cables are also chosen as the candidates for adjustment. Furthermore, the member masses are not weighted or measured against those of the analytical ones, and accordingly the densities of the main cross frame, longitudinal beams and the plate can also be selected as updated parameters.

A sensitivity analysis computes the sensitivity coefficient defined as the rate of change in a particular response quantity with respect to a change in a structural parameter. The parameters with high sensitivity can be further examined and selected for model updating, while those with low sensitivity can be eliminated. A well-conditioned updated problem necessitates the selection of those parameters which will be most effective in reducing the discrepancies between the analytical results and their identified counterparts. In this study, the sensitivity analysis is performed by calculating the derivatives of the first 12 modal frequencies with respect to various parameters. Out of 15 structural parameters to be updated, five of them show a significant influence on natural frequencies. Although the natural frequencies are not sensitive to either the Young's modulus of plate or the spring connection between the deck, towers and piers, these parameters were still selected because they may affect mode shapes due to their obvious uncertainties. Finally a total of 10 structural parameters were chosen for updating (see Table 13.3).

13.7.2 Objective Functions

The objective functions are formulated in terms of the discrepancy between analytical and experimental natural frequencies and MAC as shown below (Jaishi and Ren, 2005):

$$J(x) = \sum_{i=1}^{n} \alpha_i \left(\frac{f_{ai} - f_{ei}}{f_{ei}} \right)^2 + \sum_{i=1}^{n} \beta_i \frac{\left(1 - \sqrt{MAC_i} \right)^2}{MAC_i} \tag{13.1}$$

where f_{ai} and f_{ei} are the analytical and experimental natural frequencies of the ith mode, respectively; α_i and β_i are the weighting factors, whose values depend on the accuracy of the test. In this study, α is two times β considering the lower accuracy of mode shapes.

In each iteration step, the analytical and experimental mode shapes are matched using the MAC criterion, and the frequency discrepancies are calculated. The selected updated parameters are estimated during the iterative process. The tuning process terminates when the following tolerances are achieved:

$$0 \le \left| f_{ai} - f_{ei} \right| \le u \tag{13.2}$$

$$l \le MAC \le 1 \tag{13.3}$$

where u represents the upper limit whose value can be set as absolute error between the ith analytical frequency and experimental frequency, and l represents the lower limit of the MAC value.

13.7.3 Updating Results

Given in Table 13.2 is the final correlation of natural frequencies and mode shapes after model updating. It can be seen that the significant differences of the natural frequencies between the FE model and experiments are reduced from 17.56% to 4.55% in the first vertical mode and from 12.40% to 5.63% in the second vertical mode. Moreover, the correlation of mode shapes is also improved significantly and all MAC values exceed 0.9. For the first vertical mode, the MAC value increases from 0.55 to 0.97, and that of the second vertical mode increases from 0.54 to 0.96. As shown in Figure 13.25, the analytical mode shapes of both main cables and deck components are in good agreement with the measured results as shown in Figure 13.25.

Table 13.2 Measured and computed modal parameters after model updating.

Direction	Mode No.	f_e (Hz)	f_a (Hz)	Diff. (%)	MAC	Span
	1	3.907	3.755	-3.89	0.99	Main
	2	4.981	4.819	-3.25	0.92	Main
	3	5.371	5.024	-6.46	0.95	Main
Lateral	4	7.875	7.700	-2.22	0.99	Ma Wan
	5	7.911	7.925	0.18	0.97	Main
	6	8.643	8.422	-2.56	0.97	Main
	7	9.081	8.659	-4.65	0.99	Tsing Yi
	1	5.273	5.513	4.55	0.97	Main
Vertical	2	5.664	5.983	5.63	0.96	Main
	3	9.191	8.683	-5.53	0.99	Tsing Yi
Torsional	1	8.789	9.094	3.47	0.90	Main
	2	9.766	9.602	-1.68	0.93	Main

Table 13.3 lists the changes in the updating parameters. The initial values of the parameters in the table are the design values. It can be seen that there are slight to moderate reductions in the elastic moduli of the components except that of the plate has experienced 83% reduction, indicating that the initial guess of the moduli is relatively accurate. It can also be found that the connection stiffnesses particularly the vertical connections have significant reductions. In the scaled model, the longitudinal beams and truss beams in the deck module that is made of aluminium were connected by welding. Besides, the plate and the deck were

connected by special adhesive. Moreover, all the vertical bearings on piers T1, T2, T3, M1 and Tsing Yi tower were linked to the deck using epoxy adhesive. All these connections are quite different from the fixed ones and thus reduce the structural stiffness, especially the vertical and torsional stiffnesses.

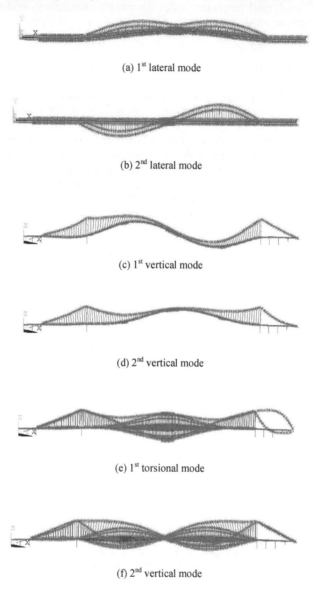

(a) 1st lateral mode

(b) 2nd lateral mode

(c) 1st vertical mode

(d) 2nd vertical mode

(e) 1st torsional mode

(f) 2nd vertical mode

Figure 13.25 The first six analytical vibration modes

Table 13.3 Changes in updated parameters.

Parameters to be updated	Initial values	Updated values	Change (%)
Elastic modulus of main cables (Pa)	9.4×10^{10}	8.85×10^{10}	-5.8
Elastic modulus of cross frames (Pa)	6.89×10^{10}	6.67×10^{10}	-3.1
Elastic modulus of longitudinal beams (Pa)	6.89×10^{10}	5.91×10^{10}	-14.2
Elastic modulus of truss beams (Pa)	6.89×10^{10}	5.79×10^{10}	-15.9
Elastic modulus of plates (Pa)	6.89×10^{10}	1.18×10^{10}	-82.8
Elastic modulus of bearings (Pa) (Ma Wan tower and deck, pier M2 and deck)	6.89×10^{10}	6.21×10^{10}	-9.9
Mass density of deck (aluminium) (kg/m^3)	2720	2825.5	3.9
Stiffness of lateral connections (N/m) (Towers and deck)	3.25×10^8	3.21×10^8	-1.2
Stiffness of vertical connections (N/m) (Tsing Yi tower and deck)	3.00×10^8	3.00×10^4	
Stiffness of vertical connections (N/m) (M1 and deck, T1-T3 and deck)	1.19×10^9	1.12×10^5	

13.8 SUMMARY

This chapter presents a preliminary study in designing and setting up a test-bed for the research on SHM of long-span suspension bridges. The test-bed encompasses the delicate physical model and the updated FE model of a long-span suspension bridge. The design and setup of the physical model are presented in detail. The geometric measurements and the modal tests were carried out on the physical bridge model to identify its geometric configuration and dynamic characteristics. The identified results indicate that the dynamic properties of the physical model fulfil the elementary design requirements. The first natural frequency of the bridge model corresponding to the lateral sway is about 3.907 Hz, which ensures the physical model is not too flexible and insensitive to environmental interruptions. The shapes and sequence of the first a few vibration modes of the bridge model also match with those of the prototype bridge. The FE modelling of the bridge model was also established, followed by a model updating in terms of the modal test results. The natural frequencies and mode shapes of the updated model are in good agreement with the measured results. The test-bed comprising the delicate physical model and the updated FE model of a long suspension bridge could serve a benchmark problem for SHM, in particular structural damage detection, of long suspension bridges.

13.9 REFERENCES

Allemang, R.J., and Brown, D.L., 1982, A correlation coefficient for modal vector analysis. In *Proceedings of the 1st International Modal Analysis Conference*, Orlando, Florida, edited by DeMichele, Dominick J., (Bethel, Connecticut: Society for Experimental Mechanics), pp. 110-116.

ANSYS, 2004, *User's Manual*, Version 9.0, (Southpointe: ANSYS Inc.).

Brownjohn, J.M.W., Xia, P.Q., Hao, H. and Xia, Y., 2001, Civil structure condition assessment by FE model updating: methodology and case studies. *Finite Elements in Analysis and Design*, **37**, pp. 761-775.

Carvalho, J.B., Datta, B.N., Lin, W.W., and Wang, C.S., 2006, Symmetry preserving eigenvalue embedding in finite-element model updating of vibrating structures. *Journal of Sound and Vibration*, **290(3-5)**, pp. 839-864.

Chang, F.K., 1997, *Structural Health Monitoring: Current Status and Perspectives*, (Boca Raton, Florida: CRC Press).

Doebling, S.W., Farrar, C.R., and Prime, M.B., 1998, A summary review of vibration-based damage identification methods. *The Shock and Vibration Digest*, **30**, pp. 91-105.

Farrar, C.R., *et al.*, 2003, *Damage Prognosis: Current Status and Future Needs*. Report LA-14051-MS, (Los Alamos: Los Alamos National Laboratory).

Glaser, S.D., Li, H., Wang, M.L., Ou, J.P., and Lynch, J., 2007, Sensor technology innovation for the advancement of structural health monitoring: a strategic program of US-China research for the next decade. *Smart Structures and Systems*, **3(2)**, pp. 221-244.

Jaishi, B., and Ren, W.X., 2005, Structural finite element model updating using ambient vibration test results. *Journal of Structural Engineering, ASCE*, **131(4)**, pp. 617-628.

Johnson, E.A., Lam, H.F., Katafygiotis, L.S., and Beck, J.L., 2004, Phase I IASC-ASCE structural health monitoring benchmark problem using simulated data. *Journal of Engineering Mechanics*, **130(1)**, pp. 3-15.

Mottershead, J.E., and Friswell, M.I., 1993, Model updating in structural dynamics: A survey. *Journal of Sound and Vibration*, **167(2)**, pp. 347-375.

Mufti, A.A., 2001, *Guidelines for Structural Health Monitoring*, (Winnipeg, Canada: Intelligent Sensing for Innovative Structures).

Simiu, E. and Scanlan, R., 1996, *Wind Effects on Structures*, 3rd ed., (New York: John Wiley).

Wong, K.Y., Man, K.L., and Chan, W.Y., 2001a, Monitoring Hong Kong's bridges real-time kinematic spans the gap. *GPS World*, **12(7)**, pp. 10-18.

Wong, K.Y., Man, K.L., and Chan W.Y., 2001b, Application of global positioning system to structural health monitoring of cable-supported bridges. In *Proceedings of Health Monitoring and Management of Civil Infrastructure Systems, SPIE*, **4337**, Newport Beach, USA, edited by Chase, S, (Bellingham, Washington: SPIE), pp. 390-401.

Xu, Y.L., Ko, J.M., and Zhang, W.S., 1997, Vibration studies of Tsing Ma suspension bridge. *Journal of Bridge Engineering*, **2(4)**, pp. 149-156.

Zhang, Q.W., Chang C.C., and Chang T.Y.P., 2000, Finite element model updating for structures with parametric constraints. *Earthquake Engineering and Structural Dynamics*, **29**, pp. 927-944.

13.10 NOTATIONS

f_{ai}, f_{ei}	Analytical and experimental natural frequencies of the ith mode.
J	Objective function.
l	Lower limit of the MAC value.
n	Number of modes.
u	Upper limit of frequency residual.
α, β	Weighting factors in objective function.

CHAPTER 14

Epilogue: Challenges and Prospects

14.1 CHALLENGES

Structural health monitoring (SHM) is a cutting-edge technology, providing better solutions for design, construction, maintenance, functionality, and safety of long-span suspension bridges. It is based on a comprehensive sensory system, a sophisticated data transmission, processing and management system, and an advanced structural computation, evaluation and decision making system. Therefore the SHM of long-span suspension bridges is integrated with various engineering disciplines including advanced structural engineering, modern information technology, latest risk assessment and bridge management. The SHM process involves the long-term observations of a bridge's surrounding environment and responses over time by an array of sensors, extraction of structural performance oriented features from these measurements, numerical computation of the bridge, and condition assessment and prediction. After extreme events occur, such as earthquakes or strong winds, the SHM is used for rapid condition screening and can provide the real-time and reliable information about the serviceability, safety, and integrity of the bridge.

In this regard, the previous chapters have detailed design of the SHM system, SHM-oriented modelling, monitoring of various loading and environmental effects, damage detection methods, SHM-based rating system, and test-bed establishment for long-span suspension bridges. Although a considerable amount of information has been covered by these chapters, there are still some important issues not being fully discussed and some challenging issues to be solved in the near future.

14.1.1 Durability and Optimization of the Sensor Network

The sensors and the sensor network used for SHM of a long-span suspension bridge are exposed to very harsh environments. The life-span of the sensor network is likely much less than that of a long-span suspension bridge, which could be more than 100 years. The accurate measurements of various loading, environmental conditions, and bridge response rely on the quality of the sensors very much. Durability of the sensors and the sensor network during their life-span needs special attention. Advanced and reliable sensors have to be developed.

The quality and completeness of the data collected by the SHM systems depend not only upon the capabilities and quality of the sensors used, but also upon where the sensors are deployed on the structure. The selection of a subset of measurement locations from a large number set of candidate locations can be viewed as a three-step decision process: (a) sensor quantity; (b) sensor placement optimization; and (c) evaluation of different sensor configurations. The problem of sensor placement optimization has been addressed by engineers and researchers in the areas of aeronautical and mechanical engineering. While a variety of performance indices have been proposed, the existing investigations on sensor network suffer several limitations for health monitoring of long-span bridges. Previous studies mainly addressed the sensor network to acquire modal information as accurately and richly as possible, which can only perform SHM at a global level. These studies mainly determined the optimal sensor locations by considering structural performance without considering the performance of the sensor network. The data quality could be heavily contaminated by noise during signal transmission over the network.

In view of these unresolved issues, great efforts must be made to optimize sensor placement based on not only structural performance but also sensor network performance. It has also become necessary to design the sensor network for long-span bridges by considering the fusion of different kinds of sensors to achieve the utmost goals of the SHM. As the SHM system is targeted for long-term applications, the service life of sensors and the ease of replacement or supplement of sensors should be taken into account in the design of the sensor network for future SHM of long-span bridges.

14.1.2 Novel Damage Detection Methods

Long-span suspension bridges may be subject to progressive deterioration in service and damage due to extreme loading events. Early damage detection and warning are vital for an effective planning of maintenance and repair work. During the past few decades, numerous methods have been developed to identify the structural damage. However, a long-span suspension bridge consists of hundreds of thousands of components, and the local damage often does not affect the global responses of the bridge significantly, making the damage detection analysis inaccurate and sometimes impossible. Successful application of damage detection to long-span suspension bridges is still very limited although some methods have been successfully applied in mechanical and aeronautical engineering. Novel damage detection algorithms, such as integration of local responses with global responses, need to be developed.

14.1.3 Advanced Computational Simulation

The number of sensors in an SHM system is always limited for a long-span suspension bridge, and the locations of structural defects may not be directly monitored by the sensors. Numerical analysis or computational simulation based on finite element models of a bridge is necessary to simulate loadings that are difficult to measure directly, predict the bridge responses and detect the damage in critical structural members/joints where the sensors may not be installed. Although some simplified finite element models are effective for global dynamic analysis, they are insufficient for local stress/strain analysis and performance assessment of long-span bridges. To facilitate an effective performance assessment of a long-span bridge, new finite element modelling techniques, such as multi-scale modelling technique, with SHM oriented shall be developed. Accordingly, new computation simulations of a long-span bridge subject to various loading or a combination of multiple loadings needs to be investigated.

One task of the SHM is to prevent the long-span bridge from catastrophic collapse. An advanced computer simulation platform is thus required to develop extreme loading models by sufficiently utilizing the measurement data from the SHM system, to model the bridge including geometric and material nonlinearities, to perform load path and redundancy analysis, to carry out progressive collapse analysis of the bridge under the extreme loadings, and finally to evolve an appropriate early warning system.

14.1.4 Bridge Rating and Life-cycle Management

Existing bridge management systems generally rely on visual inspections and NDT inspections, and the bridge condition is often assessed through a condition rating or condition index. Although SHM technologies have been developed for bridge condition assessment, they have not been effectively utilized for bridge management systems. There is no effective and scientific bridge rating system for long-span suspension bridges at present to provide a rational basis for rating risk of major bridge structural components and for determining types and frequencies of inspection and maintenance by utilizing the SHM technology. It is imperative to integrate SHM systems with advanced computer simulation platforms to form smart systems and combine quantitative monitoring results from the smart systems with the results from visual and NDT inspections and the qualitative opinions from experts to achieve our goal. From the life-cycle management point of view, how the initial cost of the SHM will result in the total cost saving and improved bridge safety needs further study.

14.2 PROSPECTS

Although there are many challenging issues ahead in the SHM of long-span suspension bridges, the implementation of SHM technologies for practical applications is being widely considered and is in process as mentioned in Chapter 1. Mutual respect and effectively coordinated collaboration are required among professional engineers, academic researchers, government agencies, contractors, managers, and owners. This is because all sectors will be benefited from the SHM technologies for improving structural reliability and operational efficiency while reducing life-cycle maintenance and operating cost.

Rapid development in sensing technology, information technology, computation simulation, and asset management will eventually overcome the remaining challenging issues to fulfil the goals and objectives of the SHM. Wireless sensors and networks, bio-inspired sensors, nano sensors, multi-function sensors and others will provide new sensing technologies for SHM. Development of knowledge discovery algorithms will allow more reliable, accurate, and efficient data processing and data explanation. In addition, more powerful computation platforms will be developed in terms of hardware and software such that the SHM can serve the associated sectors better.

Another important aspect making the SHM technology successful and sustainable is education of the young generation. It is most likely that some viewpoints introduced in this book will be overruled and some methodologies will be improved by our young generation in the near future. This is the main purpose for the authors to write this book.

Index